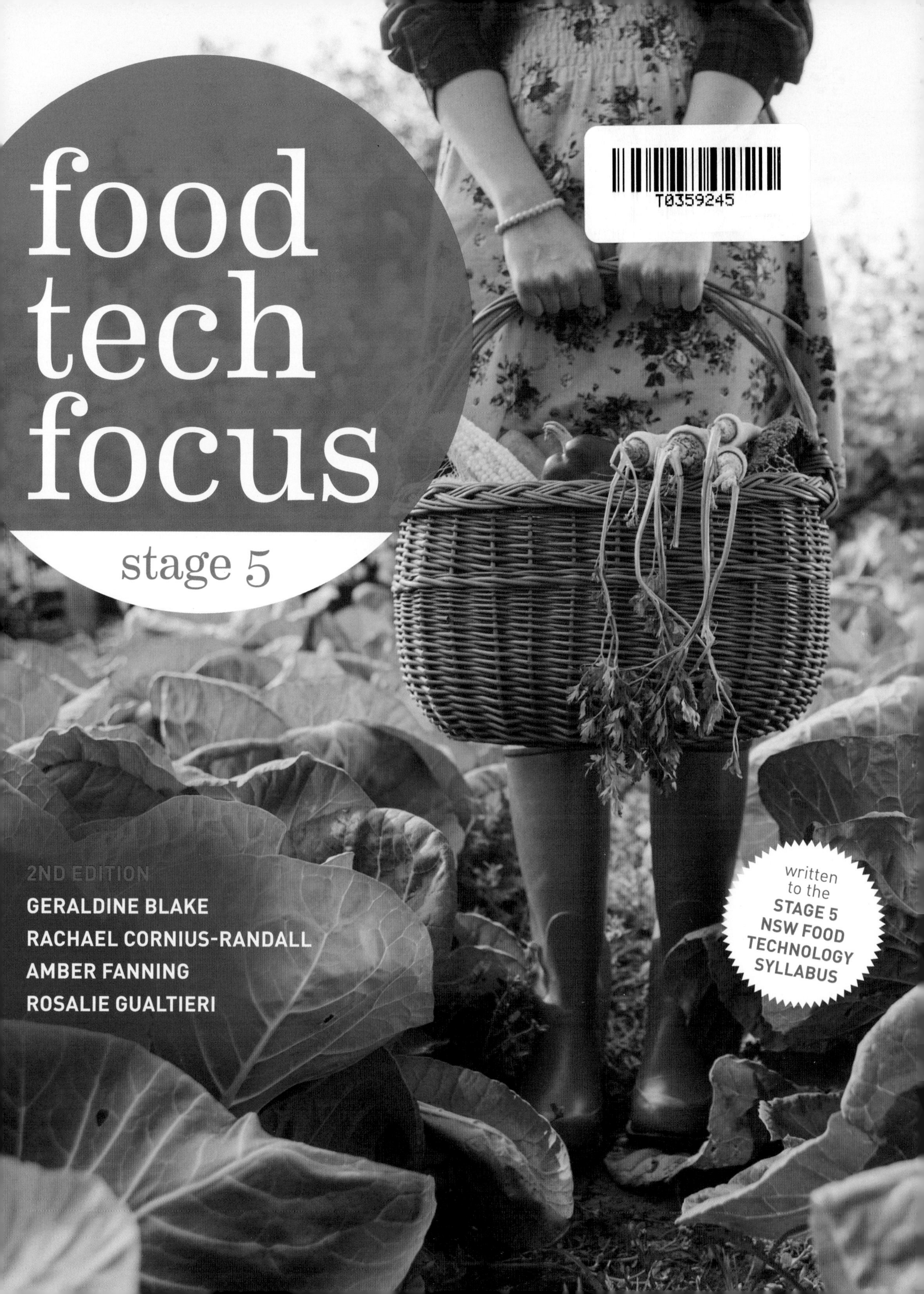
food
tech
focus
stage 5
T0359245
2ND EDITION
GERALDINE BLAKE
RACHAEL CORNIUS-RANDALL
AMBER FANNING
ROSALIE GUALTIERI
written
to the
STAGE 5
NSW FOOD
TECHNOLOGY
SYLLABUS

Food Tech Focus Stage 5
2nd Edition
Geraldine Blake
Rachael Cornius-Randall
Amber Fanning
Rosalie Gualtieri

Associate publishing editor: Sam Bonwick
Editor: Nadine Anderson-Conklin
Copy editors: Gene Anderson-Conklin and Mandy Herbet
Proofreader: Patrick O'Duffy
Indexer: Russell Brooks
Cover and text designer: Aisling Gallagher
Cover image: Stocksy United/Jovo Jovanovic
Permissions researcher: Helen Mammides
Photographer: Mark Fergus
Chef and stylist: Melanie Vandegraaff of Red Courgette
Photo shoot production: Rhonda Fergus
Tableware: Cedar Hospitality Supplies
Typesetter: Cenveo Publisher Services

Acknowledgements
Text design images:
iStockphoto/ivan-96, iStockphoto/denisk0, iStockphoto/SvetaGaintseva, iStockphoto/nicoolay, Shutterstock.com/Nikiparonak, Shutterstock.com/Morphart Creation, Shutterstock.com/Danussa, Shutterstock.com/geraria, Shutterstock.com/ostill, Shutterstock.com/Natykach Nataliia, iStockphoto/appleuzr, iStockphoto/Joe Biafore, iStockphoto/Alex Potemkin, iStockphoto/jeffbergen, iStockphoto/AndreyPopov

For product information and technology assistance,
in Australia call **1300 790 853**;
in New Zealand call **0800 449 725**

For permission to use material from this text or product, please email
aust.permissions@cengage.com

National Library of Australia Cataloguing-in-Publication Data
Blake, Geraldine, author.
Food tech focus stage 5/ Geraldine Blake, Rachael Cornius-Randall, Amber Fanning, Rosalie Gualtieri.

2nd edition.
9780170383417 (paperback)
Includes index.
For secondary school age.

Food industry and trade--Textbooks.
Food--Textbooks.

Cornius-Randall, Rachael, author.
Fanning, Amber, author.
Gualtieri, Rosalie, author.

664.0712

Cengage Learning Australia
Level 7, 80 Dorcas Street
South Melbourne, Victoria Australia 3205

Cengage Learning New Zealand
Unit 4B Rosedale Office Park
331 Rosedale Road, Albany, North Shore 0632, NZ

For learning solutions, visit **cengage.com.au**

Printed in Australia by CanPrint Communications Pty Limited.
3 4 5 6 7 8 21 20 19

9780170383417

contents

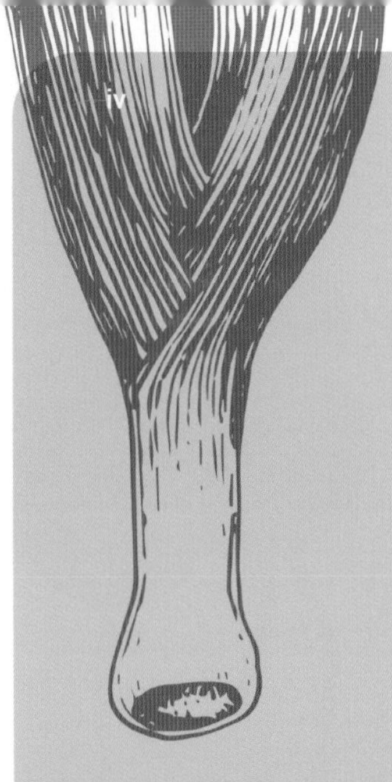

Chapter 8 Food for life

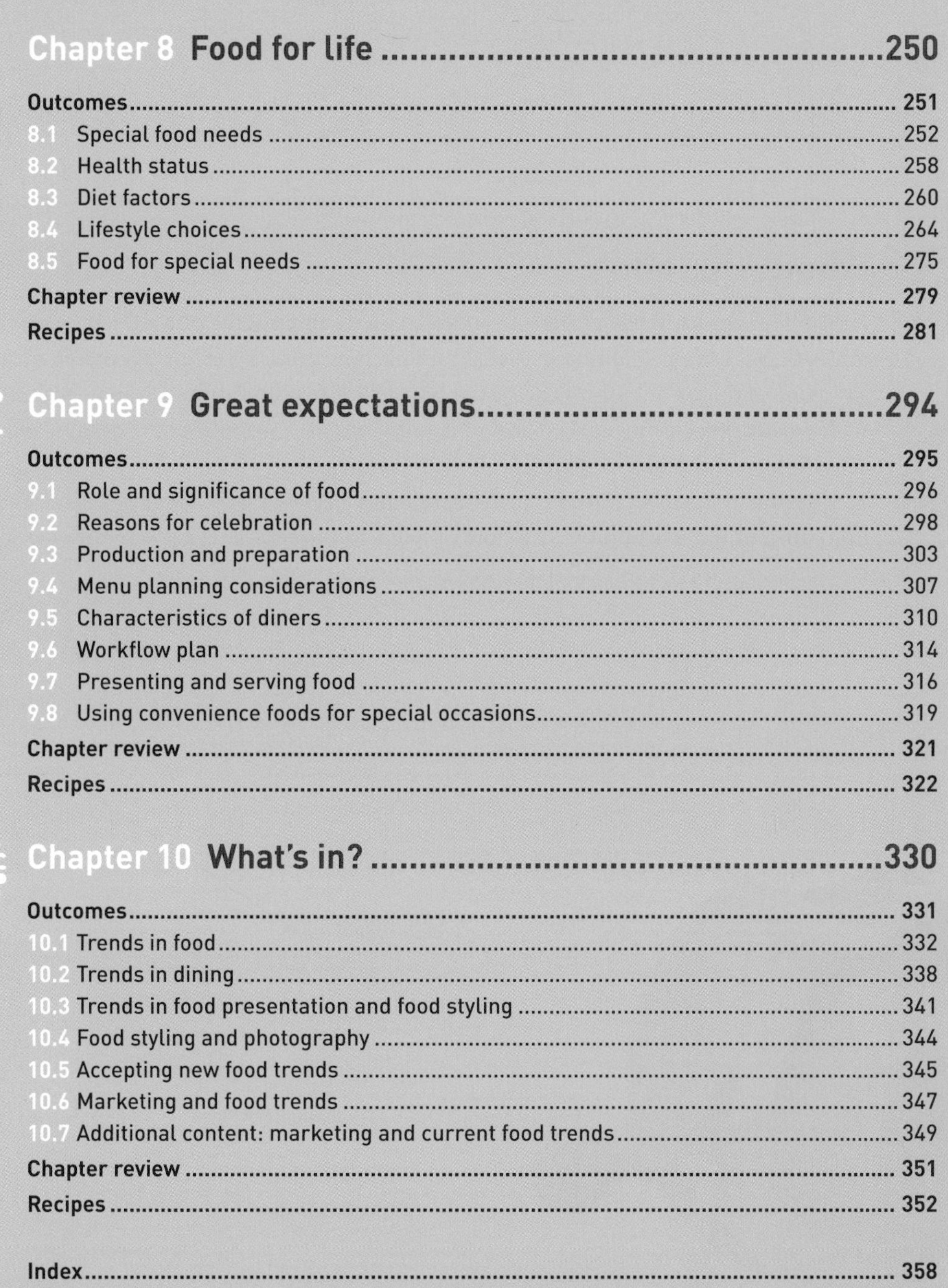

Chapter 9 Great expectations

Chapter 10 What's in?

9780170383417

introduction

About this book

Food Tech Focus has been written to meet all the requirements of the New South Wales Stage 5 Food Technology Syllabus. This new edition retains the direct, clear-cut approach and logical lesson flow of the bestselling first edition but is new and improved with comprehensively checked and updated content, a stunning new design, and activities around numeracy, literacy, ICT and nutrition.

Favourite recipes from the first edition and all new dishes from popular cuisines from around the world provide a host of tried and tested practical activities for students. Recipes now also include bespoke inspirational photography.

Recipes related to the chapter. Recipes appear at the end of the relevant chapter.

Chapter outcomes help students and teachers monitor their progress through the course of study.

CHAPTER OUTCOMES

In this chapter you will learn about:

- the use of foods native to Australia
- early European influences on food habits
- multicultural influences on food habits and methods of cooking
- the evolution of an Australian cuisine
- influences on food selection
- factors affecting current consumption patterns
- development of food production and processing from both historical and contemporary perspectives.

In this chapter you will learn to:

- investigate traditional and contemporary use of native/bush foods
- discuss the impact of early European influences on food habits
- identify the major multicultural influences on contemporary Australian diets
- discuss the defining characteristics of Australian food and prepare food that reflects the changing nature of Australian cuisine
- examine the influences on food selection and changes in eating habits
- relate changes in consumption patterns to their social, economic, nutritional and environmental impact
- investigate the development of the Australian food industry in consideration of food-related technologies that have emerged over time.

RECIPES

- Lemon myrtle chocolate crackles, p.109
- Wattleseed jam cockles, p.110
- Quandong jam, p.111
- Traditional damper, p.112
- Bush damper, p.113
- Contemporary Aussie damper, p.114
- Mem ran (North Vietnamese spring rolls), p. 115
- Chicken, minted pea and ricotta meatballs, p. 116

Bush tucker to contemporary cuisine

FOCUS AREA

FOCUS AREA: FOOD IN AUSTRALIA

Defining Australian cuisine is not easy. Is it meat and three vegetables, or bush tucker such as kangaroo and macadamia nuts, or foods from other cuisines with a local spin? This chapter looks at the effects migration has had on the food Australians eat. This chapter will examine the history of food including the traditional bush foods prepared by Indigenous Australians and the influence of early European settlers.

FOOD FACTS

Before 1770, there were more than 300 000 Aboriginal people living in 500–600 tribal groups, each consisting of between 100 and 1000 people. Each tribe had its own territory and language.

Men and women serving in the Australian army are now trained in how to survive on bush foods in case of combat in Australia.

A pound is approximately equal to 450 grams (the size of a packet of pasta); an ounce is about 30 grams (the size of a small bag of chips); and a pint is about 600 mL (the size of a bottle of coke).

Food manufacturers have responded to the growing trend towards eating out with a range of chilled, frozen and microwavable 'quick fix' meals, which are promoted as the restaurant dishes you don't have to go out for.

Recent developments in food packaging and storage include flexible and lightweight plastic packaging materials, frozen-food cabinets, refrigerated dairy cases and dual-oven heating trays. Food quality control has also had an impact on food packaging.

FOOD WORDS

coolamon elongated wooden dish used for gathering plant and animal foods

food habit pattern of eating; the way in which food is prepared, served and eaten

indigenous food (bush tucker) food native to Australia that was present before European colonisation

multiculturalism existence of different cultures within the one society

ration fixed allowance of food

satiety state of being fed or satisfied

scurvy disease caused by a lack of Vitamin C

79

Food facts are interesting facts and information related to the key content.

Food words are key glossary terms for the chapter.

Unit reviews
Chapters are divided into units. Graded activities and questions are provided at the end of each unit for students to apply the knowledge and skills learned.

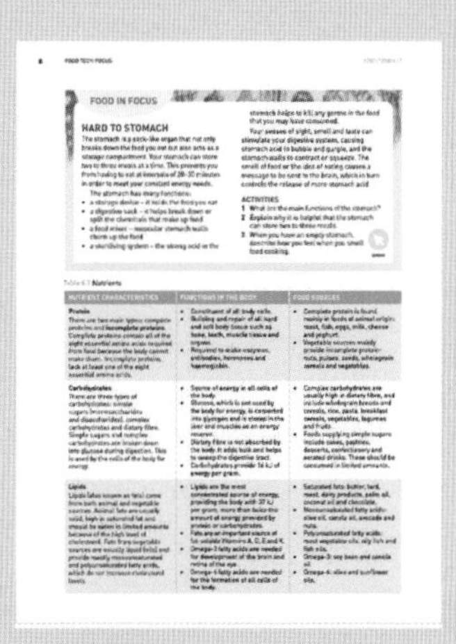

Food in Focus
Support unit content in a case study approach. Contain questions to consolidate and extend student knowledge.

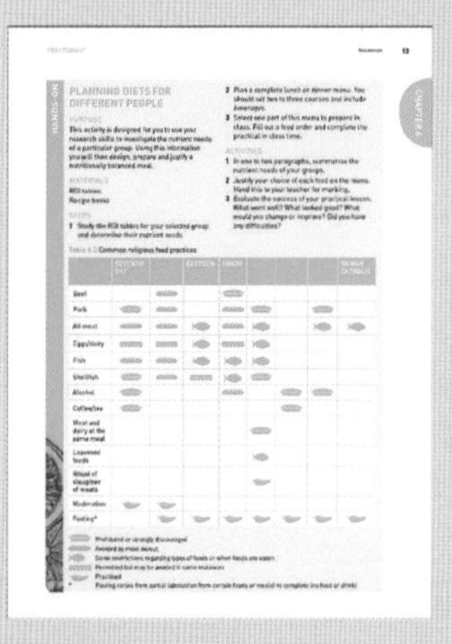

Hands-on
Structured practical tasks that deepen and broaden practical knowledge.

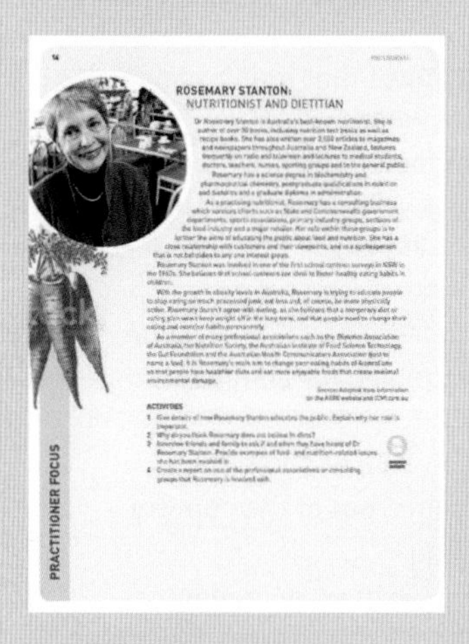

Practitioner focus
Allow students to make real-life connections between the classroom and industry.

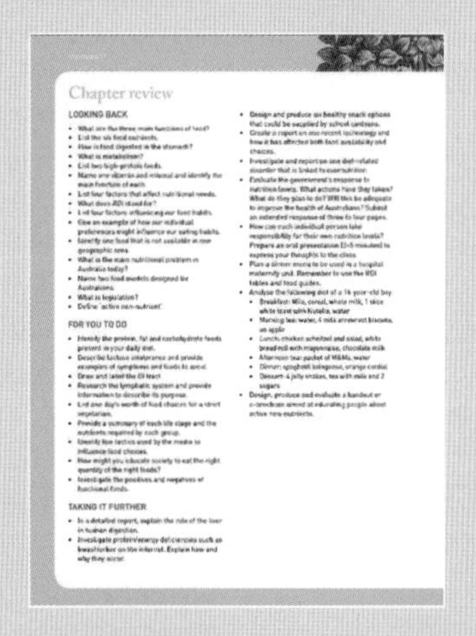

Chapter review
Extending on the unit reviews, graded activities and questions for students to apply knowledge and skills learned throughout the chapter.

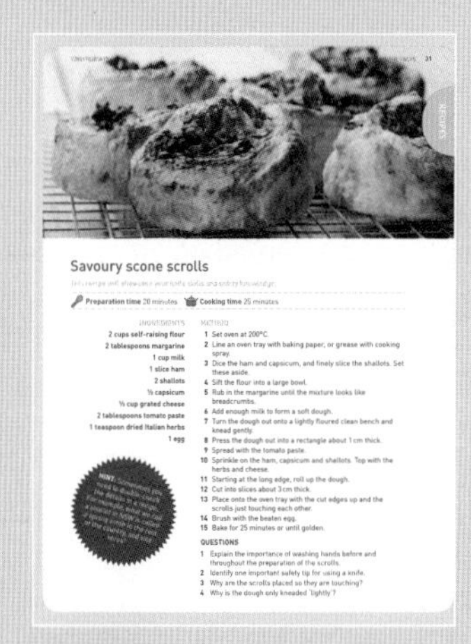

Recipes
All recipes have been tested and provide students with opportunities to develop skills in food preparation and processing.

NelsonNet

Food Tech Focus is a premium Cengage title and is fully supported by the NelsonNet platform. Access to NelsonNet also provides students with additional web-based materials, such as activity sheets and quizzes.

NelsonNetBook

The NelsonNetBook is your digital textbook. Readable online and offline on desktops, laptops and tablets, it reproduces the student text in digital form. With annotations and reviewing tools, and the ability to add and customise your book, NelsonNetBook is accessible immediately via access codes. Please note that any notations made to the NelsonNetBook will expire two years after the access code is activated.

Weblinks

Students and teachers can link directly to external websites referred to in *Food Tech Focus* via the free, unprotected weblinks site located at **http://foodtechfocus.nelsonnet.com.au.**

Disclaimer

Please note that complimentary access to NelsonNet and the NelsonNetBook is only available to teachers who use the accompanying student textbook as a core educational resource in their classroom. Contact your sales representative for information about access codes and conditions.

9780170383417

glossary

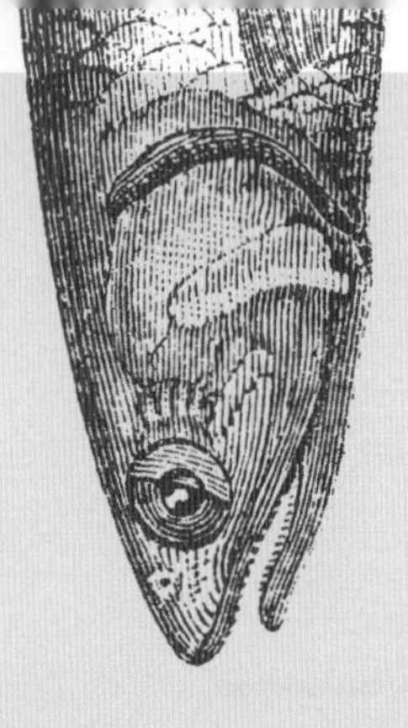

à la carte meaning 'from the card' – a menu that is written with individual prices for dishes

anaemia condition resulting from an inadequate intake of iron, causing the sufferer to feel weak, tired and fatigued

antibody protein produced by the body to fight foreign particles

balanced diet nutrition intake that provides energy, nutrients, dietary fibre and fluids in amounts appropriate for good health

biodegradable able to decompose because of the activity of living organisms such as bacteria

blanch plunge into boiling water for a short time, then into cold water

cafe informal restaurant, offering a range of hot meals

canapé small piece of toast spread with a variety of toppings and served as an appetiser

carbo-loading ingesting large amounts of carbohydrate-rich foods in one sitting, usually before an athletic event

cholesterol constituent of some fats that sticks easily to the walls of the arteries

chronic constant, continuing for a long time

clean eating a diet of whole foods that have had minimal processing

colostrum first breast liquid produced by a mother after childbirth; it is high in proteins, minerals and antibodies

commercial for profit

condiment a substance, such as a sauce, that gives a special flavour to a food

constraint restriction

consumer a person who uses a product or service

convalescence period of recovery of health and strength after injury or illness

coolamon elongated wooden dish used for gathering plant and animal foods

criteria standards to judge against

cross contamination transfer of bacteria or other micro-organisms from one food or item of equipment to another

crouton small cube of bread crisply fried or toasted

crudité raw vegetable cut into bite-sized strips and served with a dip

cuisine style of cooking

deficiency lack or absence of a particular nutrient

degustation menu appreciative tasting of various foods and focusing on the senses

dehydration lack of water in the body

developing nation country with low levels of production of goods and services and limited technology

diverticulitis inflammation of small, abnormal pouches in the wall of the digestive tract

duty of care legal obligation imposed on an individual requiring adherence to a standard of reasonable care while performing any acts that could foreseeably harm others

emaciated very lean or underweight through lack of food

employee a person working for another person or business for pay

employer a person or business that employs one or more people, especially for wages or salary

energy fuel we need from food to function and be active; energy requirements vary depending on your age, body size and physical activity

enhance improve or build on

enrich add nutrients to improve nutritional value

enterprise a company organised for commercial purposes

evaluate to determine the value of something

export send goods or produce out of the country

fast eat very little or nothing

fermentation production of acid or alcohol by micro-organisms that changes the texture, flavour or aroma of food

finger food small appetiser or sweet item that you can pick up and eat using your fingers

finite having bounds or limits; many natural resources are available in fixed amounts and once used cannot be replenished

food allergy abnormal, unpleasant or adverse reaction to food by the body's immune system

food habit pattern of eating; the way in which food is prepared, served and eaten

food intolerance abnormal, unpleasant or adverse reaction to food caused when the nerve endings in different parts of the body become irritated

herbs leafy plants, valued for their flavour and scent

hormone chemical substance made by the body to perform a specific function; for example, to stimulate growth and development

hypertension very high blood pressure, often linked to a high salt intake

immunity ability to fight or resist disease or infection

indigenous food (bush tucker) food native to Australia that was present before European colonisation

irritable bowel syndrome pain and bloating of the abdomen

julienne narrow strips that are approximately 3 mm wide
kilojoules standard unit of energy measurement
lactation production of milk for breastfeeding
leavened bread bread that has risen because of the addition of a leavening agent such as yeast
legislation law passed by a government
life cycle stages through which people pass that begin when a female egg is fertilised and end at the time of death
logistics planning and carrying out of any complex or large-scale activity
malnutrition lack of proper nutrition, caused by not having enough to eat or not eating enough of the right things
manufacturer business involved in the making, processing or production of goods
media forms of communication – radio, television, newspapers, magazines, social media, world wide web – that reach large numbers of people
micro-organism microscopic organism such as bacteria
modified changed or altered
morbidity illness
mortality death
multiculturalism existence of different cultures within the one society
neural tube defect condition that occurs when the central nervous system of the foetus is not correctly developed; commonly referred to as spina bifida
osteoporosis condition of weak and porous bones due to a lack of calcium in the diet
perishable having a short shelf life; spoiling quickly and needing careful storage
photosynthesis process whereby leaves of a plant use the energy of the sun to change carbon dioxide and oxygen into carbohydrates, which are then stored in the plant
policy management plan or strategy devised by a government or organisation
portion to measure or divide foods into predetermined quantities
portion size specific size, shape and weight of food to be served
processed foods foods that have undergone change from their original raw state by a manufacturing process
rancid unpleasant or stale smell or taste, usually affecting fats and oils
ration fixed allowance of food
refugee person who flees for safety to another country
sanitise make clean and free from germs and dirt
satiety state of being fed or satisfied
scurvy disease caused by a lack of Vitamin C
sedentary involving little exercise, a lot of sitting
self-sufficient ability to support oneself; communities that are self-sufficient need very little outside assistance to produce food
sensory relating to the senses or sensation
spices aromatic seasonings obtained from the bark, buds, fruit, roots, seeds or stems of various plants
spore organism that produces one or more organisms, similar to a seed
staple food basic food that a population eats on a regular basis and in large quantities
stock control determination of the supply of goods kept by a business
stunted restricted growth and development of the body
subsidise provide financial aid
subsistence farmer small farmer providing only enough produce to meet the family's food needs
sustenance nourishment in the form of food
tamper meddle with for the purpose of altering
tariff tax on imports so that locally produced goods become less expensive by comparison
toxin poisons generated by micro-organisms that can cause illness
tumour abnormal swelling or lump in the body
vegan people who consume no animal products of any kind, including dairy and eggs
wean introduce semi-solid foods into the diet of the infant and reduce milk intake

CHAPTER 1

Mark Fergus

Just give me the facts

FOCUS AREA

CORE: FOOD PREPARATION AND PROCESSING

The study of food is a fascinating exploration of science, industry, history, law and eating! This chapter provides information about safe practices, basic knowledge of the Food Technology room and equipment, and a quick overview of the processing of food in industry.

CHAPTER OUTCOMES

In this chapter you will learn about:

- food safety and hygiene
- causes of food spoilage and the principles of food preservation
- reasons for cooking food
- properties of food
- ingredients in food preparation and the equipment used to prepare food
- the role of technology in food preparation and its social effects
- the effects of food preparation and processing
- levels of industrial food preparation
- food presentation and service
- food packaging.

In this chapter you will learn to:

- demonstrate safe and hygienic work practices
- identify ingredients that are a high spoilage risk and investigate methods to minimise this
- apply the principles of food preservation in food preparation and storage
- appreciate the sensory properties of food
- prepare foods that demonstrate their functional properties
- create food items for different cuisines using appliances and technology
- prepare and cook foods to prevent nutrient loss
- explain the role of food additives in processed food
- discuss the impact of food processing on the environment, the economy, society and health
- experiment with food presentation and garnishing
- explore the functions and diversity of packaging.

RECIPES

FOOD FACTS

Milk, soft cheese, cream, chicken, red meat, fish, ice-cream, cooked pasta, cooked rice, coconut, pastries and gelatine have been identified as most likely to cause food poisoning.

Hot foods usually have a stronger aroma than cold foods. This is due to the heating of the chemical compounds in the food during the heating or cooking process.

Flavour is what people commonly call the 'taste' of food. It is actually a combination of smell, taste, spiciness, temperature and texture. Much of the flavour of food comes from smell, so when you are unable to smell you have lost much of your ability to experience flavour.

Why are so many culinary terms French? Historically, French chefs were very skilled. When the French Revolution started and everyone was losing their heads, the chefs to the royal families decided that now was a good time to travel to distant places. The chefs took their skills and culinary terms all over the world. French became the standard language of cooking.

Frozen vegetables first became available in Australia following the Second World War. They were some of the first mass-produced processed foods in Australia.

FOOD WORDS

à la carte meaning 'from the card' – a menu that is written with individual prices for dishes

biodegradable able to decompose because of the activity of living organisms such as bacteria

blanch plunge into boiling water for a short time, then into cold water

crouton small cube of bread crisply fried or toasted

fermentation production of acid or alcohol by micro-organisms that changes the texture, flavour or aroma of food

rancid unpleasant or stale smell or taste, usually affecting fats and oils

sanitise make clean and free from germs and dirt

spore organism that produces one or more organisms, similar to a seed

toxin poisons generated by micro-organisms that can cause illness

1.1 Food safety and hygiene practices

Safety in the kitchen is everyone's responsibility. You are responsible for the cleanliness of your hands, apron and workspace, and for your work practices. You are also expected to behave in a manner that will not harm others.

Personal hygiene

Imagine you are in a restaurant and about to eat a delicious meal. Would you eat the meal if the chef had not washed her hands after visiting the bathroom? What about if she blew her nose, then picked up a stray chip to put on your plate before it was served to you? As a customer, you would not know any of these things had happened, and the germs that may be passed on to your meal could make you *very* sick indeed.

Food hygiene

There are three principles of food hygiene:

- prevent contamination of foods
- kill the bacteria by cooking properly
- prevent the growth of food-poisoning bacteria.

Foods will be safe from contamination if correct personal hygiene is maintained, but this is not enough to keep food safe. Cross-contamination occurs when raw foods come into direct or indirect contact with cooked foods. Raw foods should be treated as if they contain food-poisoning bacteria, and handled carefully. Separate equipment should be used for raw and cooked foods; for example, chopping boards and knives used for raw meat should be washed before being used for cooked meat. Raw foods should be stored separately from cooked foods. Food should be properly covered. In the refrigerator, cooked food should be stored on higher shelves and raw food on lower shelves. Washing up must be thorough. The water in the sink should be as hot as you can handle and detergent should be used.

Remove food scraps first and wash in a logical order (from cleanest to dirtiest) glassware, cutlery, crockery then cookware. Remember to wipe down all work areas and cook tops as well.

Most food-poisoning bacteria will be killed by cooking – provided the cooking temperature is high enough. The cooking temperature depends on the type of food and the cooking method. The crucial factor is the internal or inside temperature of the food. For example, rare roast beef should reach an internal temperature of 60°C, while chicken and hamburger patties should have a minimum internal temperature of 75°C. The temperature in these cases is measured using a meat thermometer.

The growth of food-poisoning bacteria is greatest in the temperature danger zone. The danger zone is between 5°C and 60°C. All raw and cooked food must be stored outside the danger zone. This means the refrigerator must keep foods cooler than 5°C and the oven, microwave or bain-marie should hold warm foods at above 60°C.

iStock.com/purple_queue

Figure 1.1 What good personal and food hygiene practices are demonstrated here?

Shutterstock.com/Vasina Natalia

Figure 1.2 Washing up in a logical order from 'cleanest' to 'dirtiest' will save time and water!

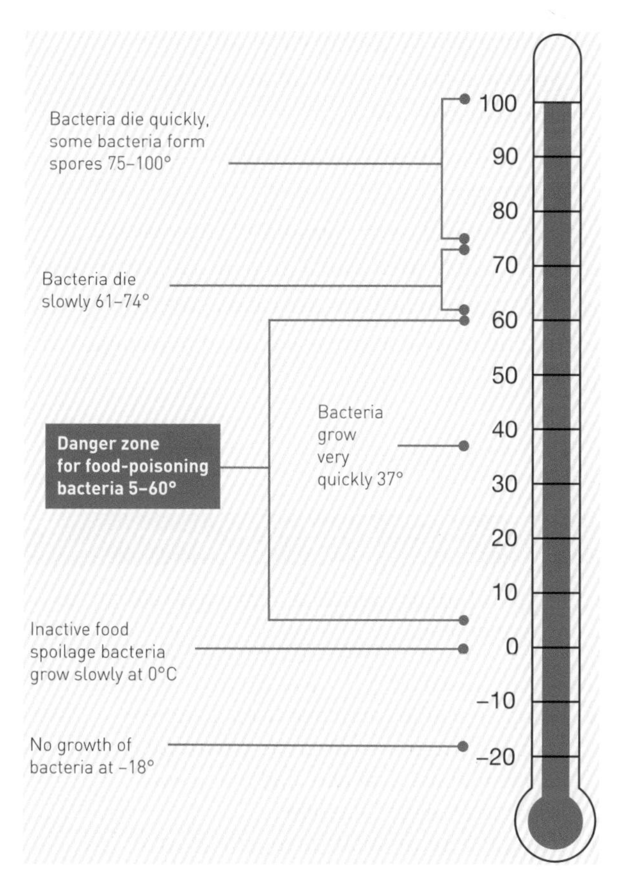

Figure 1.3 Temperatures of foods are measured on the inside of cooked food and the outside of raw foods.

Frozen foods should be thawed in the refrigerator (large food items like frozen chickens may take up to two days) or the microwave. Leftovers should be reheated quickly to a temperature of 75°C to ensure bacteria do not have time to grow when they are passing through the danger zone. This is done easily in the microwave.

Safe work practices

Generally, common sense and good judgement should determine how you behave in the classroom, but there are some special rules for the Food Technology room. These are important rules because this classroom can be dangerous.

- Beware of utensils with sharp edges. Walk with knives pointing downwards, your arm by your side and the blade facing away from you. Do not place knives into a sink of soapy water as they can be hidden from view and may injure the person washing up.
- Work efficiently. Clean as you go, close cupboard doors and drawers as you use them, mop up spills immediately so there are no slips or trips, and never run.
- Think safety. Watch out for pot handles pointing towards the edge of cook tops, knives on the edge of benches and other potentially dangerous situations.

Figure 1.4 What features of this kitchen present hazards?

- Be careful when working with steam, gas, hot oil and electricity. Do not use appliances with damaged cords. Be particularly careful with electrical appliances around water.
- Finally, follow all instructions. If in doubt, ask questions rather than guess.

Food safety legislation

Australia has a set of national Food Safety Standards designed to reduce the incidence of food-borne illness. The standards outline requirements for all food businesses including:

- training staff in food hygiene and safety
- providing facilities to store food and protect it from contamination
- providing handwashing facilities on the premises.

The standards were introduced after statistics revealed people were eating more meals not prepared at home (restaurant, take-away and convenience foods) and that the incidence of food poisoning was increasing.

UNIT REVIEW

LOOKING BACK

1 List and explain the three principles of food hygiene.
2 Explain the link between the temperature of high-risk foods, such as chicken, and food safety.
3 Suggest why staff training in food handling is required for employees in food businesses.

FOR YOU TO DO

4 Write a short article for your school newsletter advising students and parents about the temperature danger zone, making it relevant to school lunches.

TAKING IT FURTHER

5 Using an online catalogue from a kitchen supplies outlet, research the different types of knives that are available and the uses for which they are intended.
6 Using the South Australian Safe Work website, access the 'Virtual Hotel' and try the safety test!

South Australia Safe Work

1.2 Causes of food deterioration and food spoilage

Microbial activity

Micro-organisms occur naturally in the environment. While some can spoil food, others are useful in food production; for example, the manufacture of beer, cheese, yoghurt and wine.

Micro-organisms can be:

- bacteria
- yeasts
- moulds.

Bacteria

Bacteria are microscopic organisms that cause food poisoning as a result of being consumed live in food, or consuming the **toxins** that the bacteria produce once in your stomach.

Shutterstock.com/Rohatynchuk Mykola

Figure 1.5 Could this display be a food safety danger? What factors could make these foods unsafe?

The optimum conditions for bacteria are:

- food supply
- a low-acid environment
- time to grow
- warm temperature (between 5°C and 60°C: the danger zone)
- oxygen (although some of the more dangerous bacteria do not need oxygen to survive)
- a moist environment.

Under ideal conditions, bacteria double in number every 20 minutes.

Table 1.1 Bacteria most commonly associated with food poisoning

NAME	HOW FOOD IS CONTAMINATED	SOURCE OF CONTAMINATION	FOODS INVOLVED	SYMPTOMS
Salmonella	*Salmonella* causes an infection when food is eaten that contains live bacteria. It is the most common cause of food poisoning in Australia.	Found in the gut of animals, particularly chicken. Cross-contamination is frequently the cause. Flies, mice, birds and pets are other sources of contamination.	Raw meat, poultry and sausages. It is found in food outlets where cross-contamination has occurred, or where foods have not been properly thawed.	Vomiting, diarrhoea, listlessness within 6–72 hours. It can be fatal in the elderly, very young and chronically ill.
Clostridium perfringens	This bacteria causes an infection when food is eaten that contains live bacteria. The bacteria also release a toxin into the intestine once consumed.	Raw meat, vegetables coated in soil or dirt, unhygienic food handling.	Commonly associated with cold and reheated meat, especially casseroles or other foods cooked in bulk, such as large roasts, stew leftovers.	Onset occurs usually within 8–24 hours of consuming contaminated food, causing nausea, diarrhoea, abdominal cramping. Symptoms last for 12–24 hours.
Staphylococcus	*Staphylococcus* is a toxin. Food poisoning occurs when the food that contains this toxin is eaten. Cooking does not destroy the toxin.	50% of the population carry this bacteria in nasal passages, mouths and infected skin, wounds and pimples. Coughing, sneezing and contact via hands spreads the bacteria.	Foods eaten cold such as custards, cold meats, pre-prepared buffet dishes.	Vomiting, diarrhoea, abdominal cramps within 1–8 hours of eating contaminated food. Symptoms can last for 1–2 days.
***Escherichia coli* (known as *E. coli*)**	*E. coli* causes an infection when food is eaten that contains live bacteria. Its presence indicates poor food-handling practices.	*E. coli* lives in the bowels of humans and animals. It is transferred by poor hygiene in the workplace. Many raw foods contain *E. coli* so cross-contamination through infected food preparation surfaces and equipment is common.	Raw and processed meat and poultry.	Abdominal upset with severe diarrhoea. It can be fatal in babies and young children.
Listeria	*Listeria* is very resistant to extremes in temperature and can both grow in refrigerated environments and survive high temperatures of pasteurisation.	Main sources are dust, water, fish, shellfish, birds and insects.	Milk products and cheese. Soft cheeses (such as brie and camembert) are likely sources, as are manufactured meats (such as ham, corned beef and salami) and soft-serve ice-cream.	Nausea, vomiting and diarrhoea within 24 hours. It has been linked to stillbirth and miscarriage.

1.2 Causes of food deterioration and food spoilage

Yeasts

Yeasts are single-celled plant organisms that can produce slime on fruit juices and vinegar products and can cause other foods to ferment and thus spoil. The **fermentation** process produces bubbles on the surface of the liquid and gives a strong yeasty odour and taste. Foods affected by yeasts can give a tingly sensation to the tongue and a slightly acidic flavour. While these are not pleasant, they are not harmful.

Moulds

Mould is different from the other micro-organisms in that it can be seen by the naked eye. Moulds are a form of fungi and reproduce by forming **spores** on the surface of foods such as bread, fruit and cheese.

They appear as a dark cottonwool-like mass on the surface of food. The spores are carried in the air. Moulds enjoy a warm, moist and nutrient-rich environment where they grow rapidly. Some moulds found in grains and nuts produce dangerous toxins if consumed. Heating to 60°C for 10 minutes can destroy them. The appearance of mould makes the food undesirable, and while they do not cause as much illness as bacteria, the presence of mould can damage a food outlet's reputation. Moulds are used in food processing in the production of 'blue-veined' cheeses such as Stilton and the velvety rind of brie cheese.

Enzymatic changes

Enzymes are proteins that help to speed up reactions. Some enzymes found naturally in food can cause food spoilage. These natural chemicals cause foods to ripen and age. This action does not 'switch itself off' and continues past the optimum ripening stage, such as when the enzymes continue to convert starch in fruits to sugar, causing increasing softening in overripe fruit. Heating foods above 60°C or storing below 4°C slows the action of enzymes considerably.

HANDS-ON

FOODS PRODUCED WITH BACTERIA, YEASTS AND MOULDS

PURPOSE

The aim of this activity is to taste the positive effects that micro-organisms play in the production of some foods.

MATERIALS

Toothpicks
Small ramekins
Small cups
Food samples such as blue-veined cheese, soft cheese such as brie or camembert, AB (bacteria-enhanced) yoghurt, fermented milk drinks such as Yakult

STEPS

1 Your teacher will distribute samples of each product for you to try.
2 Note your reactions to the foods.

ACTIVITIES

1 List the foods you tasted that you had not tried before.
2 Explain the difference you think the addition of the bacteria, yeast or mould made to the taste, texture or smell of the product.
3 Examine the package of each product. How does the manufacturer explain the presence of the bacteria, yeast or mould?

Physical and chemical reactions

Storing food correctly is important. Correct storage reduces the opportunity for foods to be exposed to conditions that will affect their physical and chemical properties.

Non-perishable foods can be stored in the pantry for long periods of time. The pantry should be clean and well-ventilated and at room temperature (between 10°C and 20°C). It should be dry (free from moisture) to prevent mould from developing. Dry goods should be stored in airtight containers to stop contamination by weevils, insects and rodents.

Canned foods can be stored in the pantry for many months, but the quality of the goods purchased is important. Dented or swollen cans should be avoided, as they may be misshapen by gas produced due to microbial activity or chemical action.

Cold storage includes refrigerated and frozen storage areas. Refrigerators should be operating at or below 5°C, while freezers usually keep foods below −18°C.

All foods in the refrigerator should be well covered and organised so as to avoid cross-contamination. Poultry and red meats should be stored and wrapped separately and cooked foods stored on higher shelves. Foods should never be placed directly on refrigerator shelves.

Frozen foods can suffer 'freezer burn' if not packaged correctly. Removing as much air as possible from the package before freezing and sealing the food well prevents the build-up of ice crystals, which causes the drying effect of freezer burn. Many foods may be successfully frozen but, once reheated, must be eaten or the remainder thrown away. Thawed and reheated food should never be refrozen.

Environmental factors

Food can be contaminated as a result of environmental factors. These may be as obvious as dirt and dust or as a result of less obvious situations, such as:

- insect spray used in the kitchen while food was uncovered
- foods being exposed to the air
- damaged packaging
- a food handler using a gloved hand for both serving food and handling money
- rough handling
- vegetables or flowers used as a table decoration coming into contact with food
- display labels on a buffet coming into contact with food
- food stored in the temperature danger zone
- wait staff accidentally touching food on the plate while serving meals
- food placed on crockery that has been inadequately cleaned or **sanitised**.

The production and serving of food is a great responsibility and not one to be taken lightly. In Food Technology you must consider safety and hygiene factors very carefully and always think before you act.

UNIT REVIEW

LOOKING BACK

1 Which is more dangerous: bacteria, yeasts or moulds? Defend your answer.
2 Explain the significance of the temperature danger zone to bacteria.
3 List the optimum conditions for the growth of bacteria.
4 List foods where bacteria, yeasts and moulds are used in the production of the product.
5 Explain the meaning of the term 'enzyme' and give a common example of an enzymatic reaction.
6 State the optimum storage conditions for pantry, refrigerator and freezer storage.
7 List five everyday actions that could cause food contamination in your home.

FOR YOU TO DO

8 Analyse the table of bacteria on page 7. Write a newspaper article describing the type of bacteria you consider to be the greatest risk to public health – don't forget to explain why!
9 Design an experiment to demonstrate the action of enzymes and how their action may be slowed.
10 Store a slice of bread loosely wrapped in plastic in the freezer for one week. Describe the effects of freezer burn.

TAKING IT FURTHER

11 As a class, start a blog compiling news articles related to food safety and food poisoning. Comment on the issues related to each article.

1.3 Principles of food preservation and storage

For centuries, food has been preserved so that it can be stored for long periods of time at an acceptable quality. Food is preserved to:

- prevent wastage; for example, making marmalade when there are lots of oranges and lemons available
- store food against a later shortage or to improve choice
- maintain the nutritive value of the food and make food preparation easier and quicker.

There are many methods of food preservation. Each method involves controlling a number of factors.

Moisture levels

The rate of growth of micro-organisms depends on the amount of water available. Dehydration or drying is an old method of preservation. It means removing up to 90 per cent of the moisture because bacteria will not survive in these conditions. Foods that have been dried are not as heavy and, when kept in an airtight container, will have a long shelf life. Drying is a relatively simple procedure and does not necessarily need special equipment. Traditionally, drying was carried out in the open air in the sun, but it can be successfully done in a microwave or home oven on a very low temperature. Fruits and vegetables can be successfully dehydrated, and dehydrated beef commercially known as 'beef jerky' is also available.

Salt draws moisture from the cells of food. Salting is commonly done using brine (a salty water solution). This is the method used to create meats such as corned beef. The salt draws moisture from the meat and is dissolved in it. The meat then reabsorbs the salty juices, which inhibits bacterial growth. A mild brine will keep meat for several weeks. Salting was the method used to preserve meat for early settlers travelling to Australia. Many vegetables are also suitable for salting and then bottling.

Sugar is the key ingredient in preserving fruits in the form of jams, jellies, marmalades and fruit butters. Sugar has a dehydrating effect similar to that of salt. In strong concentrations it increases the food's natural sugar levels so that micro-organisms cannot grow.

Shutterstock.com/Jerome Stubbs

Figure 1.6 Preserved meat like this sells on the streets of many cities and villages across China.

It is important that the amount of sugar is just right to ensure the preservation qualities. A general rule of thumb is that equal parts of sugar and fruit (by weight) are required. The keeping qualities of the product are improved if it is refrigerated after opening.

Addition of chemicals

Traditionally, the main chemicals used to preserve foods were salt and sugar. Chemical preservatives have been developed to replace these but the principles are the same. Used in commercial food preservation and processing, they can be identified by their product labels. Food additives are identified by code numbers.

Preservatives are numbered in the 200s; for example, sorbic acid (200) is used to preserve cheese because it prevents the growth of yeasts and moulds.

Potassium nitrate (252) or sodium nitrate (251) may be added to a brine (salt) solution to enhance the colour of salted meat, as a salt solution on its own will turn the meat grey. These chemicals also help to prevent bacterial growth.

Antioxidants are in the 300s. Ascorbic acid (300) stops browning in unprocessed fruits, while butylated hydroxytoluene (321) prevents the flavour from deteriorating in fats and oils due to the action of enzymes.

Smoking food, one of the oldest methods of preservation, is used to extend the shelf life of meat and fish, eggs, nuts and cheese. While the smoking process has a drying effect, the chemicals in the wood smoke also act as a preservative. Some chemicals are poisonous to the bacteria, while other compounds stop meat from oxidising and going bad. In domestic smokers, the food is removed from a brine solution and is placed on a rack above a burner. Different woods, brine flavourings and herbs create different flavours in the preserved food. Smoke contains carcinogenic (cancer-causing) substances, so smoked foods should not be eaten in large amounts.

Temperature

Freezing prevents the growth of some bacteria and moulds, so the rate of food spoilage is reduced. Freezing does not prevent enzyme action, so foods such as vegetables must be **blanched** before freezing to deactivate the spoilage enzymes.

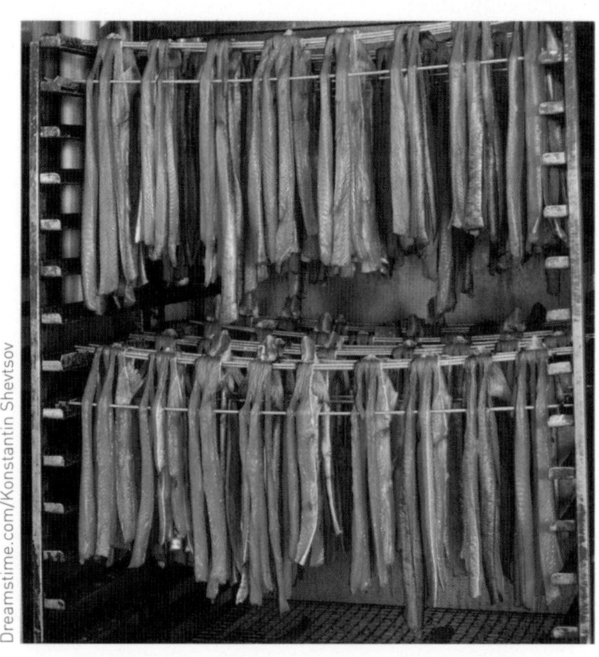

Dreamstime.com/Konstantin Shevtsov

Figure 1.7 This shows commercial production of smoked salmon. Similar results can be achieved at home on a barbecue.

A large variety of foods including fruits, vegetables, breads, meats, fish and cooked meals are suitable for freezing.

Freezing stops all the chemical activity in the cells of the food. Water crystals are formed during the freezing process and rupture some cells in the food. This explains why foods with a very high water content such as strawberries, lettuce, celery and watermelon will be damaged severely during freezing, and collapse on thawing.

To keep the quality of frozen foods high when thawed, the ice crystals must be small. This is achieved by freezing the product very quickly.

Bottling preserves food by heating it in a container, usually a glass bottle or jar. The heat kills the micro-organisms and then creates a vacuum to store the food in an airtight environment. Bottling is frequently used to preserve fruit and some vegetables. The heating process sterilises the food and kills the yeasts and moulds already present. The addition of acid to some fruits helps control the growth of micro-organisms.

The process usually involves heating the bottled food (which may or may not be already cooked) in a water bath to between 74°C and 100°C. The time for heating depends on the food, size of the bottle and the acidity of the food. Firm fruits are best preserved in pieces, whereas soft fruits are best preserved as syrups or purees.

Shutterstock.com/monticello

Figure 1.8 Have you ever wondered how fruit and vegetables were preserved? Bottled fruits and vegetables such as this feature in the district exhibits of the Sydney Royal Easter Show.

pH level

Vinegar is a product of the fermentation process. Acetic acid is produced when the alcohol from the fermentation process is converted. This process gives various vinegars their distinctive flavour. The addition of vinegar to preserve foods is commonly known as pickling. The high level of acidity prevents the growth of micro-organisms. The natural liquids in the foods are replaced by the vinegar, which also gives the food flavour. The use of vinegar to preserve may take three forms:

- packing raw food in jars then adding vinegar
- placing food in brine, draining the brine, then immersing the food in vinegar to remove excess moisture, keep vegetables crisp and prevent the formation of bacteria
- cooking the food in a solution of vinegar and flavourings, commonly known as chutneys or relishes.

Oxygen

Foods can also be spoiled by the absorption of oxygen. The effect of removing or eliminating oxygen is often used in conjunction with other preservation methods such as bottling and freezing. Expelling most of the air from a plastic bag of sausages before freezing prevents the trapped cold air from turning the fat in the sausages **rancid**.

One method of preservation involves removing oxygen only without changing the nature of the food and is known as vacuum packaging or cryovac packaging. This method is usually done in conjunction with refrigeration, and is commonly used for red meats, smallgoods such as ham and other cured meats, and cheese.

UNIT REVIEW

LOOKING BACK

1 Explain the principles of food preservation used in the production of sun-dried tomatoes, sultanas, marmalade and smoked ham.
2 What method of preservation is used to make corned meat?
3 Why should smoked foods be eaten in moderation?
4 What is cryovac packaging? How does it preserve food?
5 Use the correct terms to explain what will happen to watermelon if it is frozen.

FOR YOU TO DO

6 Design an experiment to find out the most appropriate method of making homemade sun-dried tomatoes, comparing natural methods (the heat from the sun) and technology (the microwave). Be sure to consider hygiene factors, time taken and the quality of the end product.
7 Explain how bottling effectively preserves fruit and vegetables.

TAKING IT FURTHER

8 As a class, create a set of recipe cards for preserved fruit and vegetable recipes. Each person should find their own recipe, test it and then write the title, ingredient list and method on an A5 card, adding a photo of their finished product. Laminate the card for display in your classroom, or swap and test others' recipes!
9 Investigate the additive codes listed on the Food Standards website. Identify the additives contained in a packet of flavoured chips or coloured lollies.

Food Standards Australia

1.4 Sensory properties of food

Before using food in any food preparation exercise, the sensory properties must be considered. These include colour, texture, odour and taste. Sound is also often considered in the sensory analysis of food, which is covered in more detail on page 167.

Colour

What do you see when you imagine a crusty bread roll or a juicy tomato? Chances are your image of this food will have a lot to do with its colour. The colour of food is a useful guide to judge freshness and ripeness, how well cooked it is or whether it has been spoiled or damaged.

Similarly, colour enhances enjoyment of food. Careful use of colour combinations can make a meal more visually inviting.

Sometimes colour can confuse our senses when combined with taste. Perhaps you have eaten an iced doughnut decorated in club colours to show support for a football team in the grand final, or tried a flavoured clear bottled water. These are not the flavours normally associated with these colours!

Texture

Texture is known as 'mouthfeel'. It can be visually detected by the physical structure of the food, such as crisp iceberg lettuce, or by the surface appearance of the food, such as velvety smooth chocolate mousse.

Variety in texture is important in a meal as it adds interest. Examples include crusty bread served with a creamy soup, or pappadums as a side dish for an Indian curry.

Odour

What is your favourite food smell? Is it bread baking? The coconut cream smell of a Thai restaurant? Sausages and steak on a barbecue? The smell of freshly cut pineapple? Aroma or odour is a very strong sensation that draws people's attention to food and contributes greatly to their eating enjoyment. Sense of smell is also very useful in judging whether a food is 'off'.

Taste and smell are closely linked – try holding your nose and tasting different foods. You've probably experienced how difficult it is to enjoy a tasty meal when you have a cold because your sense of smell is affected.

Taste

Do you have a sweet tooth? The tongue detects taste. The taste buds on the tongue are stimulated by different flavours which in turn register on different areas of the tongue. The specific taste sensations are sweet, sour, bitter, salty and umami (a Japanese word meaning 'pleasant savouriness').

Flavour

Flavour is said to be the combination of taste and smell, texture and appearance. Flavour gives the total impression of the dish. The flavour of a dish is usually tested before serving to determine whether any flavours are too dominant or the food too bland. A meal that is balanced in flavour will usually contain a variety of taste sensations, leading from mild to stronger, with stronger flavours paired with a blander accompaniment, such as a curry served with rice and ending with a sweet dish.

Shutterstock.com/Mat Hayward

Figure 1.9 How does colour affect our anticipation of the flavour of food?

FOOD IN FOCUS

TAKING THE TASTE TEST

The MyScience program at CSIRO is a program developed for school students, helping them develop skills in scientific experimentation.

One group developed an experiment to determine the influence of plate colour on food selection. Professional researchers have looked at how **food colour**, as well as **plate colour and shape,** influences the perception of taste. Even the type of **cutlery** can make a difference!

The students laid out three plates – one green, one red and one blue – each with a pile of coloured lollies on it. They then invited volunteers to approach the table and select a lolly. Most volunteers chose lollies from the red plate.

The second group investigated the question – which is more sour? A sour lolly or a sour fruit? The MyScience mentor demonstrated that sour lollies such as Warheads are coated in malic acid, causing the intense sour taste until the acid is dissolved in saliva. The demonstration involves dropping the Warheads into a weak solution of bicarb soda and water. The solution will fizz immediately then die down. If you remove the Warhead and drop it into a second bicarb solution, nothing will happen. The malic acid has been dissolved, and the sour taste has gone.

ACTIVITIES

1 Why do you think the red plate was the most popular for the selection of lollies?
2 What other factors may have influenced the plate choice of the volunteers?
3 Design an experiment to test the influence of food colour, plate shape or any other factor you can think of. Carry out your taste-testing experiment. Did the volunteer selections match your expectations?

UNIT REVIEW

LOOKING BACK

1 List the five senses and explain how foods can be made to appeal to each.
2 Using each sensory property as a list heading, provide 10 words that could be used to describe each.
3 Which two senses are closely linked? What can you do to test this?

FOR YOU TO DO

4 Using recipe books, design a three-course menu that incorporates all the sensory properties of food. Draw the plating or presentation of each. Label your diagrams to highlight features to note, such as colours, special accompaniments, crockery or glassware chosen.

TAKING IT FURTHER

5 Conduct an online survey of the class to determine the likes and dislikes of certain foods based on taste. Are sweets or savouries the popular choice?

Survey Monkey

1.5 Functional properties of carbohydrates, proteins and lipids

Video tutorial

The functional properties of food affect the physical and chemical changes that occur during food storage, preparation and presentation.

Carbohydrates

Carbohydrates are a component of food that is found in foods containing large amounts of sugar, fibre or starch. Cereal products, vegetables and fruit are the best sources of carbohydrates. Carbohydrates are involved in the following reactions.

Caramelisation is the browning process formed by the action of heat on sucrose (a simple carbohydrate). Foods containing sugar will brown when heated to about 160°C. A good example of this is the golden colour of toffees and the browning of gravy.

Dextrinisation is a slightly different type of browning, produced by the action of heat on starch. The best example of this is the golden appearance of bread and baked cakes.

Crystallisation occurs when sugar is dissolved in a liquid and heated. As heating continues, the liquid evaporates, leaving the sugar to clump together and starting the crystallisation process. The best examples of crystallisation are used in confectionery, such as boiled lollies and toffee apples.

Gelatinisation refers to the thickening and setting of food. Starch and protein are used to thicken food preparations. When starch is added to water and heated, the grains swell enormously and cause the product to thicken. Examples are the addition of cornflour to thicken gravy and using arrowroot in confectionery.

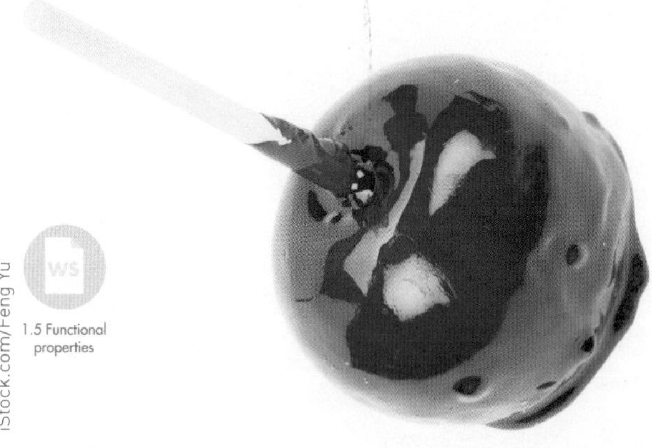

WS 1.5 Functional properties

iStock.com/Feng Yu

Figure 1.10 Toffee apples undergo caramelisation (making the toffee a golden colour) and crystallisation (causing the toffee to be brittle) in the cooking process.

Proteins

There are many different types of protein, and these allow foods to become more viscous or set. These properties have many applications in food preparation.

Two important processes are denaturation and coagulation. To denature means to change properties, and once a protein has denatured, it cannot return to its original form. An example of this is the application of heat to an egg. The second step of denaturation is coagulation, whereby protein thickens and changes into a semi-solid mass. Examples of denaturation and coagulation include scrambled eggs, baked custard and beaten egg white. The process of denaturation and coagulation is affected by the application of acid, alkalis, sugar and salt.

Syneresis occurs when the coagulation process continues due to heating. This is seen when the protein squeezes out the liquid in the product, such as curdling in custards, the separation of liquid from meat when it is cooked to well done, and the weeping of liquid from meringue on top of a lemon meringue pie.

Shutterstock.com/MaxkatelUSA

Figure 1.11 Creating meringue requires denaturation, turning runny egg white into a light, fluffy foam.

Lipids

Lipids, or fats and oils, have many properties that make them a useful ingredient in food and a useful component when frying food.

Emulsification is the dispersal of a fat or oil throughout a liquid. Oil and water do not mix, so emulsification requires the use of a third ingredient. If oil and water are beaten for a period of time, the oil is reduced to very small droplets that appear to be distributed in the water. However, after standing for a while the mixture will separate. An example of this is clear salad dressing such as French dressing. Egg yolk, salt, paprika and mustard are examples of emulsifying agents that work by wrapping around the small oil droplets, preventing them from joining back together and separating from the water on standing. Some common examples of food emulsions are milk, mayonnaise, gravy and cheese.

UNIT REVIEW

LOOKING BACK

1 Explain the functional properties demonstrated by toast, brown gravy, toffee, scrambled eggs and mayonnaise.
2 Use diagrams to explain the processes of emulsification and gelatinisation.
3 Describe the process of syneresis.

FOR YOU TO DO

4 In groups, design a series of posters to illustrate the functional properties of carbohydrates, proteins and lipids. Use recipes, magazine photographs and hand-drawn illustrations to enhance your presentation.

TAKING IT FURTHER

5 Conduct an experiment to demonstrate the denaturation of egg white. Determine the point at which the egg foam is at its maximum volume, and show what happens if the egg white continues to be beaten. Demonstrate your findings to the class.
6 Find a simple recipe for homemade mayonnaise. Test it out. Why does the oil or butter need to be added very gradually?

1.6 Basic ingredients used in food preparation

All recipes use basic ingredient groups in varying quantities. These ingredients are combined in different ways to create distinctive flavours and tastes that are unique to different cuisines.

Proteins are found in foods such as meats, milk, cheese, yoghurt, fish, eggs, soy beans, nuts, cereals and vegetables.

Carbohydrates are found in cereal products, vegetables and fruits. They can be in the form of sugar or starches.

Lipids include fats and oils, margarine, butter, lard and copha. Lipids are found in small amounts in cream, egg yolk, dairy foods, some fish and nuts.

Herbs and spices are used in cookery to add interest and excitement to meals. They include common herbs and spices such as basil, pepper, thyme, parsley, chives, cinnamon, ginger, garlic, vanilla and mint.

Australian cuisine traditions in the past reflected the British and European influences of early European settlement. Influences from Asia and recognition of indigenous heritage have contributed a rich variety of tastes, ingredients and flavours to the supermarket shelves and restaurant tables.

The difference between cuisines and flavours is the use of certain ingredients or cooking methods. For example:

- Asian cooking uses quick cooking methods such as stir-frying or steaming, herbs, spices and carbohydrate-rich noodles and rice
- Indian cooking is known for using aromatic spices
- Greek cooking utilises vegetables grown in the Mediterranean region
- Italian cuisine is famous for pasta and vegetable and meat sauces
- Japanese sushi is celebrated not only for its flavour but also its complex preparation process.

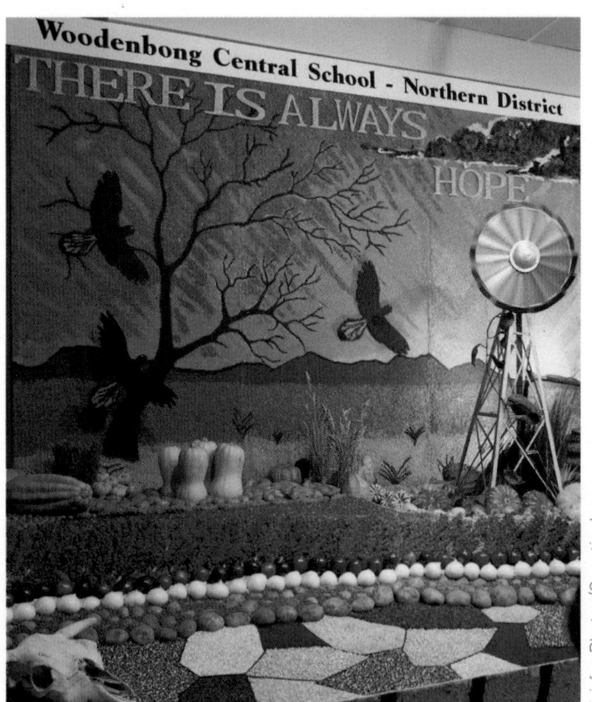

Fairfax Photos/Quentin Jones

Figure 1.12 Some areas of NSW are well known for their cuisine specialties and local produce. What is the specialty of your local area?

UNIT REVIEW

LOOKING BACK

1 Why do you think certain foods or flavours are linked with certain countries?
2 Explain the diversity of cuisines available in cities compared to country areas of Australia.
3 Give five food examples of protein foods, carbohydrate foods and lipids.

FOR YOU TO DO

4 Select and make a recipe using the ingredients familiar to a particular cultural cuisine. Write out the recipe procedure, and include a short section after the recipe showing evidence of research of the culture and how this recipe represents that culture.

TAKING IT FURTHER

5 Many people believe Asian cuisines are very similar. Choose two Asian countries and research the cuisines of these countries. Make sure you investigate the ingredients used, particularly herbs and spices, cooking methods and equipment, accompaniments and serving. Present your work as a two-page spread suitable for use in a food magazine.
6 As a group of four, prepare a mystery box of ingredients containing four ingredients typical for a dish from a cuisine of your choice. Swap your box with another group and plan a dish around the ingredients you receive.

1.7 Methods and equipment

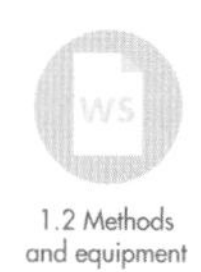

1.2 Methods and equipment

In Food Technology, just as in other specialist subjects, there are certain tools of the trade with which you should be familiar. These include utensils, cookware, small appliances and large appliances.

Utensils

Utensils are small hand-held tools designed to perform particular tasks. It is important that the right tool is selected for the job required.

- **Cutting, peeling and slicing:** knives (cook's, paring, filleting, cleaver or bread), zester, apple corer, vegetable peeler, bean slicer, poultry shears and biscuit cutters
- **Mixing:** spoons (wooden, slotted, metal or skimmer) and whisks used in bowls (ceramic, plastic or stainless steel) or saucepans
- **Sieving and straining:** colanders, sifters and salad spinners
- **Mashing and grating:** grater, potato masher, garlic crusher and mortar and pestle
- **Weighing and measuring:** measuring spoons, cups and jugs, thermometers (meat and candy) and scales.

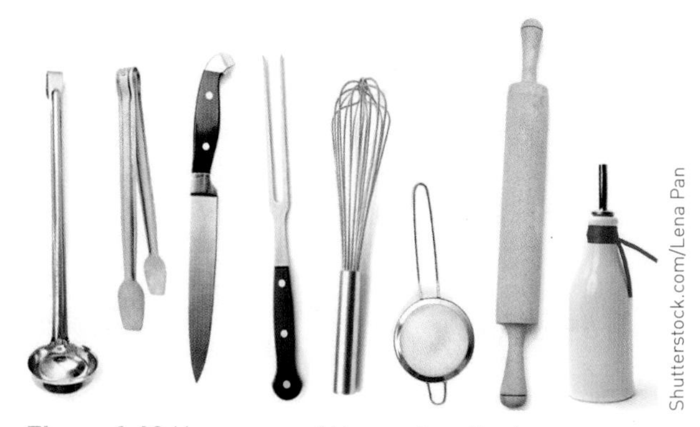

Figure 1.13 How many of these utensils do you recognise? How many have you used?

Cookware

Cookware is the equipment used for cooking or baking. Some types of cookware have specific uses, such as omelette pans, fish poachers and muffin tins. Cookware may be made from rigid metal, such as stainless steel or aluminium, or flexible latex. Cookware includes:

- saucepans, frypans, stockpots and woks
- bakeware such as cake tins of various shapes and sizes, roasting pans, casserole dishes and soufflé dishes.

Appliances

Small appliances are labour-saving devices and are generally electric, such as microwave ovens, toasters, electric beaters ('stick' blenders and fixed dual-beater mixers), food processors, deep fryers, blenders, sandwich toasters or presses, rice cookers and electric frypans, woks or skillets.

Large appliances are fixed in the Food Technology room and may be powered by electricity or gas. They include ovens and cooktops, refrigerators and freezers.

Advances in technology have meant that food preparation has become less labour-intensive. Electric bench-top mixers replaced wooden spoons in cake-making and microwave ovens have made reheating of food much faster. These advances have meant people need less skill to prepare meals and have more time to dedicate to other tasks.

Recipe terminology

Equipment and appliances are used in combination with methods of preparation and cookery to prepare food. Terminology used in the method section of a recipe indicates the action to take and the equipment that may be required. Some common terminology used to describe preparation of cookery methods include:

- **bake:** to cook food using dry heat from the oven
- **beat:** to mix an ingredient or batter vigorously to incorporate air and make it light and smooth
- **boil:** to cook in liquid to a temperature of at least 100°C
- **chop:** to cut food into very small pieces
- **dice:** to cut food into cube-shaped pieces
- **fold:** to combine a whisked ingredient carefully into another mixture
- **fry:** to cook food in hot oil by immersion (deep frying) or shallow frying on both sides
- **peel:** to remove the outer skin or shell
- **poach:** to cook in a liquid heated to 94°C or less
- **sauté:** to fry lightly in a little butter and/or oil
- **sear:** to brown the outside of meat quickly in a hot pan before baking in the oven
- **shred:** to cut food finely
- **sift:** to shake a dry ingredient through a sieve or sifter to remove lumps and incorporate air
- **steam:** to cook food using the steam of boiling water
- **stew:** to cook meat slowly at simmering point
- **whip:** to beat a mixture until it is light and fluffy.

UNIT REVIEW

LOOKING BACK

1 Give an example of a recipe and ingredients (where appropriate) that would require the use of the following utensils:
   - apple corer
   - balloon whisk
   - mortar and pestle
   - measuring cups
   - measuring jug.

2 Explain why the following appliances are described as labour-saving: toaster, microwave, blender and electric mixer.

3 Give examples of recipes that might include the following terms in the method: poach, sift, whip, dice and fold.

FOR YOU TO DO

4 Design a recipe card for Upside-down cake without using specific cooking terms. Include diagrams for each step and label the utensil or equipment used.

TAKING IT FURTHER

5 Devise a utensil and equipment list for the Upside-down cake recipe.

1.8 Physical and nutritive effects of preparation and processing

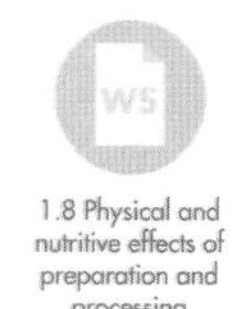

1.8 Physical and nutritive effects of preparation and processing

Some foods can be eaten raw. Others will have their flavour, texture and general appeal enhanced by the alteration of the food through processing or cooking. These processes alter the physical and nutritive properties of the food.

The shape of food can be altered in preparation through cutting, grating and shaping. The colour of foods can be enhanced by removing the skin, such as with kiwi fruit, and by blanching, such as with snow peas, which are briefly plunged into hot water and then placed in iced water to enhance the intense green colour. Flavours can be enhanced by the use of marinades and spices. The texture of food is changed by using utensils such as a meat mallet or blender, or by the process of cooking. Aroma is enhanced by various methods of cooking – just think of the smell of bread baking, compared with the bland smell of raw dough, or the aroma of onions being barbecued, compared with the smell of raw onions!

These processes can also alter the nutritive value of the food. Vitamins may be dissolved in either water or fat. Water-soluble vitamins such as Vitamin C are almost completely destroyed during cooking. Vitamin C is also affected by oxygen, so when vegetables are peeled or cut, the oxygen in the air contributes further to its destruction. B group vitamins are found in green leafy vegetables and are also water-soluble; cooking methods such as boiling will destroy them. Fibre is a type of carbohydrate that cannot be digested, but is essential to the good functioning of our gastro-intestinal tract. Fibre is commonly found in fruits and vegetables, and certain processes reduce the amount of fibre available. Peeling an apple or straining the pith from orange juice removes valuable fibre from the diet. Frying may add undesirable amounts of fat to a food as fat is absorbed during the cooking process. If the oil used to deep fry is not at a high enough heat, the absorption will be even higher.

iStock.com/alxpin

Pomegranate

Shutterstock.com/Anna Sedneva

Shutterstock.com/jiangdi

Shutterstock.com/Aedka Studio

Figure 1.14 Some foods, such as these exotic fruits, need little preparation to look interesting!

UNIT REVIEW

LOOKING BACK

1 Provide examples of the following foods where:
  a the flavour is enhanced by cooking
  b the appearance is improved by peeling
  c the shape is improved by cutting or slicing.

2 List three methods by which the texture of the food may be altered.

FOR YOU TO DO

3 The aroma of food can evoke many memories. Describe a memorable food experience based on the smell of food.

4 Fat-soluble vitamins will be found in foods that contain fat. Suggest cooking processes that may cause fat-soluble vitamins to be lost.

TAKING IT FURTHER

5 In a small saucepan, boil a piece each of pumpkin, potato and carrot until very soft. Drain the water into a glass and allow to cool. When cool, try drinking the vegetable water – it will have more water-soluble vitamins than the cooked vegetables! How does it taste? What does it look like? Use this activity to explain the nutritive value of making soup with vegetables high in water-soluble vitamins.

1.9 Industrial food processing

Most food goes through a series of treatments before it is ready to be eaten by a consumer.

Levels of processing

There are two levels of processing in industry.

Primary processing

Primary processing is the changing of raw foods after harvest or slaughter into ingredients that can be consumed individually, or further used in the production of manufactured food products. Primary processing results in very little change to the physical form of the food. Some examples of primary processing include milling of wheat into flour, packaging potatoes into 1.5 kilogram bags and refining rice to remove the outer husk.

Secondary processing

Secondary processing changes primary processed food into other food products. The physical appearance, texture and flavour of the food changes dramatically. Some examples include flour and other ingredients being made into noodles; potatoes peeled and cut before being fried as chips; peaches peeled, pitted and canned in syrup.

Shutterstock.com/FotograFFF

Figure 1.15 Primary processing requires little more than harvesting and preparing for sale.

iStock.com/Savas Keskiner

Figure 1.16 Biscuits demonstrate secondary processing. The more a product is changed from its original ingredients, the more highly processed it is.

FOOD IN FOCUS

HOW ARE TIM TAMS MADE?

Every day Grant Turner goes to work, he makes 1.5 million biscuits. It's why he has cranked up his 1 km production line that snakes around Sydney's Arnott's factory, so it churns out half a billion of the hallowed choccy biscuits every year.

And plenty of packs have been opened [in fifty years] – more than 45 million every year, enough to span Sydney to Ayers Rock and back again, if you laid every Tim Tam out in a row.

Or look at it this way – churning out 3000 Tim Tam biscuits every minute, the Arnott's Huntingwood factory could feed a biscuit to every person in a packed house at ANZ Stadium, every half-hour.

The factory that teases hungry motorists as they travel along the M5 is a wondrous riposte to those who have issued the death warrant to Australian manufacturing. Having completed an overhaul of its operations in 2007, which included the installation of state-of-the-art robotics, this is a manufacturing operation humming to a productive – and profitable – beat.

Grant Turner walks the length of Production Line 5, watching the Tim Tams' conception. Sugar, flour, colours and flavours begin the process in a gigantic hopper which mixes the ingredients for 20 minutes to create the cookie dough.

Then the conveyor belt steps up to warp speed. Biscuits cut 1 mm thick, 11 holes punched, baked for 90 m through six gas-fired ovens, cooled by freezing air, biscuits flipped and filled with cream, bathed in a pool of chocolate, another dose of freezing air, choirs of robotic arms picking and filling.

And suddenly a pack of Tim Tams is born. And a few million of its siblings.

Source: Adapted from '50 years of temptation – everything you ever wanted to know about Tim Tams' by Andrew Carswell, *The Sunday Telegraph*, Feb 18, 2014.

ACTIVITIES

1 Write a flow process diagram to explain the process of manufacturing Tim Tams.
2 What is the purpose of punching holes in the biscuits or cooling the biscuits before they are sandwiched with the cream filling?
3 Do some research to find a video of commercial biscuit production. What major differences do you notice?
4 Propose a new version of Tim Tams. Draw a flow process diagram to include the steps to make your new variety.

Additives

Food additives are ingredients that would not normally be eaten as a food by itself. Industrial food processing uses many additives for a variety of purposes:

- anti-caking agents are used to keep dry powdery products from clumping together, such as cake mixes
- artificial sweeteners make a product sweet without using sugar, such as diet soft drinks
- colours add or restore colours to foods, such as sauces, sweets and snack foods
- flavourings and flavour enhancers improve the taste or flavour of a food or restore flavour lost during processing, such as MSG (monosodium glutamate)
- humectants prevent food from drying out
- preservatives protect food against the action of micro-organisms
- thickeners and vegetable gums give a food product a uniform consistency and texture, such as some confectionery and savoury sauces.

Additives often have very long chemical names. Processed foods must be labelled with all ingredients. Additives are identified by their classification (colour, humectant, vegetable gum, etc.) and assigned a three-digit number. These coded numbers can be deciphered using the Food Additives List found on the Food Standards Australia and New Zealand website.

Food Standards Australia and New Zealand

Environmental, social and economic effects of industrial food preparation

Environmental effects

Environmental issues include litter creation, food wastage and the recycling of byproducts of the production process. Processed food generates a great deal of packaging, which in turn generates waste. Previously, much of this packaging was used as landfill. Now, recycling provides the opportunity to reuse paper, plastics and glass containers and packages. Food waste includes the leftover portions of meals and trimmings from restaurant food preparation, and from large-scale food preparation. In commercial food processing, the product left over from one process can often be turned into another. For example, apples that do not meet the criteria to be sold fresh may be processed into tins of pie apple, apple sauce or apple juice. Charities and organisations such as food banks receive donations of food that would otherwise be thrown out. Scraps of meat can be sent to pet food manufacturers or used in the production of fertiliser. Cooking oil and waste from grease traps can be collected and reprocessed into stock feed, fuel and soap.

Social effects

Industrial food preparation allows processes traditionally carried out in the home to be done in the manufacturing plant. Many families lead hectic working and social lives, so prepared or partially prepared foods suit this lifestyle well. Quick cooking and minimal preparation mean more time to spend on other things. The cost, however, is that prepared foods do not require much skill, so traditional food preparation skills may be lost. There is also less control over the consumption of additives when processed foods are eaten regularly.

Economic effects

Increasing employment in the manufacturing and hospitality sectors is the result of a booming food manufacturing industry. Processed foods are more expensive than their unprocessed counterparts, so the consumption of foods prepared outside the home is costing the householder more money.

UNIT REVIEW

LOOKING BACK

1 Explain how different methods of food preparation affect the physical characteristics of food.

2 Describe methods of cooking that maximise the nutrient value of the food. Use examples to support your answer.

3 List five roles additives play in food processing.

FOR YOU TO DO

4 Working in groups of four, design a production line process for the Savoury scone scrolls recipe on p. 31. Make sure each group member has a task.

TAKING IT FURTHER

5 Using the Food Additives List from the Food Standards Australia and New Zealand website, decipher the additive codes on a snack food packet.

Food Standards Australia and New Zealand

STEPHANIE ALEXANDER
RESTAURATEUR AND AUTHOR

Stephanie Alexander is a well-known restaurateur, and author of the very popular book *The Cook's Companion*. She was inspired by a Californian program called the Edible Garden which aimed to address the eating habits of young children. She developed the Stephanie Alexander Kitchen Garden Foundation, where children across Years 3 to 6 spend a minimum of 40 minutes a week in an extensive vegetable garden that they have helped design, build and maintain on the school grounds according to organic gardening principles. They also spend one and a half hours each week in a kitchen classroom, preparing and sharing a wonderful variety of meals created from their produce.

The aim of the Kitchen Garden Program is pleasurable food education for children. The Program develops lifelong skills in the kitchen and garden and encourages children to enjoy all the benefits of growing, harvesting, preparing and sharing.

Why have a Kitchen Garden Program?

Not all kids eat well. A disturbing number go to school each day without breakfast. Many others are overweight or obese. These symptoms of the busy world in which we live are likely to become habits of a lifetime for our children and lead to serious health issues in the future. Obesity is the public health issue of the not-too-distant future – diabetes, heart disease, strokes, joint problems, dental decay, chronic constipation, depression – to name a few consequences of an unhealthy diet and lack of exercise. The fruit and vegetable intakes of Australian children and adolescents fall well below recommendations and have continued to decline in the past 10 years. Their consumption of packaged, salty, sugary, fatty snack and convenience foods grows while their consumption of fresh, seasonally available food lessens. 'Convenience' food is too easy and is being made easier – with vending machines, 24-hour retailing and junk food outlets at every corner.

Figure 1.17 Stephanie emphasises the importance of growing and cooking with fresh ingredients.

In many cultures, eating together around a table is the centre of family life. It is the meeting place where thoughts are shared, ideas challenged, news is exchanged and where the participants leave the table restored in many ways. In the Kitchen Garden Program, equal importance is given to time around the table sharing the meal that has been prepared from produce grown in the garden.

Source: **www.kitchengardenfoundation.org.au**. Images: Top right: Getty Images/AFP/Giuseppe/Cacace; Figure 1.17 AAP Image/Supplied by DEC PR.

The Kitchen Garden Foundation

ACTIVITIES

1. Outline the features of the Kitchen Garden program you think would be appealing to primary school children.
2. List the reasons given for implementing a Kitchen Garden Program.
3. Suggest how the Kitchen Garden Program can address environmental, social and economic effects of food processing.
4. Why is equal importance given to sitting at a table sharing a meal made from the kitchen garden produce?

1.10 Presentation and service of food

The presentation of food is like a 'first impression' and is very important in food service.

Visual appeal

Food should be attractive to the eye and stimulate the palate. The senses should be considered in the presentation of food – sight, smell and taste.

Colours should be contrasting but also what is expected. For example, a creamy pasta on a white plate is boring; soup should not be blue. Colour highlights can be introduced with crockery, accompanying foods or garnishes.

Textures should be appropriate. For example, fried food should be golden and crisp; vegetables should not be soggy. A variety in texture in a menu is desirable.

Height in plating food has been a trend, with plenty of plate visible around the food. Some salads are piled into a section of wide PVC plumbing pipe on the plate and the pipe carefully removed to retain the height in presentation. Another trend is the plating of food on wooden boards or careful placement of foods on dark slate tiles.

Shutterstock.com/Matryoha

Figure 1.18 Elegant, simple presentation. What characteristics of the presentation of these canapés make it an attractive dish?

Garnishes

Garnishes are used to add interest and enhance the presentation of a dish. Some traditional garnishes can indicate an ingredient of a dish. For example, fish is often garnished with lemon. Some popular garnishes are:

- lemons: zest, wedges, slices
- tomatoes: cherry tomatoes whole or cut
- celery or shallot curls
- croutons
- fresh herbs such as basil, parsley, mint, coriander or dill
- fanned strawberries
- piped cream or flavoured butters, or sauces such as raspberry coulis
- sauces such as mayonnaise, aioli, tartare or sweet chilli
- olive oil

Styles of service

Many factors influence how food is served. These include the number of people being served, the type of establishment where the food is eaten, the type of menu, the cost or price, and the time available for the meal. Following is a description of some of the more common styles of food service.

A buffet is a type of self-service. Customers help themselves from a range of dishes from a table. This style of service is also known as smorgasbord.

À la carte means 'according to the card', which indicates dishes are ordered from a written menu or chalk board. The customer can choose the number and type of dishes and the meal may be cooked to order.

Silver service restaurants have customers sitting down at empty plates. The waiter serves all the food to each customer. The vegetables are served from a separate container and served using a fork and spoon in a manner similar to tongs. The meat may be carved at the table and placed on plates by the waiter. Each customer is asked individually what he or she would like, and this is placed attractively on the plate. The waiter is responsible for the presentation of the food on each plate.

Tapas is a popular style of service. It is Spanish in origin and often involves small pieces of food served on a platter to share at the table. Multiple platters are ordered to form a meal.

UNIT REVIEW

LOOKING BACK

1 Explain the role of garnishes.
2 Describe three styles of service.
3 Speculate on the role of restaurant staff for each style of service.

FOR YOU TO DO

4 Design a menu card for an à la carte restaurant. List three entrees, six mains and four desserts. Include prices and a short description of each dish.

TAKING IT FURTHER

5 Observe a demonstration of silver service dining. Practise laying a table for three courses and serve each other using silver service techniques.

HANDS-ON

USING GARNISHES TO ADD VALUE TO A FOOD PRODUCT

PURPOSE

Even the plainest chocolate cake can be something special when given deluxe treatment. This activity gives you the opportunity to be creative with garnishes.

MATERIALS

Paring knives
Piping bags and nozzles
Paper to cut stencils
Shakers for icing sugar and cocoa
Skewers
Slice of un-iced chocolate cake (from a round cake)
Selection of fruit such as strawberries, passionfruit and kiwi fruit
Cream
Chocolate
Raspberry topping
Cocoa
Icing sugar

STEPS

1 Place the slice of cake on a plate.
2 Garnish using fanned strawberries, runny cream and raspberry topping to create spider webs, hearts or wave effects, curls made from melted chocolate, cocoa or icing sugar dusted onto the plate or cake through a paper stencil.
3 Present to the class.

ACTIVITIES

1 Which slice of cake looked the best? Why?
2 Which effects were easy to create?
3 Explain why a piece of chocolate cake might sell in a cafe for more than $5 per slice.

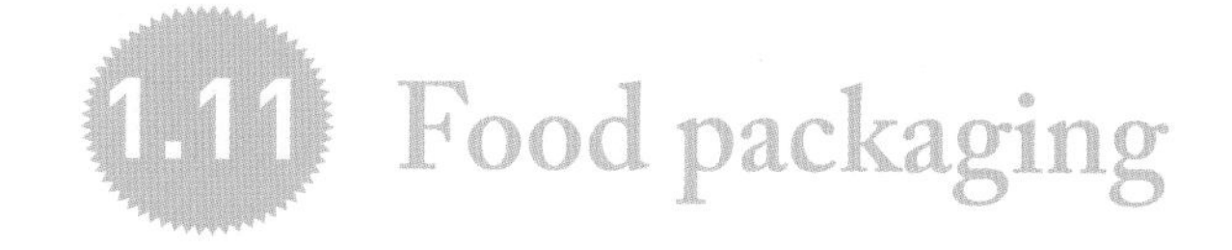

1.11 Food packaging

Forms, materials and functions

There is a variety of traditional forms of packaging. See Table 1.2 for details.

Shutterstock.com/Evgeny Karandaev

Figure 1.19 Food packaging performs numerous functions and is an essential part of food processing.

Table 1.2 Packaging characteristics and disadvantages

PACKAGING TYPE	DESCRIPTION	CHARACTERISTICS	DISADVANTAGES
Paper and cardboard	Paper is one of the most widely used forms of packaging in the world.	• Easily folded into different shapes and sizes • Lightweight but relatively strong when made into various thicknesses or laminated • Able to be recycled • A renewable resource (timber plantations) • Easy to print on • Very cheap to produce.	Lack of resistance to moisture, odours, insects and rats, and the high possibility of damage by crushing or squashing during transport.
Glass	Glass packages are very popular for various reasons.	• Resealable • Able to be printed on • Able to be recycled and reused • Transparent so food can be seen easily • Able to be coloured or tinted • Non-reactive with other foods • Very strong and durable • Able to be heated to high temperatures so it is hygienic and safe.	Can be cracked or broken, may be heavy to transport and is relatively costly to recycle.
Metal	Standard cans are made from aluminium, tinplate and steel. Aluminium can also be rolled into a fine sheet and sold as aluminium foil. Aluminium foil can be used to coat or line packages like muesli bar wrappers or potato crisp bags.	• Sterile and safe • Easily opened with a ring pull or can opener • Relatively lightweight • Protects against moisture, light and air • Strong and durable • Easy to stack • Heat resistant • Easily labelled • Versatile.	May rust when poorly handled, may dent if dropped causing damage to the seal. Not a renewable resource, although it can be reused and recycled, but can be difficult to recycle (especially when aluminium is mixed with other metals, such as tinplate, and both metals must be separated).
Plastic	Plastic packages can be either: • flexible like those used for cakes, biscuits and cling wraps, which are thin plastic films that are made into a reel or sheet • rigid like those used for ice-cream containers, margarine, yoghurt and sauce bottles, which are solid, moulded plastics that are very strong and heat-resistant.	• Able to be formed into many different shapes • Safe • Easy to carry • Relatively cheap • Strong and stable • Sometimes able to be recycled (such as PET plastics) • Lightweight • Resealable • Easily labelled.	May be made from a non-renewable resource (coal, oil and gas) and in many cases is not recyclable.

Technological developments

As technology develops, so too does packaging technology. Many new and innovative forms of packaging have been designed to meet the needs of consumers. These new designs have proved to be very useful in the food industry.

Modified atmosphere packaging (MAP)

This type of packaging has had the air inside the package either changed or removed completely.

Barrier packaging

Some foods, like fruit and vegetables, give off gases that help to ripen the food. However, manufacturers don't want food to ripen too quickly, especially during transport and storage, so they have developed a special plastic that lets out ripening gases and lets in other helpful gases. Fruit and vegetable refrigerator storage bags are a type of barrier packaging.

Active packaging

When foods are packaged, oxygen is still used up and carbon dioxide and other gases produced. Manufacturers can now place small sachets called 'scavengers' into food packages. These help to absorb moisture and other gases that may spoil food. They are used in wet or dry noodle packs.

Modified gas packaging

Certain gases like oxygen contribute to spoiling in food. Manufacturers have discovered how to remove oxygen from a package and replace it with another gas, such as carbon dioxide or nitrogen, that helps to slow food decomposition. Fresh pasta can be packaged in a moulded plastic tray with a peel-back lid using modified gas packaging.

Vacuum packages

Oxygen contributes to spoiling in food. Manufacturers can remove all air by sucking it out with a special type of vacuum. Meat like bacon can be packaged in vacuum packages.

Figure 1.20 How many types of packaging feature in a typical gift box of chocolates? Which elements are recyclable?

HANDS-ON

CAN YOU TASTE THE DIFFERENCE?

PURPOSE

Soft drinks are packaged for convenience in different forms. Compare the same product from different packages in this activity. Do they taste different?

MATERIALS

Blindfolds

Glasses for each student

Soft drink from three different packages: glass, plastic, aluminium can

STEPS

1. Your teacher will have three samples of a soft drink for you to taste. One will have come from a glass bottle, another from a plastic PET bottle, and the third from a can.
2. Taste each one and try to identify the package that each sample came from.

ACTIVITIES

1. Compare your answers to others in your class. Did they agree? What led you to these conclusions?
2. When your teacher reveals the sources, were you right? Were you surprised?

Environmental impact

Many consumers are concerned with the quantity of packaging that is produced, then thrown out as waste. Governments, together with food manufacturers, are trying to do what they can to minimise the impact that packaging has on the environment. The main issues include:

- **biodegradable** versus non-biodegradable packaging – some packages break down quickly, while other take years or may never completely break down
- reusing and recycling packaging – some packages can be recycled and reused and not just taken to the rubbish dump
- reducing unnecessary packaging – many food companies make packages inside packages; this may be convenient but it puts great pressure on the environment
- pollution from packaging – discarded packaging is a common feature of streets and waterways, and wildlife easily become suffocated or strangled by dumped packaging
- wastage of resources (especially non-renewable resources such as coal, oil and gas) – packages made from plastic require non-renewable resources to make them and eventually supplies will be depleted
- reduction in space available for landfill – as rubbish dumps and tips fill up and more land is needed for building homes and workplaces, communities must consider how waste can be reduced.

Food packaging waste will always have an impact on the environment. It is up to governments, food companies and consumers to do what they can to improve the situation.

Labelling/legal requirements

There is a legal requirement that aspects of the contents of a food package must be stated on the label.

However, not all food has to be labelled. Some of the best foods may be unlabelled. These include:

- unpackaged foods; for example meat, fruit, vegetables or restaurant meals
- food made on the premises where it is sold; for example, bread from a bakery
- food packaged in the presence of a consumer or at the request of a customer; for example, take-away food or pizza
- food sold at a fundraising event such as a school fete.

UNIT REVIEW

LOOKING BACK

1. **a** List five reasons for packaging food.
   b For each reason, give an example of a packaged food.
2. How does packaging encourage you to buy a product?
3. Define the terms 'barrier', 'modified atmosphere packaging' and 'active packaging'.

TAKING IT FURTHER

4. Interview an elderly neighbour or relative about what grocery shopping was like when they were children. What changes do they notice now?
5. Devise an experiment to test the protective properties of an egg carton.
6. Find a package you suspect to be made from a variety of materials or may have many layers or coatings (such as a juice box). Slice the package in half and try to separate the layers. Sketch what you find, try to identify each layer and suggest why it was necessary.

FOR YOU TO DO

7. **a** Collect a photograph of a packaged food product from a magazine or newspaper advertisement, or draw your own. Label the picture, identifying the reasons the particular food has been packaged in that way.
   b Compile an advantages and disadvantages list for the packaging illustrated for your product.
8. In your own words, explain what you think 'over-packaging' is.

1.11 Food packaging

Food Standards Australia

Food Labels
What do they mean?

1 Nutrition information panel.

Most packaged foods must have a nutrition information panel. The information must be presented in a standard format which shows the amount per serve and per 100g (or 100ml if liquid) of the food. Examples of a nutrition information panel and the nutrients that have to be listed in the nutrition information have been outlined below.

There are a few exceptions to requiring a nutrition information panel such as:

- very small packages which are about the size of a larger chewing gum packet
- foods with no significant nutritional value (such as a single herb or spice), tea, and coffee
- foods sold unpackaged (unless a nutrition claim is made)
- foods made and packaged at the point of sale, for example bread made in a local bakery.

NUTRITION INFORMATION

Servings per package: 3
Serving Size: 150g

	Quantity per Serving	Quantity per 100g
Energy	608kJ	405kJ
Protein	4.2g	2.8g
Fat, total	7.4g	4.9g
– Saturated	4.5g	3.0g
Carbohydrate, total	18.6g	12.4g
– Sugars	18.6g	12.4g
Sodium	90mg	60mg

*Percentage of recommended dietary intake

Ingredients: Whole milk, contentrated skim milk, sugar, banana (8%), strawberry (6%), grape (4%), peach (2%), pineapple (2%), gelatine, culture, thickener (1442).

All quantities above are averages

2 Percentage labelling.

Packaged foods have to carry labels which show the percentage of the key or characterising ingredients or components in the food product. This will enable you to compare similar products. The characterising ingredient for this fruit salad yoghurt is fruit and you can see from the ingredient list that it is banana 8%, strawberry 8%, grape 4%, peach 2%, and pineapple 2%. An example of a percentage of a characterising component would be the amount of cocoa solids in chocolate. Some foods, such as 'white bread' or 'cheese', have no characterising ingredients.

3 Name or description of the food.

Foods must be labelled with an accurate name or description, for example fruit yoghurt must contain fruit. If it were to contain fruit flavouring rather than real fruit, the label would need to say 'fruit flavoured yoghurt'.

4 Food recall information.

Considering the number of foods available, recalls of unsafe or unsuitable foods are uncommon. Food labels must have the name and business address in Australia or New Zealand of the manufacturer or importer, as well as the lot identification of the food (or date coding). This makes food recalls, on the rare occasion that they are necessary, more efficient and effective. In Australia each year there are about 70 food recalls, most of which are precautionary and due to the food manufacturer identifying a problem from their own testing.

Details of Australian recalls are on the Food Standards Australia New Zealand website at **www.foodstandards.gov.au**. New Zealand recalls are on the New Zealand Food Safety Authority website **www.nzfsa.govt.nz**.

5 Information for allergy sufferers.

Some foods, food ingredients or components of an ingredient can cause severe allergic reactions in some people – this is known as anaphylaxis. Foods such as peanuts, tree nuts (e.g. cashews, almonds, walnuts), shellfish, finned fish, milk, eggs, sesame and soybeans and their products, when present in food, may cause severe allergic reactions and must be declared on the label however small the amount. Gluten is also included in this list but the caution is more for those with Coeliac Disease rather than allergy. Those who are wheat allergic must stay away from all wheat including gluten.

In addition, foods containing sulphite preservatives must be labelled as containing sulphites if they have 10 milligrams per kilogram or more of added sulphites. This is the level that may trigger asthma attacks in some asthmatics.

For more information on food allergies see the Anaphylaxis Australia website **www.allergyfacts.org.au** or Allergy New Zealand **www.allergy.org.nz**.

3 Name or description of the food
1 Nutrition information panel
7 Ingredient list
2 Percentage labelling
9 Food additives
12 Country of origin
4 Food recall information
6 Date marking
8 Labels must tell the truth
10 Legibility requirements
5 Information for allergy sufferers
11 Directions for use and storage

6 Date marking.

Foods with a shelf life of less than two years must have a 'best before' date. It may still be safe to eat those foods after the best before date but they may have lost quality and some nutritional value. Those foods that should not be consumed after a certain date for health and safety reasons must have a 'use by' date. An exception is bread which can be labelled with a 'baked on' or 'baked for' date if its shelf life is less than seven days.

7 Ingredient list.

You will usually find the ingredient list on the back of the product. Ingredients must be listed in descending order (by ingoing weight). This means that when the food was manufactured the first ingredient listed contributed the largest amount and the last ingredient listed contributed the least, compared to the other ingredients. So, if fat, sugar or salt are listed near the start of the list the product contains a greater proportion of these ingredients.

8 Labels must tell the truth.

Suppliers must label food products with accurate weights and measures information. Weights and measures declarations are regulated by Australian State and Territory and New Zealand Government fair trading agencies.

Fair trading laws and food laws in Australia and New Zealand require that labels do not misinform through false, misleading or deceptive representations. For example, a food with a picture of strawberries on the label must contain strawberries.

9 Food additives.

Food additives have many different purposes, including making processed food easier to use or ensuring food is preserved safely. They may come from a synthetic or a natural source. For example, emulsifiers prevent salad dressings from separating into layers and preservatives help to keep food safe or fresh longer. All food additives must have a specific use, must have been assessed and approved by FSANZ for safety and must be used in the lowest possible quantity that will achieve their purpose.

Food additives must be identified, usually by a number, and included in the ingredients list. This allows those people that may be sensitive to food additives to aviod them. A thickener has been used in this yoghurt - its additive number is 1442. A full list of numbers and additives can be obtained from the FSANZ website. Some additives are derived from food allergens which must be identified, for example lecithin (soy).

10 Legibility requirements.

Any labelling requirement legally required in the Food Standards Code must be legible, prominent, and distinct from the background and in English. The size of the type in warning statements must be at least 3mm high, except on very small packages

11 Directions for use and storage.

Where specific storage conditions are required in order for a product to keep until its 'best before' or 'use by' date, manufacturers must include this information on the label. For example, 'This yoghurt should be kept refrigerated at or below 4°C'.

12 Country of origin.

Australia and New Zealand have different country of origin labelling requirements. In Australia, packaged, and some unpackaged, foods must state the country where the food was made or produced. This could just be identifying the country where the food was packaged for retail sale and, if any of the ingredients do not originate from that country, a statement that the food is made from imported or local and imported ingredients. Australian legislation also lays down rules about 'Product of Australia', which means it must be made in Australia from Australian ingredients, and 'Made in Australia', which means it is made in Australia with significant imported ingredients.

In New Zealand, country of origin requirements only apply to wines.

For more information

There is more information about food labelling on the FSANZ website **www.foodstandards.gov.au** or in the book *Choosing the Right Stuff- the official shoppers' guide to food additives and labels, kilojoules and fat content* published by Murdoch Books and available at all good bookshops. For expert nutrition and dietary advice contact your family doctor or an accredited practising dietitian.

You can find a dietitian in a number of ways:

In Australia:

Contact Nutrition Australia at **www.nutritionaustralia.org**

Visit the 'Find a dietitian section' of the Dietitians Association of Australia's website **www.daa.asn.au**, check the Yellow Pages or call 1800 812 942 to find an Accredited Practising Dietitian near you.

In New Zealand:

Contact the New Zealand Nutrition Foundation on (09) 489 3417, email nznf@nutrition.org.nz or website **www.foodworks.co.nz/nutritionfoundation**

Visit the 'Find a Dietitian' section of the New Zealand Dietetic Association's website at **www.dietitians.org.nz** or check the Yellow Pages.

Disclaimer: This poster has been produced as a guide to consumers only. Industry and enforcement agencies should refer to the Food Standards Code.

Figure 1.21 Food labels – what do they mean? As a consumer, do you know how to decode them?

Chapter review

LOOKING BACK

1 Define 'cross-contamination'.
2 State the temperature range for the danger zone.
3 Suggest two ways frozen food may be safely thawed.
4 List five common food-poisoning bacteria and some of the foods associated with them.
5 With reference to the danger zone, explain how bottling and freezing are effective preservation methods.
6 List four reactions that occur when carbohydrates are heated.
7 With reference to mayonnaise, explain the process of emulsification.
8 Identify four elements crucial to a recipe.
9 Define the following terms: boil, fold, sauté, sear and stew.
10 Explain the significance of cooking foods containing water-soluble and/or fat-soluble vitamins.
11 Suggest two food preparation methods to increase fibre in the diet.
12 Describe the two levels of food processing. What do you imagine tertiary processing to be?
13 Provide two negative environmental effects of food processing.
14 Explain the two ways food processing has economic impacts on families.
15 List four types of food packaging materials. What are the advantages of each?
16 List 12 pieces of information found on a food label.

FOR YOU TO DO

17 Distinguish between food hygiene and personal hygiene. Use photographs from magazines to illustrate your answer.
18 Using movie maker software, create a 60-second television commercial informing consumers of some simple food safety issues in the home.
19 Design a menu for a two-day camping trip. Try to incorporate a range of foods processed by different methods.
20 Conduct a detailed survey to determine the most and least liked foods in the class. Include questions that determine which sensory properties lead to each preference.
21 Using digital photography, create an information poster explaining denaturation and coagulation of egg protein.
22 Write and record a radio advertisement encouraging listeners to take up cooking. Explain why using fresh ingredients is important.
23 Create a class collection of take-away menus. Which ones could use improvement in presentation? Develop a new version using Microsoft Publisher or a similar program.
24 Collect a food package such as a label from a can or a plastic wrapper from a packet of biscuits. Create a poster identifying all the required information.

TAKING IT FURTHER

25 Research the use of colour-coded chopping boards. Why were certain colours allocated to certain foods?
26 Draw the floor plan of your Food Technology classroom. On it, label the location of firefighting equipment, fire blankets and first aid kits. Highlight any potential hazards.
27 Investigate the origins of an unusual ingredient found in your local supermarket. Use the internet to help.
28 Design a production line for the mass production of vegetarian sushi. Allocate jobs for efficiency. How many people would be involved?
29 Investigate an organisation that accepts unsold food or donated non-perishable foods such as OzHarvest or Foodbank NSW. Write a one-page report on how this organisation works.
30 Examine the Clean Up Australia Day website and locate the latest statistics on litter from Clean Up Australia Day. How much of each type of food packaging is represented?

Clean Up Australia Day

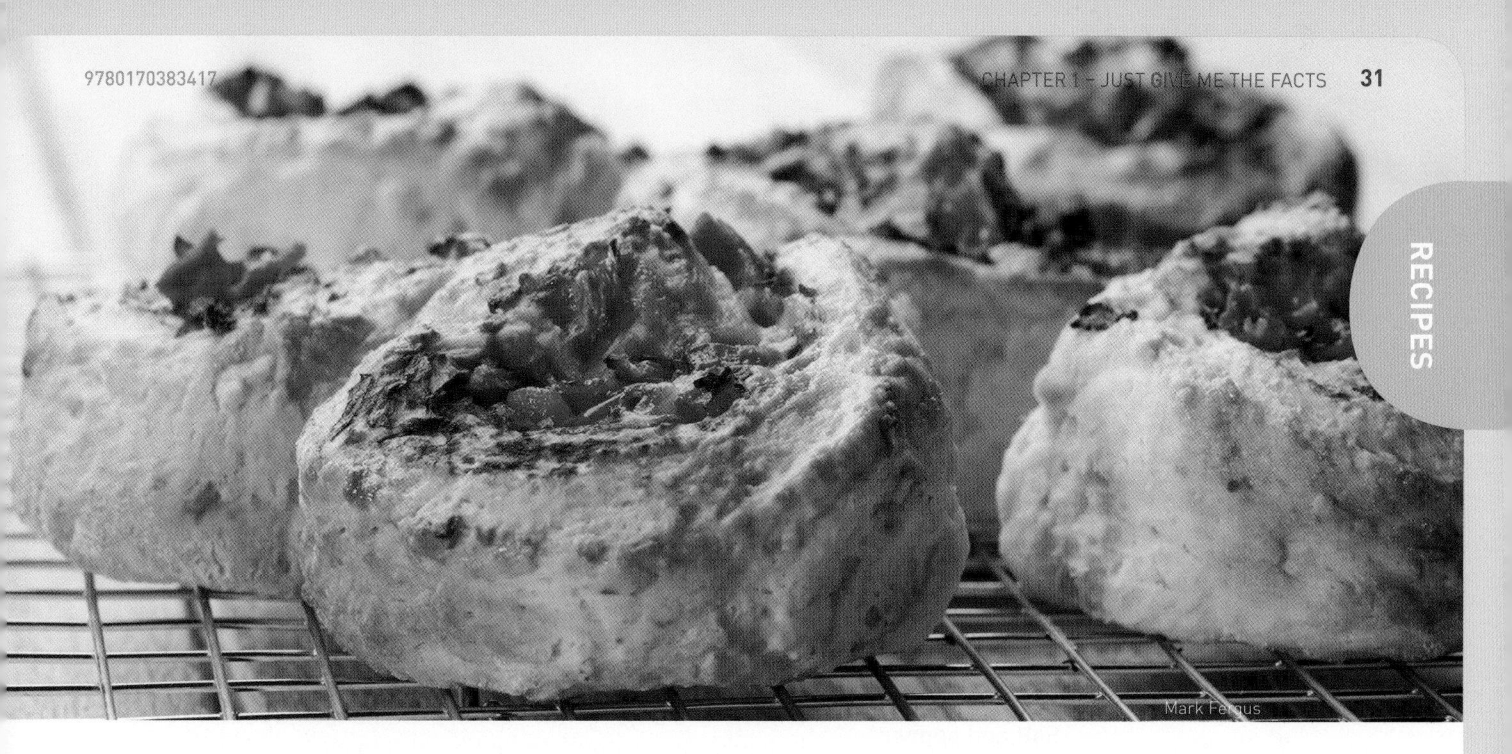
Mark Fergus

RECIPES

Savoury scone scrolls

This recipe will showcase your knife skills and safety knowledge.

 **Preparation time** 20 minutes  **Cooking time** 25 minutes

INGREDIENTS

- 2 cups self-raising flour
- 2 tablespoons margarine
- 1 cup milk
- 1 slice ham
- 2 shallots
- ½ capsicum
- ⅓ cup grated cheese
- 2 tablespoons tomato paste
- 1 teaspoon dried Italian herbs
- 1 egg

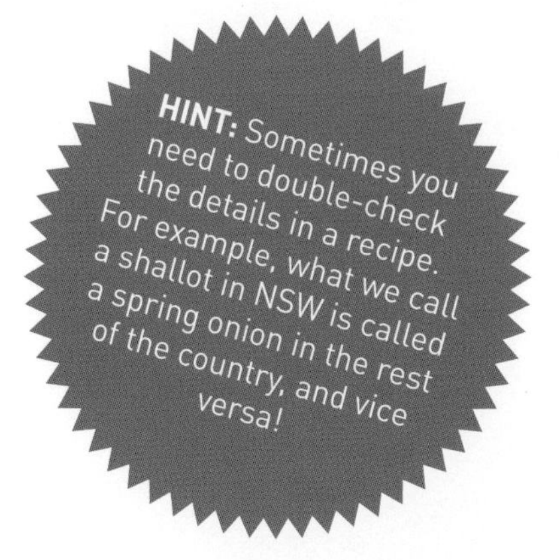

METHOD

1. Set oven at 200°C.
2. Line an oven tray with baking paper, or grease with cooking spray.
3. Dice the ham and capsicum, and finely slice the shallots. Set these aside.
4. Sift the flour into a large bowl.
5. Rub in the margarine until the mixture looks like breadcrumbs.
6. Add enough milk to form a soft dough.
7. Turn the dough out onto a lightly floured clean bench and knead gently.
8. Press the dough out into a rectangle about 1 cm thick.
9. Spread with the tomato paste.
10. Sprinkle on the ham, capsicum and shallots. Top with the herbs and cheese.
11. Starting at the long edge, roll up the dough.
12. Cut into slices about 3 cm thick.
13. Place onto the oven tray with the cut edges up and the scrolls just touching each other.
14. Brush with the beaten egg.
15. Bake for 25 minutes or until golden.

QUESTIONS

1. Explain the importance of washing hands before and throughout the preparation of the scrolls.
2. Identify one important safety tip for using a knife.
3. Why are the scrolls placed so they are touching?
4. Why is the dough only kneaded 'lightly'?

Greek yoghurt labneh

This is a type of cheese made from yoghurt. Its shelf life can be extended further by storing in oil, which removes exposure to air.

 **Preparation time** 10 minutes plus standing/finishing time  **Serves** 4

INGREDIENTS

2 cups full fat Greek yoghurt

½ cup olive oil

3 tablespoons finely chopped fresh herbs (such as parsley, chives or thyme)

½ teaspoon finely grated lemon zest

Salt and pepper

SPECIAL EQUIPMENT

Cheesecloth and cooking twine

METHOD

1. Line a large wire sieve or colander with the cheesecloth. Make sure the edges hang over the top of the sieve/colander.
2. Place the yoghurt in the sieve.
3. Gather together the edges of the cheesecloth and tie in a bundle.
4. Place the sieve over a bowl (make sure the sieve doesn't touch the bottom of the bowl).
5. Place in the refrigerator for 2–3 days.
6. After 2–3 days, gently squeeze the bundle to remove excess liquid - the yoghurt should be very thick.
7. Untie the string and roll the yoghurt into 3 cm balls.
8. Place the balls into a clean glass jar.
9. Whisk the oil, chopped herbs, lemon zest and a pinch of salt and pepper in a small bowl.
10. Pour over the labneh balls in the jar. Put the lid on and refrigerate.
11. Marinate for 8 hours before eating with crackers.

QUESTIONS

1. What is the name given to the liquid that is drained from the yoghurt?
2. What other herbs could be added to the marinade? Can you think of another presentation idea?
3. Did you like the taste of the labneh?

Mark Fergus

9780170383417

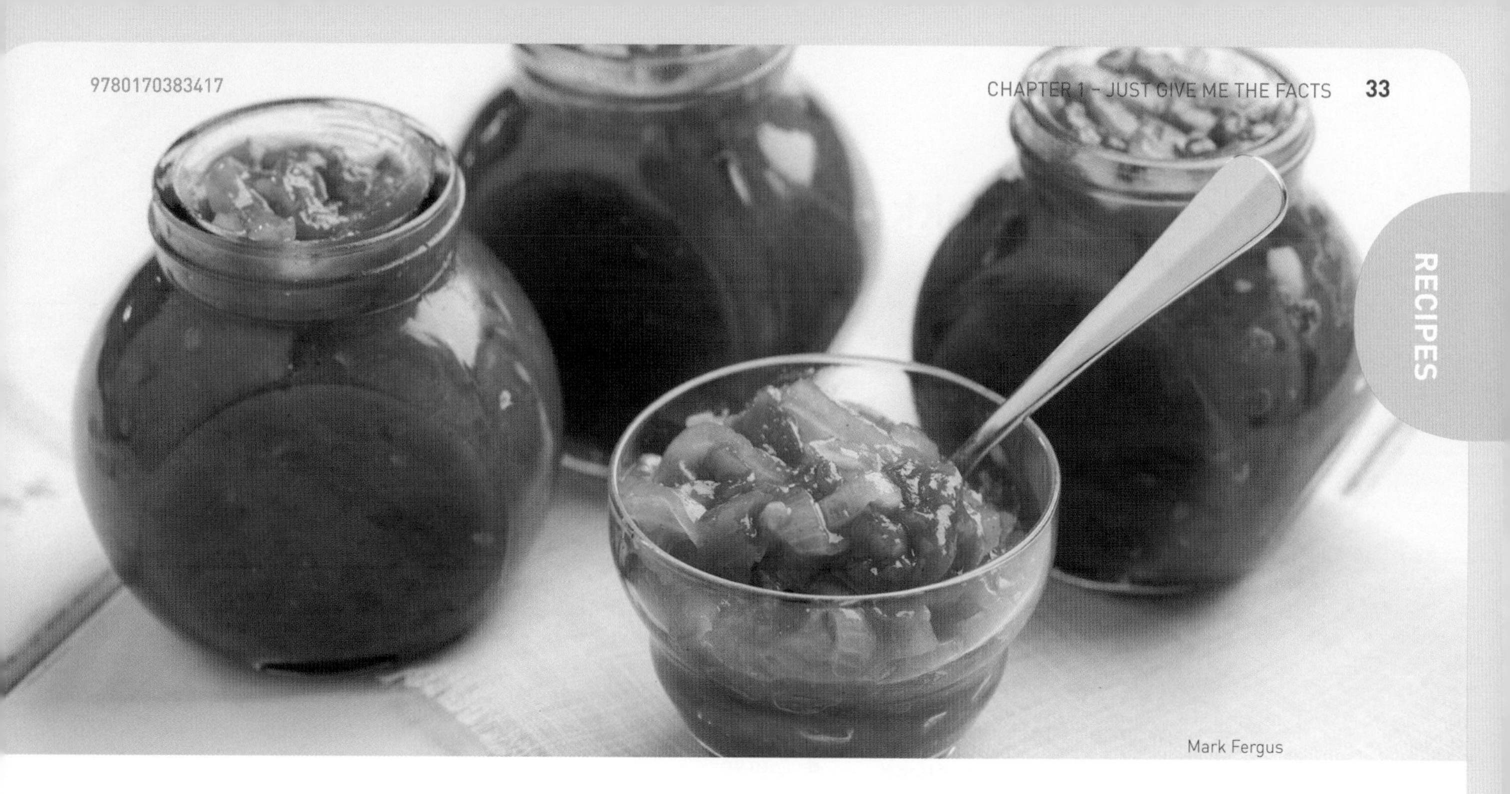
Mark Fergus

Tomato relish

This recipe uses the alteration of the pH level by the addition of vinegar as the preservation method for tomatoes.

 **Preparation time** 20 minutes  **Cooking time** 40 minutes (less if using a microwave)

 **Makes** 4 small jars

INGREDIENTS

750 grams ripe tomatoes
500 grams onions
1 cup sugar
1¼ cups malt vinegar
2 teaspoons flour
2 teaspoons curry powder
pinch cayenne pepper
2 teaspoons mustard powder
2 teaspoons salt

HINT: To peel tomatoes, carefully cut out the core (the part at the top where the stalk joins the tomato) and mark a cross on the bottom with a knife. Place the tomatoes into a saucepan of boiling water for 10–15 seconds and remove using tongs. Plunge into a bowl of iced water. The skin should now pull away easily, using a paring knife to lift the skin.

METHOD

1 Peel and chop the tomatoes.
2 Slice the onions.
3 Place the tomatoes and onions into a large saucepan with the sugar. Add the vinegar and simmer uncovered until thick (approximately 30 minutes).
4 Drain off ½ cup of the juice and blend remaining ingredients into this. Add this mixture to the saucepan.
5 Stir until boiling, then simmer for 5 minutes.
6 Bottle in sterilised jars and seal when cool.

QUESTIONS

1 Why are the remaining ingredients not mixed into the hot tomato mixture?
2 Suggest a meal the relish may accompany.
3 Using the requirements outlined in the labelling section of this chapter (pages 28–9), use a word processing package to make a label for your jars of relish.

Chicken and hokkien noodle stir-fry

This recipe has it all – flavour, texture, colour and aroma!

 **Preparation time** 10 minutes  **Cooking time** 15 minutes  **Serves** 4

INGREDIENTS

- 2 tablespoons oil
- 1 clove garlic, crushed
- 1 teaspoon fresh ginger, grated
- ½ teaspoon bottled chopped chillies (optional)
- 1 chicken breast, thinly sliced
- ¼ red capsicum, sliced
- ¼ green capsicum, sliced
- 1 stalk celery, sliced
- 1 carrot, thinly sliced
- 3 shallots, sliced diagonally
- 8 cobs baby corn
- 125 grams hokkien noodles
- 1 tablespoon soy sauce mixed with 2 tablespoons water

METHOD

1. Prepare all ingredients as described in the ingredient list.
2. Heat the oil in a wok.
3. Add the garlic, ginger and chillies and heat until aromatic.
4. Add the chicken and stir-fry until golden.
5. Add all vegetables and stir-fry for 4–5 minutes.
6. Prepare the noodles by placing in a bowl of boiling water and separating.
7. Add the noodles and soy sauce mixture to wok. Stir to combine, reduce heat and cover for 2 minutes.
8. Serve.

QUESTIONS

1. Write a recipe review of this dish that would be suitable for a food magazine. Be sure to mention the texture, colour, aroma and taste.
2. Explain three alterations you could make to the ingredients or serving of this dish.
3. Suggest a dessert that would be suitable after such a meal.

Mark Fergus

RECIPES

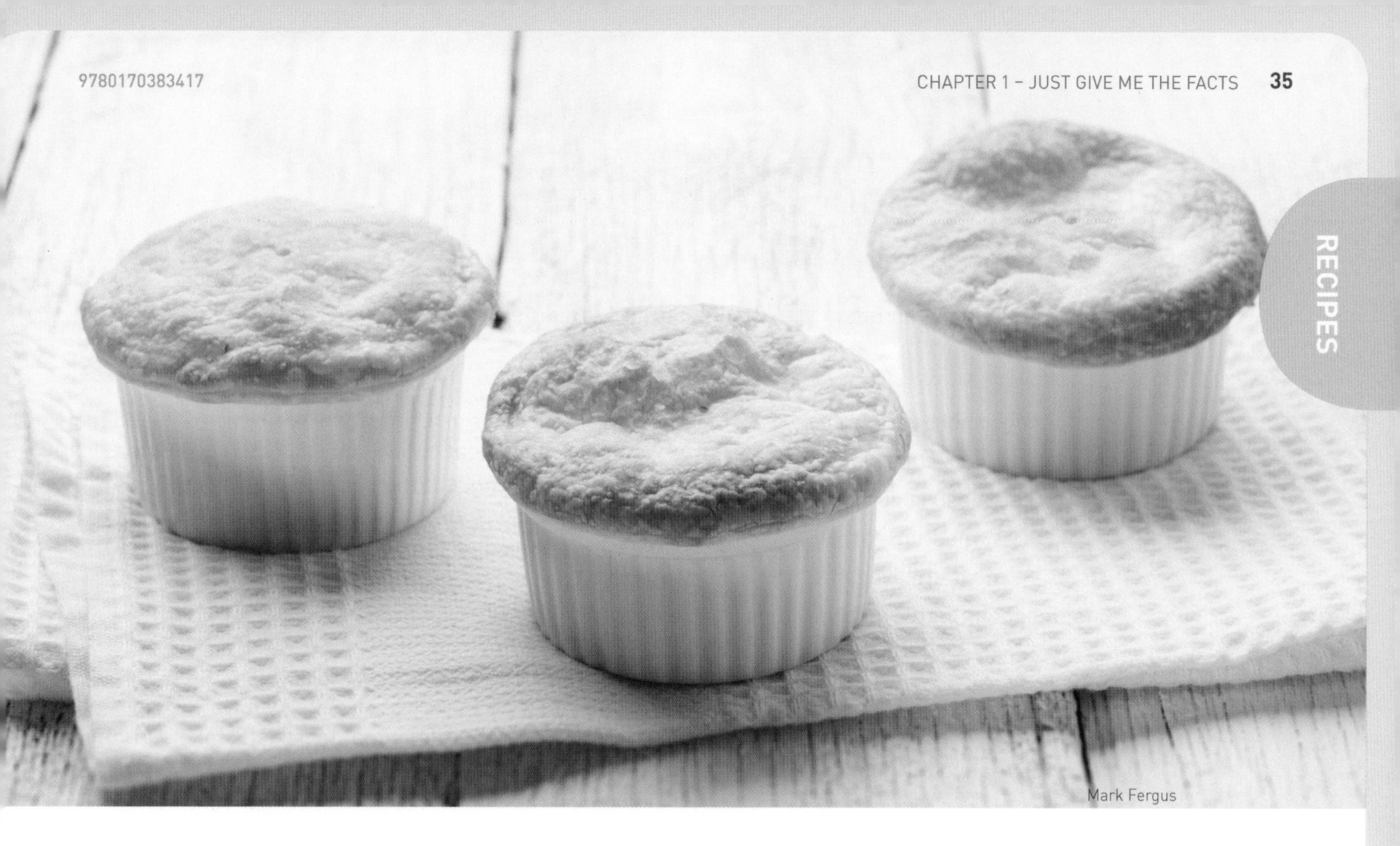
Mark Fergus

Chicken pot pie

This recipe demonstrates gelatinisation and dextrinisation.

 **Preparation time** 30 minutes  **Cooking time** 20 minutes  **Serves** 2

INGREDIENTS

1 cup cooked chicken, diced
½ onion, finely diced
1 stick celery, finely diced
1 small carrot, finely diced
2 teaspoons oil
⅓ cup frozen peas
1 tablespoon butter or margarine
2 tablespoons plain flour
¾ cup milk
1 sheet frozen puff pastry

METHOD

1. Preheat oven to 200°C. Lightly spray a medium casserole dish, or four small ramekins.
2. Dice the chicken, onion, celery and carrot.
3. Heat the oil in a small saucepan on medium-high and lightly fry the onion, celery and carrot for 3 minutes.
4. Remove the vegetables to a clean bowl.
5. In the same saucepan, add the butter or margarine and melt.
6. Add the flour, stir with a wooden spoon over the heat and cook for 1 minute.
7. Take the pan off the heat and add the milk very gradually, stirring constantly.
8. Once all the milk is added, return to the heat and stir over medium-high heat until it is thickened.
9. Add the diced chicken, cooked vegetables and frozen peas to the saucepan.
10. Mix together and spoon evenly into the casserole dish.
11. Place the puff pastry on top and trim to the shape of the casserole dish.
12. Bake in the oven for 20 minutes or until pastry is golden.

QUESTIONS

1. Which element of the recipe demonstrates gelatinisation?
2. Which element of the recipe demonstrates dextrinisation?
3. Suggest two other flavour/ingredient substitutions for this recipe.

Vegetarian sushi

This is a tricky dish with the reward in the presentation.

 **Preparation time** 30 minutes  **Cooking time** 15 minutes  **Serves** 2

INGREDIENTS

- 2 cups short grain rice
- 2 cups water
- 1 Lebanese cucumber
- ½ medium carrot
- ½ medium avocado
- 3 tablespoons water (extra)
- 3 tablespoons rice vinegar
- 1 tablespoon sugar
- 5 sheets toasted nori
- Soy sauce for dipping

METHOD

1. Place the rice and water in a saucepan and bring to the boil. Cover with lid, reduce heat to low and cook for 10 minutes.
2. While the rice is cooking, cut the cucumber, carrot and avocado into matchstick slices.
3. When the rice is cooked and has cooled slightly, mix in the extra water, rice vinegar and sugar.
4. Place 1 sheet of nori on a sushi mat or use a clean, firm cloth placemat. Spread half of the rice evenly over the nori.
5. Layer 1 slice each of the vegetables at 1 short end of the nori.
6. Carefully roll the nori, using the mat to assist. Make sure the vegetables are in the centre of the roll.
7. Wrap in plastic wrap and chill before cutting into 8 rounds. Repeat the process with other sheets of nori.
8. Serve with soy sauce to dip.

QUESTIONS

1. What method of cooking rice has been used in this recipe?
2. What is the origin of nori?
3. What other fillings could be appropriate for this classic Japanese dish?
4. Suggest innovative ways of presenting the sushi slices.

Mark Fergus

Mark Fergus

Upside-down cake

This recipe requires you to read carefully, and demonstrate your food preparation skills and safe use of equipment.

 **Preparation time** 40 minutes  **Cooking time** 30 minutes  **Serves** 4

INGREDIENTS

CARAMEL:

⅓ cup sugar

1½ tablespoons water

FRUIT:

½ Granny Smith apple, peeled, cored and thinly sliced

CAKE:

60 grams butter

⅓ cup brown sugar

1 egg

¼ cup sour cream

⅔ cup self-raising flour

¼ teaspoon baking powder

¼ teaspoon ground ginger

¼ teaspoon cinnamon

METHOD

1. Preheat oven to 190°C. Grease a 21 x 9 cm cake tin and line with baking paper.
2. Prepare the caramel topping by cooking the sugar and water in a small saucepan over a medium heat, stirring until sugar dissolves. Bring to the boil without stirring until the mixture turns golden. (Watch carefully that the mixture does not burn.)
3. Quickly pour the caramel into the cake tin.
4. Carefully arrange the sliced apple over the top of the caramel – the caramel will be HOT.
5. Using electric beaters, cream the butter and sugar until light and fluffy.
6. Beat in the egg and sour cream until combined.
7. Sift the flour and spices together. Fold into the beaten mixture, using a wooden spoon.
8. Carefully pour the cake mixture over the caramel/apple topping in the tin.
9. Bake for 25–30 minutes or until golden.
10. Remove from oven and stand in tin for a few minutes before turning out onto a serving plate.

QUESTIONS

1. Could this recipe be made without an electric mixer? What would the differences be?
2. Suggest alterations to the fruit used for this recipe to align with the seasons.

9780170383417

RECIPES

Pasta

Pasta making is one of the easiest methods of food processing you can replicate at home!

 **Preparation time** 40 minutes  **Cooking time** 10 minutes  **Serves** 4

INGREDIENTS

490 grams plain flour, or 400 grams fine Italian '00' flour

4 large or extra-large eggs

METHOD

1. On a clean bench, pour the flour into a mound. Make a well in the centre (like a volcano).
2. Add the eggs into the well.
3. Gently whisk with a fork, using your free hand to secure the walls of the mound. Draw in the flour gradually as you whisk.
4. Bring the dough together in a ball. Knead for 5 minutes until elastic and smooth. Sprinkle the ball with flour, then wrap in plastic wrap and set aside to rest for 30 minutes.
5. Divide the dough into four portions.
6. Set the pasta machine on the widest setting and flour the machine and the dough. Feed the first flattened portion through the machine.
7. Repeat six more times, folding and feeding it through the machine, narrowing the setting of the machine until the pasta is approximately 1 mm thick. Cut into fettucine or spaghetti.
8. Settle pasta into 'nests' and store in a sealed container until ready to cook.

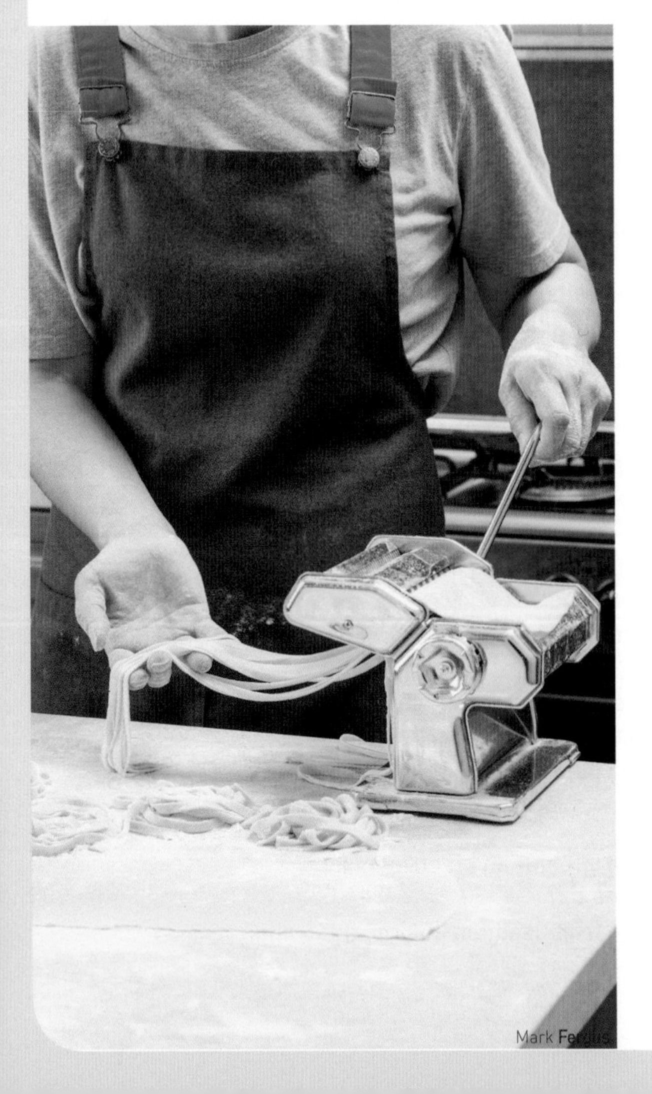
Mark Fergus

QUESTIONS

1. Research the difference between regular plain flour and fine '00' flour.
2. In industry, what types of additives might be added to this recipe before packaging for sale?
3. Suggest a simple, tasty sauce to serve with this homemade pasta.
4. Would the making of this recipe be possible without a pasta machine? How would the resulting product be different?

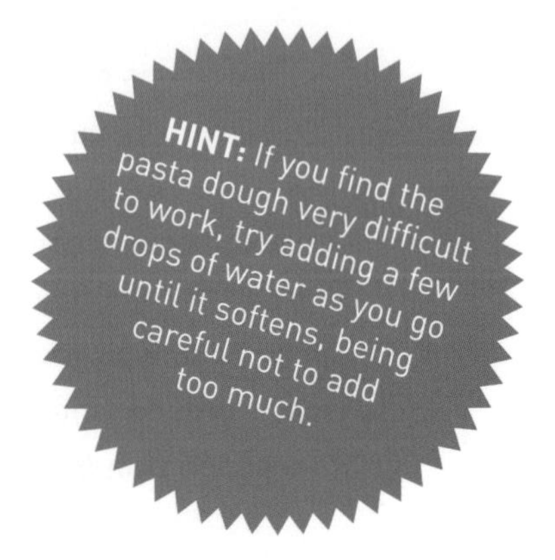

RECIPES

Mark Fergus

Easy-as apple pie

A recipe using processed foods packaged for convenience.

 **Preparation time** 20 minutes  **Cooking time** 15 minutes  **Serves** 2

INGREDIENTS

2 sheets frozen puff pastry, thawed
1 x 425 gram can pie apple
½ cup rolled oats
2 tablespoons brown sugar
2 teaspoons cinnamon

METHOD

1. Preheat oven to 200°C.
2. Spray two baking trays with cooking spray and place 1 sheet of pastry on each. Cut each sheet into 4 squares.
3. Place 1 heaped tablespoonful of apple in the middle of each square, and sprinkle with oats, sugar and cinnamon.
4. Moisten edges with water and fold in the four sides to make a parcel or envelope.
5. Bake for 15 minutes or until the pastry is golden and crisp.

QUESTIONS

1. Identify the packaging types used in this recipe.
2. This is known as a free-form pie. Why?

9780170383417

Three-ingredient party pies

Quick and easy pies using a recipe that can be adapted to produce dozens of variations!

 **Preparation time** 20 minutes  **Cooking time** 30 minutes  **Serves** 4

INGREDIENTS

400 grams lamb mince
535 grams can rogan josh sauce
5 sheets frozen shortcrust pastry

METHOD

1 Preheat oven to 200°C. Grease a 12 cup muffin tray.
2 Heat a frypan over medium heat and brown the lamb, stirring to break up lumps.
3 Add the rogan josh sauce. Bring to the boil, then reduce heat and simmer for 3 minutes.
4 Use a 10 cm pastry cutter to cut 12 rounds from 3 sheets of pastry and line the tins.
5 Use an 8 cm cutter to cut 12 rounds from the remaining pastry. These will be the pie lids.
6 Spoon filling into the prepared pie cases. Top with the 8 cm rounds and press edges to seal.
7 Using a skewer, make a small hole in the top of each pie.
8 Bake for 20 minutes or until golden on top.

QUESTIONS

1 Explain the importance of pressing the edges of the pastry when adding the tops.
2 Why are the tops pierced with a skewer? What would happen if they were not?
3 With only three ingredients, what variations could you suggest to create other pies with different flavours using prepared foods?

Mark Fergus

Mark Fergus

RECIPES

Vietnamese rice paper rolls

A food in an edible wrapper that requires no label!

 **Preparation time** 25 minutes  **Cooking time** Nil  **Serves** 4

INGREDIENTS

100 grams dried rice paper noodles
500 grams cooked, peeled prawns (optional)
2 cups shredded barbecue chicken
1 cup fresh mint
100 grams bean sprouts
1 large carrot
375 grams packet rice paper wrappers

METHOD

1. Soak the noodles following packet directions. Drain.
2. Finely slice the prawns, peel and grate the carrot, rinse and separate the leaves of mint and trim the ends of the bean sprouts. Place each prepared ingredient in a separate small bowl.
3. Half fill a shallow dish with warm water. Dip 1 wrapper into water, then place on a clean chopping board for 30 seconds or until soft enough to roll without splitting.
4. Place a small amount of prawns, chicken, noodles, mint, sprouts and carrot in the centre of the wrapper. Do not overfill or the rolls will split. Roll up, folding edges in as you go. Cover with a cloth and repeat with remaining ingredients.
5. Serve with a dipping sauce of your choice.

QUESTIONS

1. These rice paper rolls are suitable as an appetiser. Suggest three other hot appetisers that could also be served at a pre-formal gathering in the early evening.
2. Brainstorm five other foods or meals that are 'wrapped' in their own edible packages.
3. Suggest two dipping sauces that may be bought or made to accompany this dish.

CHAPTER 2

Mark Fergus

Eat well, live well

FOCUS AREA

CORE: NUTRITION AND CONSUMPTION

There is an Italian saying, '*mangia bene, vivi bene*', meaning 'eat well, live well'. Food supplies nutrients, which allow the body to function and develop from infancy to adulthood. Poor food selection may result in problems that can affect your way of life. Making good food choices can be difficult, as there is an abundance of food available. In this chapter you will examine the factors that contribute to good nutrition and learn to make healthy food selections.

CHAPTER OUTCOMES

In this chapter you will learn about:

- food nutrient groups and fibre
- foods developed to enhance health
- over- and under-nutrition, including anorexia
- food consumption in Australia
- influences on food selection
- national guidelines for healthy eating, including nutrition and labelling
- nutrition throughout the life cycle
- nutritious foods.

In this chapter you will learn to:

- explain the role of food and fibre in the body
- discuss nutritionally modified foods
- outline conditions of over- and under-nutrition, including anorexia
- conduct internet searches to identify food consumption trends and their effects on health
- outline how dietary disorders can be prevented and managed
- identify guidelines for healthy eating, including labelling symbols
- analyse the nutritive content of foods
- outline requirements of different life cycle stages
- design, plan and prepare diets for the different stages
- tabulate and analyse data using spreadsheets and graphs.

RECIPES

FOOD FACTS

Sulphur is an element found on some protein molecules. It contributes to the smell of rotten eggs.

The mineral fluoride, which helps strengthen teeth, is added to public drinking water in most parts of Australia

A single can of soft drink contains 11 teaspoons of sugar. To burn up the energy from a can of soft drink you would have to walk 3 kilometres.

Carrots really can help you see in the dark! Vitamin A is known to prevent 'night blindness' and carrots are loaded with Vitamin A.

You need around 25 grams of fibre a day. Three dried figs provide 10 grams of fibre, an apple 4 grams, half a cup of peas 9 grams, a cup of dried beans 19 grams. Make sure to drink plenty of water when you include fibre in your diet – otherwise you may get stomach pains.

Human breast milk is a probiotic food as it contains good bacteria to help prevent digestion problems in infants

Low-fat cream will not whip successfully as it does not have enough fat content, but you may use it for recipes in which whipped cream is not required.

Diabetes is the world's fastest growing chronic disease.

Gluten is found not only in breads and cereals but also in many additives in processed foods, such as thickeners.

Have a tomato with your burger! When a source of Vitamin C, such as oranges, capsicum or tomato, is eaten with iron-rich food like meat, Vitamin C helps to absorb more iron into the body.

Temperature can affect your appetite – you're more likely to be hungry if you are cold!

FOOD WORDS

chronic constant, continuing for a long time

dehydration lack of water in the body

diverticulitis inflammation of small, abnormal pouches in the wall of the digestive tract

energy fuel we need from food to function and be active; energy requirements vary depending on your age, body size and physical activity

enrich add nutrients to improve nutritional value

irritable bowel syndrome pain and bloating of the abdomen

kilojoules standard unit of energy measurement

neural tube defect condition that occurs when the central nervous system of the foetus is not correctly developed; commonly referred to as spina bifida

photosynthesis process whereby leaves of a plant use the energy of the sun to change carbon dioxide and oxygen into carbohydrates, which are then stored in the plant

processed foods foods that have undergone change from their original raw state by a manufacturing process

sedentary involving little exercise, a lot of sitting

tumour abnormal swelling or lump in the body

vegan people who consume no animal products of any kind, including dairy and eggs

2.1 The food nutrient groups

Nutrients are chemical substances found in food. There are six nutrients: proteins, carbohydrates, lipids, vitamins, minerals and water. Each nutrient has a specific role in the body.

Figure 2.1 No single food has all the nutrients in the quantities required by the body, so a variety of foods should be consumed daily.

Proteins

Proteins are found in every body cell and perform a number of functions.

- They allow the body to grow and to repair. As our cells wear down, protein is used to build new cells such as hair, nerves and skin.
- They form the basis of many of the body's chemical substances, such as enzymes, hormones and haemoglobin.
- They provide **energy** (17 **kilojoules** per gram). The body will use protein only if there are no other sources of energy. The use of protein for energy is at the expense of growth and repair.

Proteins are complex substances made from carbon, hydrogen, oxygen and nitrogen, which combine to form amino acids. There are 22 amino acids that combine in various ways to form different proteins. Each day, food must supply nine of these amino acids (10 for children), as the body cannot manufacture them. These are called essential amino acids. Proteins that contain all the essential amino acids are called complete proteins, while those lacking one or more essential amino acids are classed as incomplete proteins.

When some incomplete proteins are combined they may produce a complete protein if the amino acid lacking in one is present in the other. An example of this is baked beans on toast, which is useful for vegetarians and **vegans**.

One or two servings of complete protein foods are required daily. When you consume more protein than is required, the remainder is converted to body fat.

Table 2.1 Sources of protein

COMPLETE PROTEINS	INCOMPLETE PROTEINS
Milk Yoghurt Cheese Eggs Meat Poultry Fish Shell fish Soy beans Soy milk Tofu Textured vegetable protein (TVP), a meat substitute	Legumes, such as chick peas and kidney beans Pulses, such as split peas and lentils Nuts Seeds, such as sesame Cereals, such as rice, wheat, oats, rye and corn Cereal products, such as flour, bread and pasta

Carbohydrates

Thorugh the process of **photosynthesis**, plants produce and store carbohydrates. There are two types of carbohydrates: sugars and complex carbohydrates.

Sugars

Sugars or simple carbohydrates are usually sweet, and found in a variety of fresh and **processed foods**.

Table 2.2 Food sources of sugars

SUGAR	FOOD SOURCES
Glucose	Fruits, honey
Fructose	Fruits, honey
Sucrose	Sugar cane
Lactose	Milk
Maltose	Malt

Complex carbohydrates

Complex carbohydrates or starches have longer structures and therefore take longer to digest. They are found in grains, such as rice, corn, wheat, rye and oats, and cereal products such as bread, pasta and breakfast cereals. They are also present in fruits, vegetables, legumes and pulses.

The function of carbohydrates is to supply energy to the body. After digestion, all sugars and starches are broken down to the simplest carbohydrate: glucose. The body converts glucose to energy, which can then assist the body to undertake work. Even when you are asleep your body is busy working and using energy. Each gram of glucose can provide 17 kilojoules of energy. If the body receives more energy from food than it needs, the excess is converted to body fat and stored until required again for energy.

Lipids

Lipids are also known as fats or oils. Fats are solid at room temperature, while oils are liquid. Lipids help the body to function by providing:

- a concentrated source of energy (37 kilojoules per gram)
- warmth, as most fat is found under the skin
- protection from injury for bones and other organs
- transport for Vitamins A, D, E and K around the body
- fatty acids, which are used to help form the walls of body cells.

Mark Fergus

Figure 2.2 Butter, margarine, cream and oil are added to foods to enhance the flavour of food. Think about why we spread butter on bread.

Lipids are also found in foods such as lean meats, nuts, poultry and eggs. These foods supply other valuable nutrients as well as fats, and so make a sensible choice as a means of including lipids in the diet.

There are three different types of lipids:

- Saturated fats – found in animal products, such as butter, cream, full-cream milk, cheese and eggs, as well as coconut and many processed foods that use hardened vegetable oil. They are linked to high levels of cholesterol in the blood, which can thicken blood vessels and increase the risk of heart attack.
- Monounsaturated fats – found in olive oil, canola oil, olives, peanuts and avocadoes. These are less likely to raise blood cholesterol levels.
- Polyunsaturated fats – found in vegetable oil, margarine and fish oils. They are less likely to raise cholesterol levels in the blood and some food sources have omega-3 fatty acid, which can reduce blood cholesterol levels.

Importantly, when more lipids of any source are consumed than the body needs for energy, weight gain results.

Vitamins

Only small amounts of vitamins are required daily to protect and regulate the body. Vitamins can be referred to by their alphabetical or chemical names.

There are two types of vitamins:

- Fat-soluble vitamins A, D, E and K, which accumulate in fat stores and the liver, and can be toxic if consumed in excess.
- Water-soluble vitamins C, B complex and folate, which are absorbed quickly and need to be consumed daily. Water-soluble vitamins are easily destroyed by cooking. The body excretes any excess amounts of these vitamins.

Table 2.3 Sources and roles of vitamins

VITAMIN	ROLE IN BODY	RICH FOOD SOURCES
A	Necessary for growth, healthy skin and eyes.	Dairy foods, margarine, eggs, oily fish, green and orange fruits, and vegetables.
B1, B2, B3 (thiamine, riboflavin and niacin)	Helps the body to obtain energy from food.	Bread, flour, breakfast cereals, liver, kidney, meat and potatoes.
Folate	Contributes to the healthy development of babies in pregnancy.	Bread, flour, breakfast cereals, liver, kidney, meat and potatoes.
C (ascorbic acid)	Keeps the skin, bones and muscles healthy. Increases resistance to infection and helps the body absorb iron.	Green vegetables and fruit, especially citrus fruits.
D	Needed to form strong bones and teeth. Helps the body absorb calcium.	Butter and margarine, milk, eggs and oily fish. The action of sunlight on the skin also forms Vitamin D in the body.

Minerals

The function of minerals is to protect the body against disease and regulate body processes. The body requires over twenty mineral elements daily in very small amounts. These include minerals such as magnesium, zinc and potassium; however, the following minerals need particular consideration.

- Calcium works with other minerals, such as phosphorus, to strengthen bones and teeth. The best sources of calcium are milk, yoghurt, cheese, canned fish with bones, and nuts.
- Iron is required to form haemoglobin, which transports oxygen in the blood. Red meats are rich in iron, while poultry, fish, legumes, wholemeal bread, green leafy vegetables, breakfast cereals and eggs are also good sources.
- Sodium helps to regulate the amount of fluid in and around the body cells. It also helps the contraction and relaxation of muscles. Salt is a combination of sodium and the mineral chloride. Most Australians consume too much sodium, not just because salt is used in cooking, but because it is used in many processed foods.

Water

The human body is approximately 70 per cent water. While humans can survive without food for a few weeks, they will die of **dehydration** if they are without water for a few days.

The body requires about two litres of water a day to carry out its functions and prevent problems such as constipation and kidney stones. Good sources of water are fresh drinking water, fruits, vegetables, milk and juices. It is best to limit intake of drinks that are high in sugar, minerals and caffeine. Fresh water has no kilojoules, sugar or fat, and can contain fluoride, so it is an excellent drink choice.

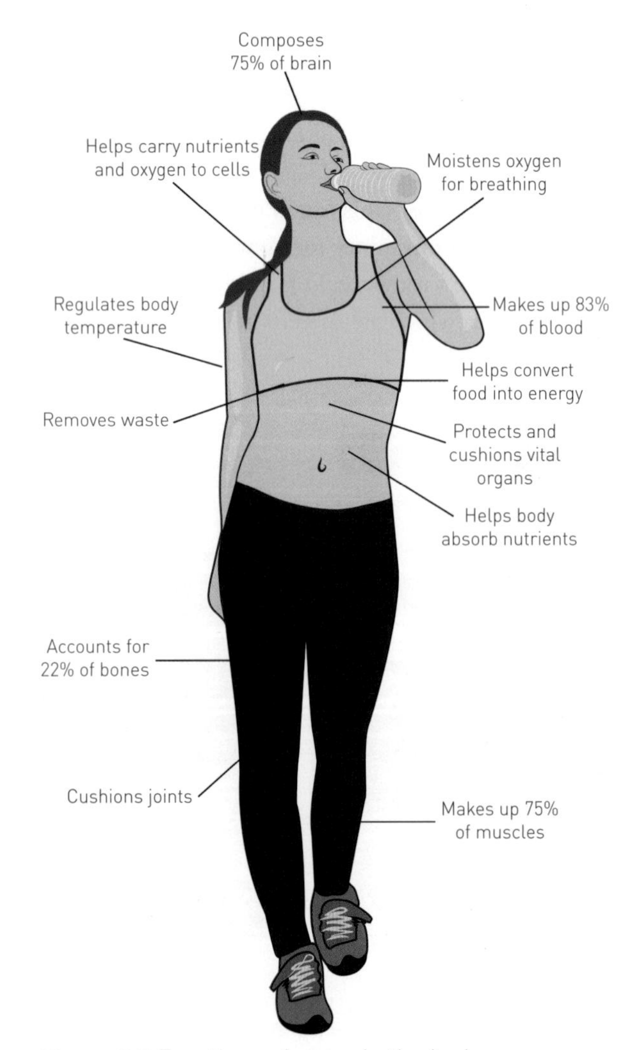

Figure 2.3 Functions of water in the body

HANDS-ON

SMOOTHIE TIME

PURPOSE

To design your own healthy smoothie.

MATERIALS

½ cup frozen fruits (saves using ice)
½ cup milk or juice
½ cup yoghurt
blender or 'bullet'

STEPS

1 Place the milk or juice in the blender and then add the yoghurt and fruit.
2 Blend until smooth. Serve.

ACTIVITIES

1 Identify one ingredient in a smoothie that is rich in the following nutrients:
- protein
- vitamins
- minerals
- water.

2 What fruits can be used in smoothies?
3 Research the method for freezing fruit at home.
4 Find a recipe for a green smoothie.

Shutterstock.com/Sergey Molchenko

HANDS-ON

MAKE RICOTTA CHEESE

PURPOSE

This fresh soft cheese is easy to make at home. It can be eaten on its own or used in savoury or sweet dishes. It makes curds and whey.

MATERIALS

1 litre milk
125 mL cream (not thickened)
60 mL white vinegar or lemon juice
Pinch of salt
Heavy-based saucepan
Thermometer
Slotted spoon
Ricotta hoop or colander lined with cheesecloth
Bowl

STEPS

1 Heat milk and cream slowly in the saucepan until it reaches 93°C. Ensure that it does not boil.
2 Turn off heat. Add the vinegar or lemon juice and quickly stir once only with a slotted spoon.
3 Allow to rest for 20 minutes to form a raft of curds on the surface.
4 Using a slotted spoon, gently lift the curds and place in the ricotta hoop to drain over a bowl.
5 Strain for 20 minutes before consuming.

ACTIVITY

1 What nutrients are found in milk?
2 What nutrients are found in cream?
3 What factors cause the milk proteins to coagulate into a curd?
4 Find a recipe that uses ricotta cheese.
5 Research how the leftover whey can be used.

Shutterstock.com/vanillaechoes

UNIT REVIEW

LOOKING BACK

1 Answer true or false to each of the statements below.
  - a Proteins are found in all body cells.
  - b Amino acids make up carbohydrates.
  - c When the body consumes more protein than it needs, the excess turns to body fat.
  - d Starches take less time to digest than sugars.
  - e Fats are liquid at room temperature.

2 How many kilojoules (energy value) per gram do each of these nutrients provide?
  - Proteins
  - Carbohydrates
  - Lipids
  - Water

FOR YOU TO DO

3 Examine the following ingredient list from a packet cake mix: sugar, wheat flour, milk solids, vegetable oils, glucose syrup, egg powder, salt, baking powder, oats and coconut cream.
  - a Circle the ingredients rich in protein.
  - b Underline the ingredients rich in carbohydrates.
  - c Box the ingredients rich in lipids.
  - d Cross the ingredients rich in sodium.

4 Complete the following poem to encourage people to drink water.

W ater is important
A
T
E
R

TAKE IT FURTHER

5 Prepare a three-minute speech covering the following points about a vitamin or mineral allocated to you by your teacher:
  - the role of the vitamin or mineral in the body
  - rich food sources of the nutrient
  - what occurs if you have an inadequate intake of the nutrient.

You may use visual aids to enhance your presentation.

2.2 The role of fibre in the diet

Fibre is a type of complex carbohydrate found in cereals, legumes and pulses, fruits and vegetables. Fibre cannot be digested but is important in maintaining a healthy digestive system and preventing problems, such as:

- constipation
- haemorrhoids
- **diverticulitis**
- **irritable bowel syndrome**
- obesity
- coronary heart disease
- diabetes
- colon cancer.

Most Australians do not eat enough fibre. While fibre supplements are available, it is best to obtain fibre from food sources that also provide other valuable nutrients.

There are three types of fibre.

Table 2.4 Types of fibre

TYPES OF FIBRE	BENEFITS	FOOD SOURCES
Soluble fibre Pectins and gums	• Slows stomach emptying so you feel full longer • Helps to control weight • Lowers blood sugar levels • Lowers cholesterol levels	Fruits Vegetables Oat bran Barley Seed husks Dried beans Lentils Soy milk

TYPES OF FIBRE	BENEFITS	FOOD SOURCES
Insoluble fibre Cellulose, hemicelluloses and lignin	• Absorbs water • Helps to move food through digestive system • Prevents constipation	Bran – rice, corn, wheat Skins of fruit and vegetables Nuts Seeds Dried beans Wholegrain foods
Resistant starch	• Maintains a healthy digestive system	Firm bananas Chick peas Potatoes Lentils Unprocessed grains, such as brown rice

Alamy Stock Photo/age fotostock

Figure 2.4 Fibre-rich foods

HANDS-ON

FIBRE-RICH BRAN

PURPOSE

Bran is the indigestible outside husk of grains such as wheat and oats. Observe the role of bran as a dietary fibre.

MATERIALS

¼ cup bran

½ cup water

STEPS

1 Soak the bran in the water.
2 Wait up to 10 minutes.
3 Observe the results.

ACTIVITIES

1 What happened to the bran?
2 How does this result resemble what occurs in the digestive system?

HANDS-ON

INDIVIDUAL MICROWAVE POPCORN

PURPOSE

Popcorn is high in fibre. It is a good snack and easy to make at home.

MATERIALS

¼ cup popcorn

Microwave-proof bowl and plate

Salt

STEPS

1 Place popcorn in a bowl and then cover with plate.
2 Microwave for 30 seconds at a time until your popcorn pops. Time will depend on your microwave power. Then add salt.
   Be careful when you remove the plate because the steam can burn you.

ACTIVITIES

Assess the advantages and disadvantages of this method to a packet of microwave popcorn you can buy from a store.

Mark Fergus

UNIT REVIEW

LOOKING BACK

1 What is dietary fibre?
2 List five health problems that can be prevented if your diet is rich in dietary fibre.
3 Name three foods rich in insoluble fibre.
4 Answer true or false to each of the statements below.
   - Fibre is found in animal foods.
   - White rice has more fibre than brown rice.
   - Lignin is a dietary fibre.
   - You need to drink plenty of water when on a high-fibre diet.
   - Dietary fibre has no kilojoules because it is mainly indigestible.
   - Popcorn is higher in fibre than chocolate.

FOR YOU TO DO

5 Change the following breakfast menu so that it is higher in fibre: sugar-frosted cereal with milk, toasted white bread with jam and orange juice.
6 Plan a high-fibre lunch for children to take to school.
7 Fresh fruits and vegetables are rich sources of fibre. For each letter of the alphabet, think of a fruit or vegetable that starts with that letter.

TAKING IT FURTHER

8 Make a Friendship Fruit Salad. In groups, organise for each team member to bring in a piece of fruit suitable for fruit salad. Prepare the fruit in class and enjoy it with some yoghurt or ice-cream supplied by the teacher.

2.3 Foods developed to enhance health

Probiotics

Probiotics are live micro-organisms, such as bacteria. The intestines contain both good and bad bugs. The good bugs help make Vitamin K, Vitamin B12 and folate. Eating a variety of healthy foods aids in keeping a balance of good and bad bugs in the digestive system. Poor diets, stress, ageing or antibiotics, however, can lead to a situation where there are more bad bugs than good bugs, causing problems such as diarrhoea and stomach ulcers. Probiotic foods therefore aim to introduce good bacteria to ensure a healthy digestive system.

Probiotic foods include yoghurts and fermented milks, such as kefir and buttermilk, that have good bacteria, such as lactobacillus and bifidobacteria. Fermented soybean products, such as miso soup and traditionally made sourdough bread, also contain probiotics.

Functional foods

Functional foods are modified or **enriched** to provide extra health benefits beyond the traditional nutrients they already naturally contain. Such products are similar in appearance to conventional foods, but their modifications provide added health benefits.

Table 2.5 Examples of functional foods and their suggested benefits

FUNCTIONAL FOODS	BENEFITS
Fermented milk drinks Yoghurts with healthy bacteria cultures Products with extra dietary fibre Gluten-free products	Improve the immune response and promote intestinal health
Margarine with plant sterols Products containing omega-3 oils	Reduce cholesterol levels, protecting against heart disease
Milks and fruit juices enriched with calcium and iron Cereals enriched with vitamins and minerals such as Vitamin B	Reduce the risk of osteoporosis and anaemia
Artificially sweetened products and diet products Low-fat, low-sugar/carbohydrate, high-protein, low-salt products	Reduce the risk of disorders such as obesity, heart disease, type 2 diabetes and hypertension

Shutterstock.com/Bananaboy

Figure 2.5 Which of these milks are functional foods?

FOOD IN FOCUS

MEADOWLEA® MARGARINE

MeadowLea® margarine was first produced in Australia in 1939. The margarine is made from Australian grown, non-genetically modified, crushed cholesterol-free canola and sunflower seeds. The company also promotes that their margarine has no artificial colours and flavours and contains 65 per cent less saturated fat than butter.

Over the years, the MeadowLea® range has developed to meet changing consumer needs. Today it provides a selection of product sizes and is available in a range of varieties, including:

- Original
- Salt Reduced
- Canola Omega 3
- Light
- Skinny
- Gluten, dairy and salt-free

Source: Adapted from information on the Goodman Fielder website

ACTIVITIES

1 What is margarine made from?
2 Of the product range listed, which margarine would be most suitable for someone who is lactose intolerant?
3 Which product(s) would be most suitable for someone who wants to cut down their fat intake?
4 Research a benefit of reducing salt intake.

Mark Fergus

UNIT REVIEW

LOOKING BACK

1 Define the term 'probiotic food'.
2 Give two benefits of consuming probiotic foods.
3 Define the term 'functional food'.
4 Give an example of a functional food and describe the health benefits of this food.

FOR YOU TO DO

5 Identify the functional and probiotic foods in the following story.
David sits down to breakfast. He pours a bowl of cereal containing added fibre to aid his digestion and hopefully reduce the risk of colon cancer and heart disease. Over the cereal he pours low-fat, high-calcium milk to control his weight and prevent osteoporosis. He spreads his toast with margarine containing plant sterols, which the packaging claims will reduce cholesterol, and before he leaves for work he picks up his strawberry yoghurt containing millions of live bacteria for bowel health.

TAKING IT FURTHER

6 Examine a package for a functional or probiotic food and indicate the benefits claimed on the package.
7 Find a recipe for making yoghurt at home.

2.4 Under- and over-nutrition and diet-related disorders

There is an abundance of food in Australia, and advances in technology have meant that our lifestyles are less active than those of past generations. Increasingly, Australians are suffering diet-related disorders that indicate over-nutrition or under-nutrition. Over-nutrition results from consumption of an excess of nutrients, while under-nutrition is the result of nutrient deficiency.

Type 2 diabetes

A person who has diabetes has too much glucose in the blood. This occurs because the person does not produce enough of the hormone insulin, or the insulin does not work well. Insulin moves glucose from the blood into the body cells, where the glucose is converted to energy. As a result, diabetes can be life-threatening and can contribute to other problems, such as heart disease, kidney disease, eye disease and infections.

About 10 per cent of diabetics have type 1 diabetes, where for some unknown reason the pancreas stops making insulin. Type 1 diabetes usually affects young people. They need daily insulin injections in addition to regular blood glucose tests and a controlled eating and exercise plan.

The most common form of diabetes in Australia is type 2 diabetes. It generally occurs in people over the age of 40 but is now increasingly affecting younger people. The pancreas produces less insulin or the insulin is blocked because of excess fat. Unlike type 1 diabetes, type 2 can be avoided. A healthy body weight, regular meals and a balanced diet help. It is also best to choose low-GI foods that slowly release energy, such as wholegrain breads and cereals, fruit, vegetables, legumes and pulses, and limit sugary foods that release glucose quickly into the blood.

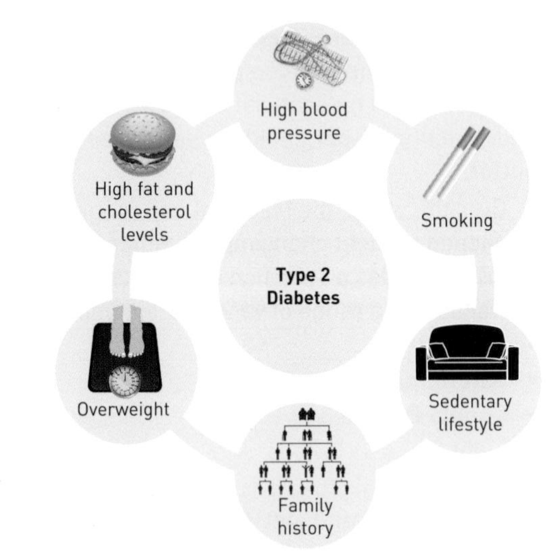

Figure 2.6 Risk factors for type 2 diabetes

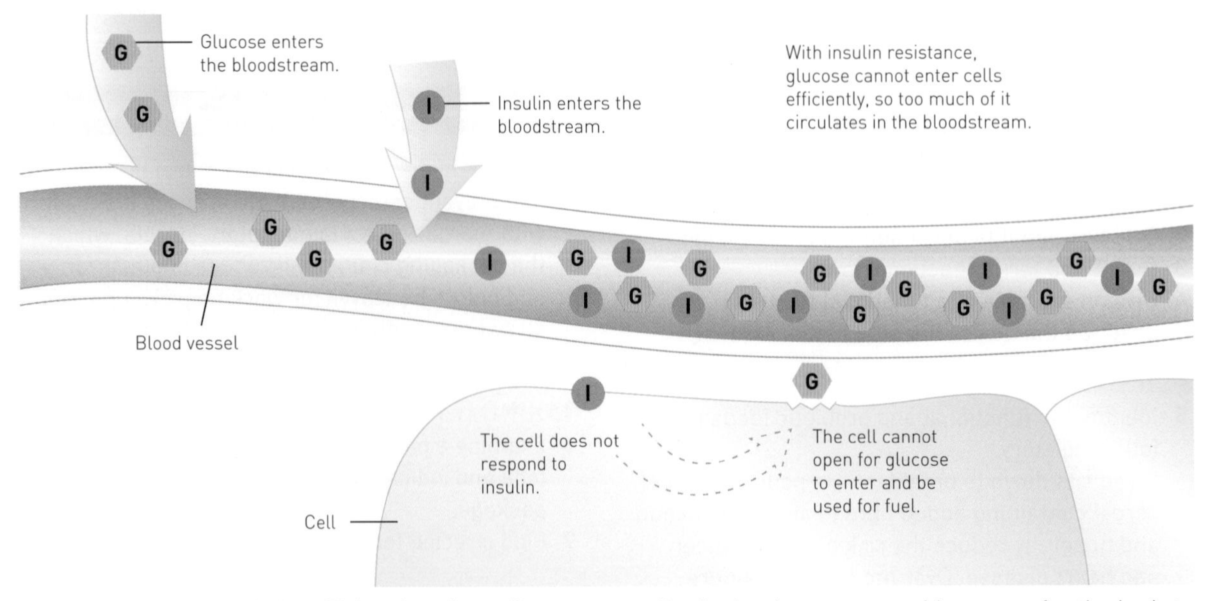

Figure 2.7 When there is insufficient insulin, cells cannot easily obtain glucose to provide energy for the body to work efficiently.

Coeliac disease

People who suffer coeliac disease are sensitive to the protein gluten, found in wheat. The gluten damages the lining of the small intestines, making it harder to absorb nutrients from food and commonly causing stomach pains.

A person with coeliac disease needs to avoid foods containing wheat and often rye, barley and oats as well. This can be difficult because gluten is present in many products, such as bread, biscuits, cakes, pastries, breakfast cereals, pasta, sausages, soups, sauces and puddings. Potato, rice, soy bean and some cornflours are good substitutes for wheat flour. These are commonly used in the gluten-free products that are available commercially.

Obesity

A person is considered obese when the body is 20 per cent over its ideal weight. Obesity is the most common nutritional disorder in Australia and can contribute to other conditions, such as heart disease, diabetes and gall bladder disease.

While there can be medical and genetic reasons for obesity, it usually results from consuming more energy from food than the body needs. The extra energy is converted to fat and stored on the body, resulting in weight gain. To lose weight, the body needs to take in fewer kilojoules than it uses for energy.

To prevent obesity, it is best to:

- eat more fresh fruit and vegetables, because they are low in energy and filling
- cut down on foods rich in sugar and fat, because they are high in kilojoules
- drink more water, because it has no kilojoules
- eat more wholegrain and high-fibre products, because they make you feel full
- keep active to use up energy.

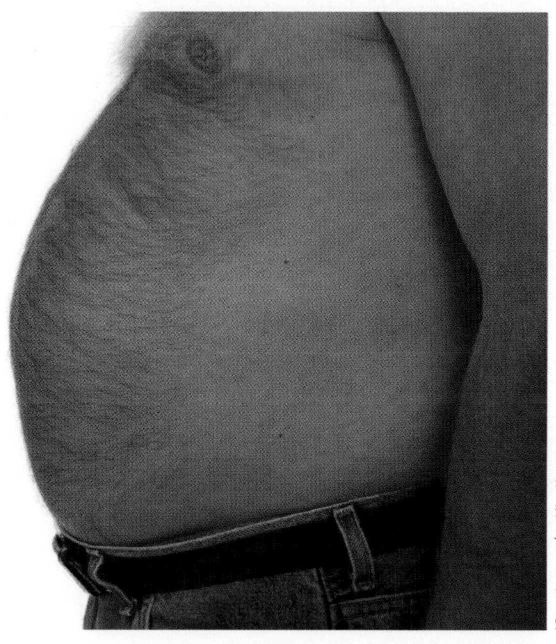

iStock.com/golfladi

Figure 2.8 One gram of alcohol provides 29 kilojoules of energy, hence the term 'beer belly'. Lack of exercise and too many kilojoules from food may also contribute to excess weight.

FOOD IN FOCUS

GET HEALTHY INFORMATION AND COACHING SERVICE

Get Healthy Information and Coaching Service® is a free information and health coaching service developed by the NSW Government. The service targets NSW adults at risk of developing chronic disease such as type 2 diabetes and heart disease due to having one or more of the following risk factors:

- not meeting healthy eating guidelines
- inadequate physical activity
- being overweight.

Coaching participants receive up to 10 free individually tailored telephone coaching sessions over six months. The health coach can help identify health goals, create action plans, maintain motivation and discuss options for a beneficial lifestyle change.

Participants who have completed the program have reported significant improvements, including:

- an increase in physical activity levels
- an average weight loss of 3.9 kg
- a reduction in their Body Mass Index (BMI)
- an average reduction in waist circumference of 5 centimetres

>

- an increase in fruit and vegetable consumption.

The development of the service is based on the Quitline, which helped thousands of smokers quit the habit and start leading healthier lives. Get Healthy at Work is another government initiative. It helps businesses identify health issues facing their workplace by providing brief health checks for employees and then setting up a Workplace Health Program for businesses to undertake.

ACTIVITIES

1 What two chronic conditions may be prevented by getting healthy?
2 Identify three changes that were report by participants after completing the program.
3 The Get Healthy program was based on what model?
4 What do you think would be the benefits for businesses to undertake the Get Healthy at Work program?

Get Healthy NSW

Anaemia

In red blood cells, iron combines with protein to form haemoglobin, which carries oxygen in the blood. A lack of iron therefore reduces the amount of oxygen carried in the blood. A chronic shortage of iron in the diet can lead to anaemia. The symptoms include:

- pale skin
- fatigue
- breathlessness
- irritability
- always feeling cold
- lower attention span.

Anaemia is more common in women, particularly those between 11 and 50 years of age, because of menstruation, and because they often restrict their food intake of iron-rich foods such as red meat. It is also common in the aged. Demand for iron increases during pregnancy, growth stages and participation in endurance sports.

Osteoporosis

Osteoporosis is the loss of calcium from bone, resulting in:

- fragile bones
- loss of height
- pain
- curvature of the spine.

In Australia, osteoporosis affects one in two females and one in three males over the age of 60. The greater the bone mass achieved during childhood and adolescence when the body is building bones, the better the chance of preventing osteoporosis later in life. But eating enough calcium is essential throughout life to maintain strong bones and to help slow down the bone loss that occurs with ageing. Regular physical activity throughout life also helps to keep bones strong and healthy, as do adequate supplies of Vitamin D and phosphorus, which help calcium to work.

Dairy foods, particularly milk, cheese and yoghurt, are the best dietary source of calcium. Canned fish eaten with the bones, green leafy vegetables (except spinach), almonds, cereals and legumes also contribute some calcium to the diet.

Coronary heart disease

Coronary heart disease occurs when the arteries become narrow from fatty deposits that cling to the artery walls. This can trigger a blood clot in the artery, which may cause a heart attack.

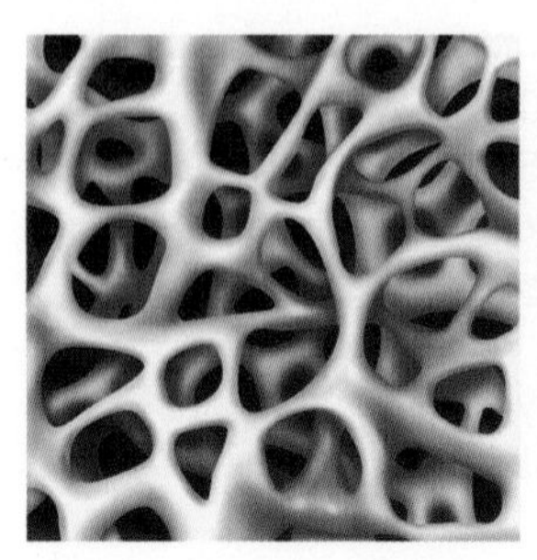

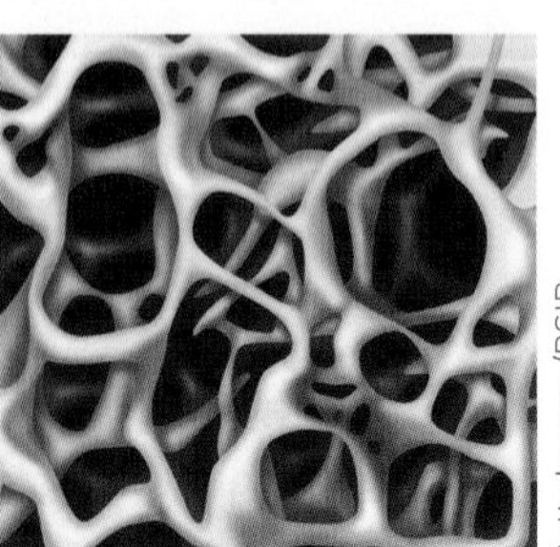

Getty Images/BSIP

Figure 2.9 Normal bone (left) and osteoporotic bone (right)

The following factors increase the risk of heart disease:

- cigarette smoking
- a **sedentary** lifestyle
- family history of the disease
- a diet high in saturated fats
- obesity
- high blood pressure
- diabetes.

Saturated fats are linked to the disease, so it is wise to limit your intake of foods containing this type of fat, as well as limiting the total fat content of your diet. Recommendations include:

- avoid fried foods and high-fat foods, such as pastries and chocolate
- switch to low-fat products
- choose lean meats and remove the skin from chicken
- read labels for their fat content
- cut down on butter, and substitute vegetable oils when cooking
- eat more plant-based foods, particularly those that are fibre-rich
- eat oily fish at least once per week (e.g. sardines, tuna and salmon) because the omega-3 fatty acids can help to lower levels of fat in the blood.

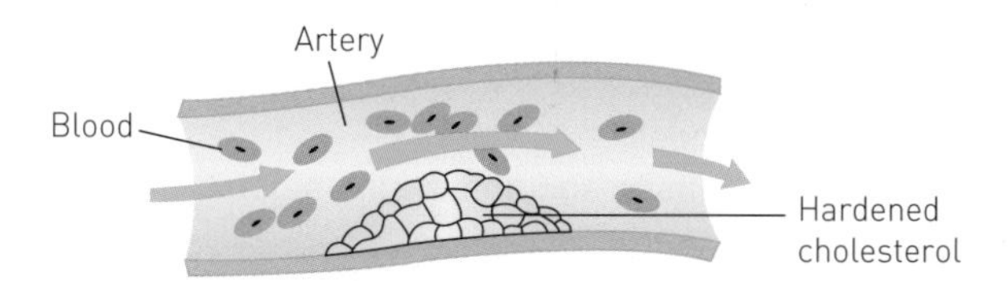

Figure 2.10 Hardened cholesterol deposits in the arteries limit blood flow to the heart. A blood clot could seal off the artery and cause a heart attack.

Hypertension

High blood pressure, or hypertension, means that the blood pumps harder than normal through the arteries. This is a risk factor for heart attacks and strokes.

Too much salt in the diet, as well as family history of the disease, can contribute to hypertension. So can obesity and smoking. A low-salt and low-fat diet, as well as regular exercise, is recommended to avoid hypertension.

To avoid a high salt intake, it is best to:

- read food labels to establish their sodium content
- limit the intake of salty foods such as take-aways, snack foods and packet soups.
- substitute herbs and spices for salt
- remove the salt shaker from the table.

Colon cancer

Bowel or colon cancer is a common cancer, primarily affecting adults. Abnormal cells on the wall of the large intestines (colon) multiply to form a **tumour**, which can spread to other parts of the body.

High intakes of alcohol and red meat appear to increase the risk of colon cancer. Maintaining a healthy weight and consuming a diet that includes a variety of vegetables, fruit, legumes, grains and some oily fish with omega-3 fatty acids appears to protect against the cancer.

Figure 2.11 Choose a physical activity you enjoy so that it becomes part of your healthy lifestyle.

DIETITIAN, NUTRITIONIST OR NATUROPATH?

Visit a nutritionist if you have weight or health concerns and want to get the right food plan for you. Concerns like digestion, improved immunity, recovery from surgery, chronic disease management, skin and hair health, allergies, kids' health, general energy and even getting a good night's sleep can often benefit from adjusting your diet.

Dietitians cover everything nutritionists do but are further specialised in clinical nutrition, medical nutrition therapy and food service management. This basically means they are more qualified to work in hospitals and with disease management, as well as recovery from eating disorders, severe allergies and helping decode food law for those in the food service business.

Both require university degrees in nutrition, but dietitians require more study and practical experience before they graduate. Only dietitians are recognised by Medicare, meaning you can get a rebate for their service, but private health insurance companies usually recognise nutritionists. Both nutritionists and dietitians may work in health promotion, private practice and community health.

Naturopathy is a holistic approach to wellness. Practitioners believe the body has the inherent ability to heal itself, with a little helping hand from nature in the form of herbal medicine, flower essences and natural supplements. Naturopaths may also provide dietary advice and detoxification techniques. Dietary strategies are not always implemented and interventions may not always be based on scientific evidence. Naturopaths often work in private practice. While naturopaths can have bachelor degrees, diplomas or other certificates, they do not have to be registered in Australia, unlike dietitians and nutritionists.

ACTIVITIES

1. What are four health concerns a nutritionist or dietitian can help to address?
2. What areas are dietitians specialised in?
3. What do naturopaths believe in relation to wellness?
4. Using the internet, investigate a job seekers website and find a current job opportunity as a nutritionist or dietitian.
5. Identify the attributes required for the job and the responsibilities of the nutritionist or dietitian in this position.

UNIT REVIEW

LOOKING BACK

1 Complete the following table in your notebook.

DISORDER	DIETARY CAUSE	ONE PREVENTION OPTION
	Too much glucose in blood	
Obesity		
		Increase red meat consumption
	Lack of calcium	
		Cut down on saturated fats
Colon cancer		

FOR YOU TO DO

2 Reply to these reader questions from a magazine.

a I have been told I have hypertension and I need to watch my diet. I love salted snack foods but can you suggest some alternatives?

b My family all have a sweet tooth but my husband has type 2 diabetes. What suggestions do you have for desserts?

c Can you suggest some interesting lunches for my teenage grandson, who has coeliac disease?

TAKING IT FURTHER

3 You have been commissioned by the government to produce material to inform the public about common diet-related disorders in Australia. Research one diet-related disorder and include in your material:

- the symptoms of the disorder
- the causes of the disorder
- how the disorder can be prevented and/or controlled
- support groups and agencies that deal with the disorder
- a recipe that can be used to help prevent or control the disorder.

2.5 Anorexia and restrained eating

Society pressures people to be slim. For some, dieting can get out of control and lead to eating disorders.

Anorexia

Anorexia nervosa is a psychological illness with serious physical consequences. Sufferers typically have an obsessive fear of gaining weight and also have a distorted body image. This leads to the sufferer depriving him or herself of food, which often leads to the characteristic low body weight. Though most common in adolescent females, the illness can occur in both males and adults as well.

Anorexia places strain on organs, which can lead to life-threatening conditions such as kidney failure, dehydration and seizures. Many teenage girls also stop menstruating because their hormone levels are affected. Some sufferers even experience growth of lanugo, a fine layer of hair all over the body, which promotes warmth.

Behaviours associated with anorexia include:

- excessive exercising
- obsession with food and kilojoules
- excuses for not eating
- a belief they are 'too fat' even when thin
- frequent weighing
- use of laxatives and diet pills
- irritability or mood swings
- wearing baggy clothes to hide weight loss.

Why some people become anorexic is difficult to say. It may result from a stressful situation, from losing control after some initial weight loss or from low self-esteem. People with anorexia are often perfectionists. Treatment for anorexia involves counselling and family support, and hospitalisation is commonly required.

Restrained eating

A restrained eater will stop eating before they are full. Unfortunately, this means they may be hungry later and are then likely to overeat, leading to weight gain. Many restrained eaters are constantly on diets.

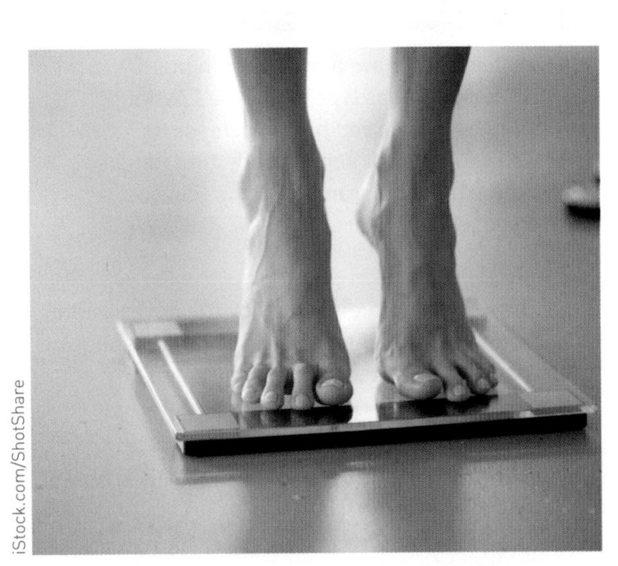
iStock.com/ShotShare

Figure 2.12 Anorexia is a form of malnutrition.

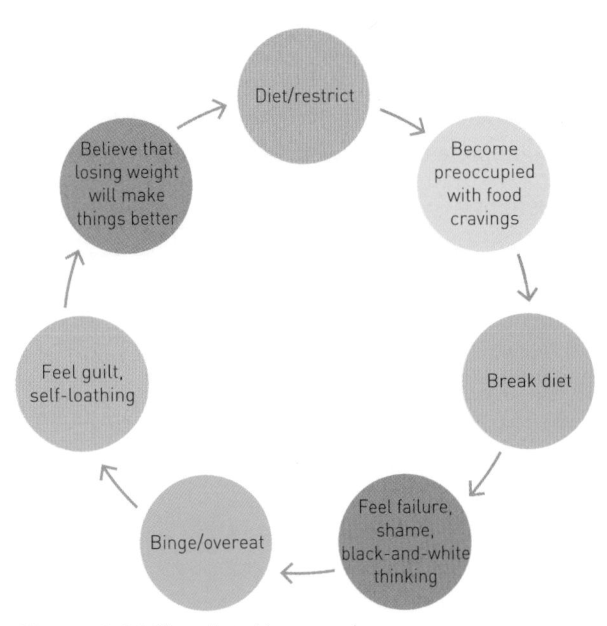

Figure 2.13 The diet–binge cycle

HOW TO BOOST YOUR BODY IMAGE

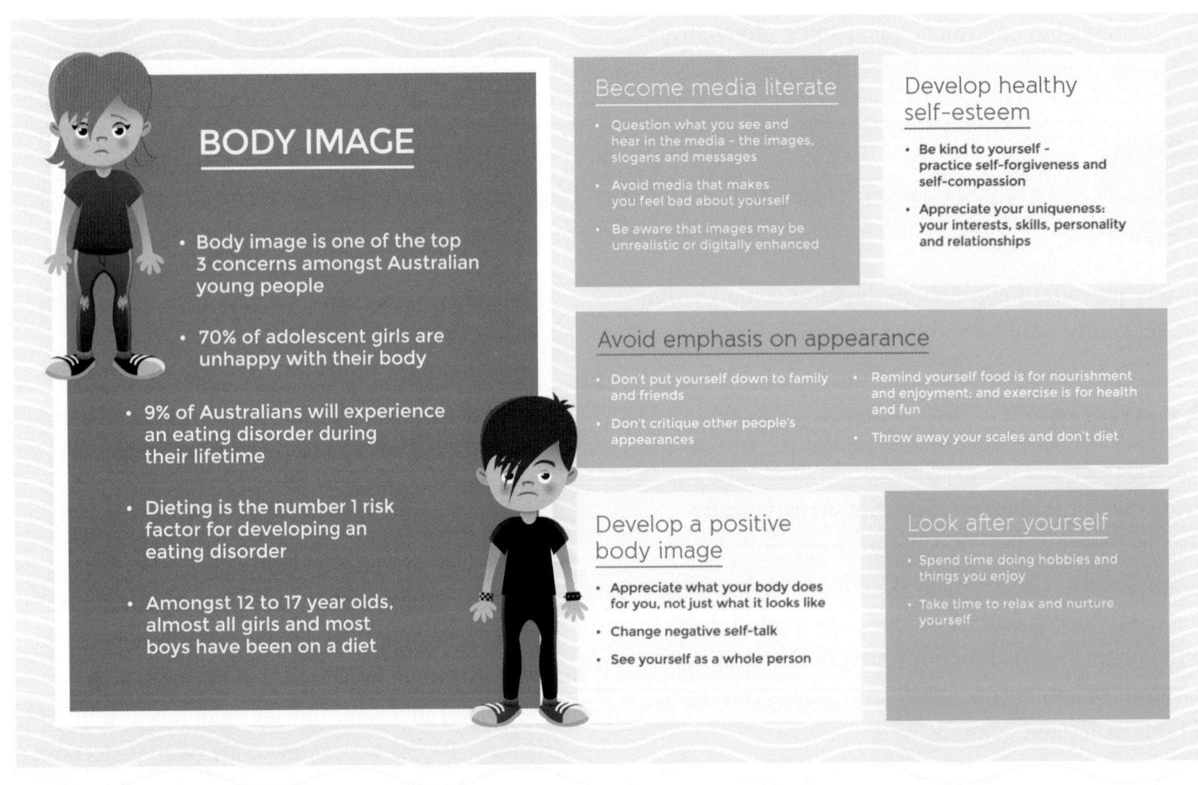

www.eatingdisorders.org.au

Figure 2.14 Advice from the Victorian eating disorder support and information service Eating Disorders Victoria

Bulimia nervosa is a severe form of restrained eating. It involves recurrent binge-eating episodes, which are followed by either purging (by vomiting or laxatives), fasting or overexercising. During a bingeing episode, a sufferer will usually eat large amounts of kilojoule-rich foods. After bingeing, a sufferer will usually feel guilt, shame and that he or she has lost control, which tends to trigger the purging behaviour. Between episodes, sufferers may follow strict diet and exercise regimens. This vicious binge-purge-exercise cycle dominates the sufferer's life. The bulimia sufferer's body suffers too, through chemical imbalances, eroded dental enamel from vomiting and the possibility of a ruptured stomach. Many bulimia sufferers maintain an average weight, or are slightly above or below the average weight for their height, which helps their illness to go undetected for a longer period of time.

Like anorexia, bulimia affects more females than males and treatment *must* involve counselling. Poor self-esteem is considered a major contributor to bulimia and restrained eating.

UNIT REVIEW

LOOKING BACK

1 What is anorexia?
2 Identify four common behaviours associated with anorexia.

FOR YOU TO DO

3 What psychological attributes do you think are needed to be a successful model in today's society?
4 An article shared on social media has the heading 'Your dieting is making you fat'. Explain why this may be so.
5 Reply to the following questions from an internet forum: 'I think my sister is out of control with her dieting. She is now too skinny. Do you think I am just jealous or could she have a problem? If she does have a problem, what can I do?'

TAKING IT FURTHER

6 Watch a movie or documentary on anorexia to assess the causes and dangers of the illness.

Food consumption in Australia

2.6 Food consumption in Australia

Changes in consumption patterns

What we eat, when we eat, where we eat and with whom we eat have all changed greatly over the last few decades. Changes in the patterns of our food consumption have occurred for a number of reasons.

- The influences of different cultures: For instance, Asian cuisine commonly uses rice, poultry, seafood and pork. European and American cultures also drink more coffee. We are eating not only more of these foods but also more varieties of these foods.
- Health concerns: We now eat less red meat, eggs, white sugar and butter because of concerns about disorders such as heart disease and diabetes. Meals have become lighter, with more emphasis on salads, chicken, seafood, pasta, stir-fries and vegetarian dishes. There has also been a trend to consume more high-protein foods and to reduce the intake of carbohydrate-rich foods.
- Changes in family structures and roles: Families are smaller and commonly both parents work. This has meant more eating of convenience foods, eating out more and more snacking as families spend less time together eating as a family. Many convenience, take-away and snack foods are high in sugar, salt and fat and low in fibre.
- Technology: This has increased the range of packaged foods available. Today there are many more forms of chilled, frozen, dried, canned and long-life products available for consumers to purchase. Technology has also made it easier to obtain and store food; for example, you can order a pizza via an app and have it delivered.

Table 2.6 Food consumption trends

FALLEN	RISEN
Bread Oats Sugar for home cookery Beef Lamb Eggs Milk Cheese Butter Beer Tea	Packaged breakfast foods Rice Sugar in processed foods Chicken Pork and bacon Seafood Soft drink Coffee Wine Oil Margarine Fruit Vegetables Bottled water

Impact on health

Diet-related disorders are on the rise in Australia, because we now consume:

- too much energy (kilojoules) as fat, alcohol and sugar
- too much salt
- not enough complex carbohydrates and fibre
- not enough of some vitamins and minerals, including calcium and iron.

Heart disease and diabetes are among the top 10 causes of death in Australia. Obesity and hypertension are contributing factors to heart disease, while childhood obesity has become a major concern for health authorities in Australia.

Influences on food selection and subsequent effects on health

Shutterstock.com/Ariwasabi

Figure 2.15 What factors may have influenced the choice of this meal?

Table 2.7 Food selection and health

INFLUENCES	EFFECTS ON HEALTH
Physiological: what food does to the body	Adults encourage children to drink milk for strong bones, but often forget the importance of milk in preventing osteoporosis themselves. Some foods are avoided because of allergies that affect the body.
Psychological: beliefs about food	Eating chocolate makes you feel good, but can lead to obesity if excessive.
Geographical: location and climate	Urban areas often have an abundance of food available that is high in fat, sugar and salt. In summer you eat more salads to cool you.
Social: friends, family, cultural or other	Eating in a group is fun, but often party and snack foods can be high in salt, sugar and fat.
Technological: technology and media	Processed foods are everywhere, but many processed foods are high in salt, sugar and fat and it is these foods that are often heavily advertised in the media. The media often encourages us to try new foods and gives us health information.
Economic: money and time	Poor people may miss out on some important nutrients because of limited income. When we have extra money or lack time we tend to treat ourselves to foods that are often high in sugar, salt and fat.

UNIT REVIEW

LOOKING BACK

1 List four foods that have decreased in consumption.
2 List four foods that have increased in consumption.
3 What nutrients do Australians consume in excess?
4 What nutrients do Australians consume too little of?
5 List the six influences on food selection.

FOR YOU TO DO

6 Australians have increased their consumption of seafood and coffee in the last few decades.
   a List three different forms of seafood that you can purchase and give an example of a recipe or dish each can be used for.

>

b List three different types of coffee you can purchase in a cafe.

7 Plan a school lunch for a 10-year-old. Outline what influenced your food selection.

8 Plan a low-cost dinner for a student who has just moved out of home to attend university. Outline what influenced your food selection.

TAKING IT FURTHER

9 Australia's oat producers want to increase their sales by placing a recipe that appeals to teenagers on social media.

a In small groups, research some recipes using oats and then produce a recipe you feel teenagers would like.

b Evaluate the recipe by asking the opinions of your peers.

10 'Eating the crust of the bread makes your hair curly.' List other beliefs people have about food.

2.7 National guidelines for healthy eating

The Australian Dietary Guidelines give advice on eating for health and wellbeing. Based on the latest scientific evidence, they describe the best approach to eating for a long and healthy life. Each Guideline is considered to be equally important in terms of public health outcomes.

2.7 National guidelines for healthy eating

Table 2.8 The Australian Dietary Guidelines

GUIDELINE 1
To achieve and maintain a healthy weight, be physically active and choose amounts of nutritious food and drinks to meet your energy needs • Children and adolescents should eat sufficient nutritious foods to grow and develop normally. They should be physically active every day and their growth should be checked regularly. • Older people should eat nutritious foods and keep physically active to help maintain muscle strength and a healthy weight.
GUIDELINE 2
Enjoy a wide variety of nutritious foods from these five groups every day: • Plenty of vegetables, including different types and colours, and legumes/beans • Fruit • Grain (cereal) foods, mostly wholegrain and/or high cereal fibre varieties, such as breads, cereals, rice, pasta, noodles, polenta, couscous, oats, quinoa and barley • Lean meats and poultry, fish, eggs, tofu, nuts and seeds, and legumes/beans • Milk, yoghurt, cheese and/or their alternatives, mostly reduced fat (reduced fat milks are not suitable for children under the age of 2 years) And drink plenty of water.
GUIDELINE 3
Limit intake of foods containing saturated fat, added salt, added sugars and alcohol a Limit intake of foods high in saturated fat such as many biscuits, cakes, pastries, pies, processed meats, commercial burgers, pizza, fried foods, potato chips, crisps and other savoury snacks. Replace high fat foods which contain predominantly saturated fats such as butter, cream, cooking margarine, coconut and palm oil with foods which contain predominantly polyunsaturated and monounsaturated fats such as oils, spreads, nut butters/pastes and avocado. Low fat diets are not suitable for children under the age of 2 years.
b Limit intake of foods and drinks containing added salt. Read labels to choose lower sodium options among similar foods. Do not add salt to foods in cooking or at the table.

c Limit intake of foods and drinks containing added sugars such as confectionary, sugar-sweetened soft drinks and cordials, fruit drinks, vitamin waters, energy and sports drinks.

d If you choose to drink alcohol, limit intake. For women who are pregnant, planning a pregnancy or breastfeeding, not drinking alcohol is the safest option.

GUIDELINE 4

Encourage, support and promote breastfeeding

GUIDELINE 5

Care for your food; prepare and store it safely

Source: National Health and Medical Research Council

STEPHANIE ALEXANDER KITCHEN GARDEN PROGRAM

Stephanie Alexander is an Australian cook, restaurateur and food writer. The Stephanie Alexander Kitchen Garden Program is run in over 600 schools across Australia. The program aims to provide positive and memorable food experiences to primary school children by teaching them how to grow, harvest, prepare and share fresh seasonal food. Students work in teams to grow the food organically and they cook dishes which use the vegetables, fruit and herbs grown in their gardens. The students are educated about the health benefits of food they have grown and the dishes reflect the Australian Dietary Guidelines. See page 23 for more information on Stephanie Alexander's Kitchen Garden Program.

ACTIVITY

List five things school children learn from participating in the Kitchen Garden Program.

Stephanie Alexander Kitchen Garden Foundation

Nutrition labelling

Food packages are designed to attract consumer attention, but under Australian and New Zealand food standards there are strict rules that govern how nutritional information about a food is labelled.

NUTRITION INFORMATION (Average)

Serving Size: 30g (2 biscuits) Servings Per Pack: 24

	PER SERVE	PER 100g
Energy (kJ)	447	1490
(Cal)	107	355
Protein (g)	3.7	12.4
Fat, Total (g)	0.4	1.3
- Saturated Fat (g)	0.1	0.3
- Trans Fat (g)	0.0	0.0
- Polyunsaturated Fat (g)	0.2	0.8
- Monounsaturated Fat (g)	0.1	0.2
Carbohydrate, Total (g)	20.1	67.0
- Sugars (g)	0.9	2.9
Dietary Fibre (g)	3.3	11.0
Sodium (mg)	81	270
Potassium (mg)	102	340
Magnesium (mg)	32 (10% RDI)*	107

*Percentage of Recommended Dietary Intake (RDI).

Ingredients: Organic wholegrain **wheat** (97%), organic sugar, salt, **barley** malt extract.
Contains cereals containing gluten.

Mark Fergus

Figure 2.16 Nutrition information panel

Health claims

Nutrition content and health claims are voluntary statements made by food businesses on labels and in advertising about a food. Claims such as 'low in fat' or 'good source of calcium' are examples of nutrition claims, while claims such as 'calcium is good for bones and teeth' or 'phytosterols may reduce blood cholesterol' are examples of health claims. Nutrition content and health claims are regulated by Food Standards Australia and New Zealand.

Remember, some of the healthiest foods may not have labels or health claims (e.g. fresh fruit and vegetables, nuts, lentils, beans, fresh meat and fish). Also check the nutritional information panel, because a label may claim that the product is low in fat, but it may still be high in sugar.

Legal requirements

By law, all manufactured packaged foods must carry a nutrition information panel.

The nutrition panel states the amount of energy (kilojoules), protein, fat, sodium (salt) and carbohydrate in a food. Saturated fats are included on the nutrition information panels because of their link to diet-related disorders such as heart disease, obesity and diabetes. Other nutrients, such as fibre, calcium and iron, must be included on the panel if there is a claim about the nutrient on the label.

The quantity of each nutrient is listed under two columns in the panel – per serve and per 100 grams. The panel helps you to compare foods so you can choose foods with less fat, kilojoules, salt and sugars.

Table 2.9 Voluntary labelling symbols

Commonwealth Department of Health

Health Star Rating system

This front-of-pack labelling system rates the nutritional profile of packaged food from half a star to 5 stars, making it quicker and easier for shoppers to compare similar packaged food. The more stars, the healthier the choice. The number of stars is based on a product's:

- energy (kilojoules)
- risk nutrients – saturated fat, sodium (salt) and sugars
- positive nutrients – dietary fibre, protein and the proportion of fruit, vegetable, nut and legume content.

Glycemic Index Foundation: www.gisymbol.com

The glycaemic index

The glycaemic index (GI) is a ranking from 0 to 100 of carbohydrate-containing foods based on how quickly they make blood glucose rise. The lower the GI rating, the more slowly the food is absorbed and metabolised, so blood glucose levels rise less dramatically. Low-GI foods are useful in controlling weight as they delay hunger as well as preventing diabetes and cardiovascular disease. It is difficult to predict a food's GI from its ingredient list or nutrition information. The GI symbol can be found on foods that meet strict nutrient criteria and have been tested by an accredited laboratory.

dlibrary.com.au

A food product displaying the Heart Tick

Food products displaying the Tick have been approved by the Australian Heart Foundation as being relatively low in saturated fat, kilojoules and salt.

2.8 The life cycle

As a person progresses through life, the body undergoes a lot of change. Each stage of the life cycle has special nutritional requirements.

Pregnancy and lactation

A balanced diet before and during pregnancy is essential to allow for the development of the baby. In this stage there is a need for:

- protein for the growth of new tissue
- calcium for the baby's bones and teeth

Shutterstock.com/Samuel Borges Photography

Figure 2.17 Pregnancy

- iron and Vitamins C and B for the increased blood supply.
- folate to reduce the risk of the child having a **neural tube defect.**

While a weight gain of 10–13 kilograms occurs during pregnancy, it is important to choose foods that are dense with nutrients rather than just kilojoules. This will help prevent excessive weight gain that may be difficult to lose after pregnancy.

After birth, many mothers breast-feed. The nutritional requirements of a breastfeeding (lactating) woman are similar to those of a pregnant woman. Alcohol and smoking should be avoided during pregnancy, as they can affect the baby's development.

Infancy (0–2 years)

iStockphoto/manley099

Figure 2.18 Infancy

In this period, babies grow and develop rapidly. Breast milk (or infant formula) contains all the nutrients required by the baby in the first 4–6 months, as it is high in calcium and protein needed for growth. Breast milk is preferred to formula as it contains valuable antibodies to protect against disease and is also easily digested.

After this period, milk is still important, but the baby must also be given foods that contain more iron and Vitamin C than are available from milk. In addition, more energy foods are required as the baby becomes more active. Iron-enriched rice baby cereal is usually the first food introduced to babies, followed by puréed fruit and vegetables and later puréed meat mixtures.

Childhood (2–11 years)

During this period, children grow steadily and they become more active. As a result, children require:

- energy-giving foods and foods containing Vitamin B to help release the energy
- protein for growth
- calcium and phosphorus for the strengthening of bones and teeth
- water to prevent dehydration.

It is important that children learn to appreciate a variety of foods during this stage, as this is when many food likes and dislikes are established. It is also best to avoid foods rich in fat and sugars, which could contribute to tooth decay and childhood obesity. Low-fat milks should also be avoided for children under 2 years of age because low-fat milks lack the essential fatty acids required for the growth and development of the child.

Left: iStockphoto/LSOphoto; Right: iStockphoto/nickp37

Figure 2.19 Childhood **Figure 2.20** Adolescence

Adolescence (12–18 years)

Teenagers are very active and more independent, but also very self-conscious. Sometimes poor food habits such as dieting and skipping meals are established during this period.

Generally, adolescents require:

- energy-giving foods
- protein for growth
- calcium for strong bones
- more iron, particularly for girls because of menstruation.

Adulthood

As the body has stopped growing by this stage, the requirements for some nutrients decrease in adults. A reduction in kilojoule intake is also commonly required, as adults tend to be less active because of family and work commitments. A well-balanced diet is essential during this stage to avoid the onset of diet-related disorders such as obesity and heart disease.

Figure 2.21 Adulthood

Figure 2.22 Late adulthood

Left: Shutterstock.com/wavebreakmedia; Right: Shutterstock.com/Andrey_Popov

Late adulthood

As adults get older, they tend to move more slowly and become less active, particularly after retirement. They also tend to lose their appetite. It is also at this stage that the effects of years of poor food habits can become evident through nutritional disorders.

It is recommended that older people continue to be active and choose foods that are:

- nutrient-dense rather than energy-dense
- high in fibre, to encourage bowel health
- high in calcium, to prevent osteoporosis.

HANDS-ON

FUNNY FACES FOOD ART

PURPOSE

Eating healthy can be fun and creative for children of all ages.

MATERIALS

Rice cakes or slices of bread

Suggested toppings:

- olives
- red capsicum
- cream cheese spread
- gherkins
- carrots
- cherry tomatoes
- alfalfa sprouts
- cheese slices or grated cheese.

grating and cutting tools

serving plate

STEPS

1 Use a rice cracker as the base for the face.
2 Spread the cracker with some cheese spread (optional).
3 Using some of the suggested toppings, cut or grate ingredients and then arrange them to form a face.

You can make a sweet version by using pikelets or pancakes as a base and toppings of fruit.

Mark Fergus

ACTIVITIES

Conduct an image search on food art and make a collage on food art projects that would encourage children to eat fruits and vegetables.

UNIT REVIEW

LOOKING BACK

1 List the different stages of the life cycle.
2 Why is an adequate intake of folate important during pregnancy?
3 Why is breast milk preferred for infants?
4 What are the extra nutritional requirements of adolescents?

FOR YOU TO DO

5 The Grech family consists of Vanessa, 3; Cameron, 15; Dianna, 38; David, 42; and Nan, 72. Determine which member or members have the greatest need for:
- iron
- energy
- calcium
- protein.

TAKING IT FURTHER

6 Organise a visit to the local preschool and/or retirement village. Treat the students or elderly people to some special but healthy and safe foods. Be sure to discuss their requirements with the director of the centre.
7 Select a stage in the life cycle. Produce a leaflet explaining the nutritional needs and tips for healthy eating.

2.9 Selecting nutritious foods

The Australian Guide to Healthy Eating and the Healthy Eating Pyramid are two tools that can help you to select nutritious foods.

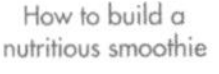
How to build a nutritious smoothie

Australian Guide to Healthy Eating poster

Healthy Eating Pyramid poster

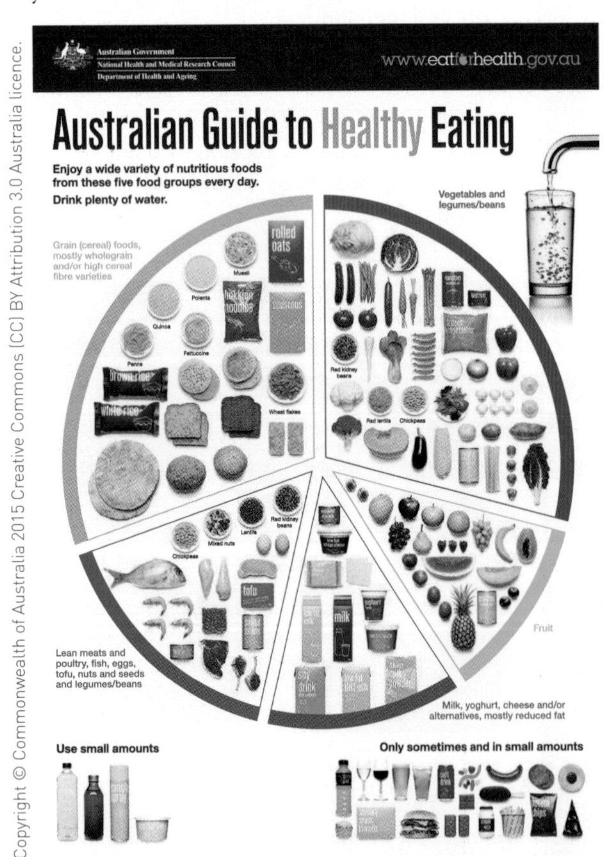

Figure 2.23 The Australian Guide to Healthy Eating is a food selection guide that visually represents the proportion of the five food groups recommended for consumption each day.

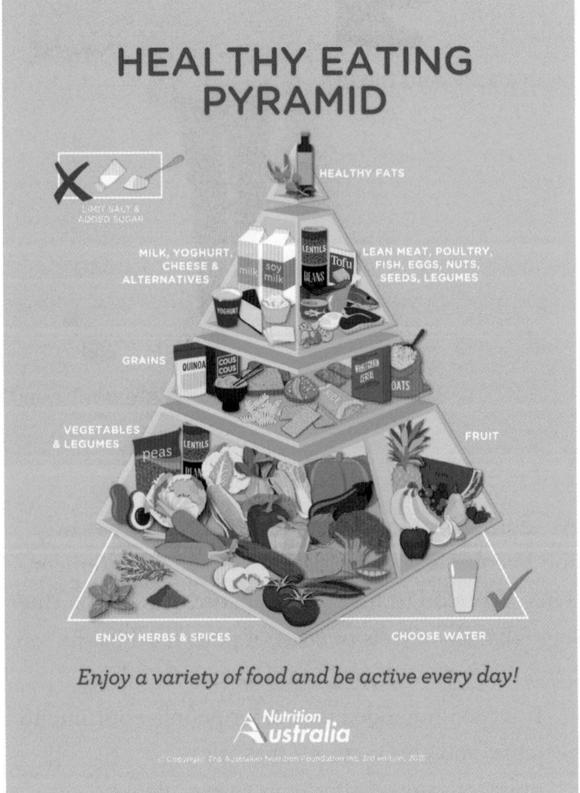

Figure 2.24 The Healthy Eating Pyramid

2.10 Changes in consumption patterns

Protein versus carbohydrates

As noted previously, the consumption of protein foods such as beef, lamb and eggs has decreased while the consumption of poultry, pork and seafood has increased. There are many reasons for this: seafood and poultry are considered relatively low-fat; salmon and other oily fish in particular are rich in omega-3 fatty acids, which may help to prevent heart disease. Soybean products have become more popular among the growing number of vegetarians, and different cultural groups have introduced Australia to new ways of preparing poultry and seafood. It is also believed that protein can delay hunger and as a result, high-protein diets have also become popular.

The consumption of carbohydrate foods such as bread and sugar for cooking has decreased, but Australians are consuming hidden sugar in many processed foods such as soft drinks and breakfast foods. As a result of Australians eating more processed foods, fibre consumption has decreased.

Over recent years, there have been many diets promoted, ranging from-low fat, high-protein, low-carb to high-fibre diets. Dietitians recommend moderation and following the Australian Guide to Healthy Eating.

Processed and unprocessed food

As food technology has improved, many more processed food products have become available. As a result, there is an abundance of chilled, frozen, dried, canned and long-life products available today. At the same time, there has been a recent trend towards consuming foods that are less processed and have undergone little change from their original state. Many farmers' or growers' markets have been established to cater for demand for fresh produce. They sell a wide range of fresh foods from across the state, ranging from fresh herbs to fresh breads and cheese. Fish markets have also become popular again with more people enjoying more varieties of seafood.

Both processed and unprocessed foods have their advantages and disadvantages.

Table 2.10 Processed and unprocessed foods

FEATURES OF PROCESSED AND UNPROCESSED FOODS	
Processed foods	**Unprocessed foods**
• Convenient – saves time and energy • Good for emergencies • Provides variety to the diet • Longer storage life than fresh • Usually more expensive • May have preservatives and other additives • Some additives may trigger allergies • Tend to be higher in fats, sugar and salt • Some have been designed for health benefits such as cholesterol-free milk • Fruit and vegetables have less fibre • Usually quantity is set – for example, 'serves 2'.	• Satisfaction in preparing food from scratch • Foods in season has more flavour and is cheaper • More time-consuming to prepare • Can control the amount of fat, sugar and salt added to the food • Usually higher in fibre, which makes the food more filling • May be higher in vitamins as processing can affect some vitamins • Environmentally friendly as it uses less packaging and energy to produce • Consumer can decide on quality and quantity of product required.

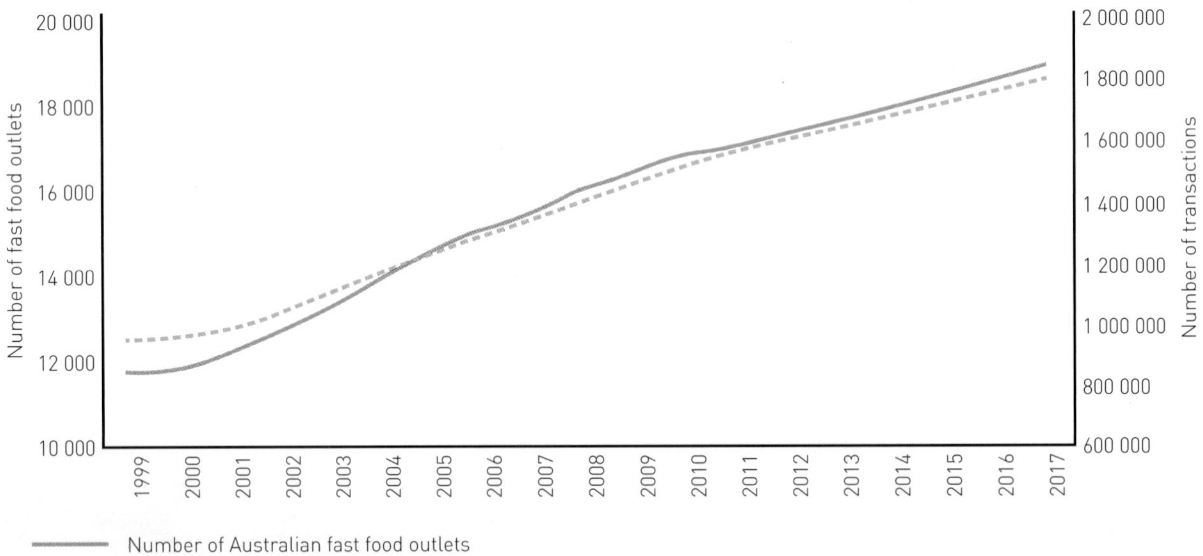

Figure 2.25 What two pieces of data does this graph indicate about fast foods? Assess what the implications of this data could be on the health of Australians.

Source: © Euromonitor International 2011

NEIL RUSSELL
EXERCISE PHYSIOLOGIST AND PERSONAL TRAINER

Neil started working as a gym instructor in 2001 while studying Human Movement Studies at the University of Technology. He graduated with 1st Class Honours in Exercise Science and lectured at both the University of New South Wales and the Australian College of Physical Education. Today, he has a very successful personal training business where he trains clients such as celebrities, athletes, models, business people and postnatal mothers. He is also an expert exercise consultant for Weight Watchers and writes many articles on health and fitness for various magazines.

Neil feels that being healthy is about achieving life balance and feeling good. He says 'focus your attention on having the healthiest food, a good night's sleep, comfortable clothes and access to a range of healthy lifestyle activities. This way you are giving your mind and body the best chance to be healthy'.

His diet is largely made up of fish, chicken and lean meats with fruit, vegetables, seeds, nuts and some grains. He also tries to stay really well hydrated. He believes in starting the day with a healthy breakfast, either quinoa porridge with banana, blueberries, chopped almonds, shredded coconut and cinnamon or a shake with similar ingredients.

When it comes to exercise, Neil undertakes activities such as sports, hiking, surfing and yoga. Neil says 'try to do things you enjoy'. Often after breakfast he does 20 minutes of yoga to help him stretch, and he loves to surf at least four times a week.

When it comes to training, Neil says you need to train specifically for what it is you want to achieve, such as weight loss, strength or speed. Personally, he trains four to five times a week, with two to three gym sessions focusing on full body strength and stability. Then he goes running or soft sand running twice a week for cardiovascular endurance.

He believes stress can be avoided by being rested, undertaking hobbies and activities that distract you from your stress and having someone you can talk to about what is stressing you. Writing music or playing an instrument can also help deal with stress in a creative way and of course, exercise helps release feel-good hormones.

Source: Neil Russell (atleta.com.au). Text: with permission of Neil Russell

ACTIVITIES

1 In two sentences, write what you think personal trainers do.
2 What four factors does Neil believe help to achieve good health?
3 What do you think are three unhealthy lifestyle activities?
4 What forms of physical activity does Neil undertake in his life?
5 List the foods that make up Neil's diet.
6 Plan a dinner menu that would suit Neil's diet.

UNIT REVIEW

LOOKING BACK

1 Which beverage is encouraged on the Australian Guide to Healthy Eating?
2 Name four foods that you can have occasionally, according to the Australian Guide to Healthy Eating.
3 To which section of the Healthy Eating Pyramid do these foods belong?
- Potatoes
- Plain, natural yoghurt
- Tuna
- Noodles
- Baked beans
- Chicken salt

FOR YOU TO DO

4 Research indicates that adults do not drink enough water a day, preferring to have soft drinks, coffee and alcohol as beverages. Develop strategies for increasing the amount of water that adults drink per day.
5 As a class, draw the Australian Guide to Healthy Eating on the playground with some chalk. Each student draws a food that belongs to a particular segment of the guide.

TAKING IT FURTHER

6 Evidence suggests that teenagers want healthier food choices.
   a Design and prepare a bread-based lunch option for teenagers that could be sold at a canteen. Think about using one of the many varieties of breads available today.
   b Explain how you have used the Australian Guide to Healthy Eating or the Healthy Eating Pyramid in selecting your lunch menu.

Chapter review

LOOKING BACK

Select the correct answer in the following multiple-choice questions.

1 What are proteins made from?
- A Fatty acids
- B Amino acids
- C Sodium
- D Glucose

2 Which vitamin is mostly found in citrus fruits such as oranges?
- A Vitamin C
- B Vitamin D
- C Vitamin A
- D Folate

3 Which is not a function of fibre?
- A Aid digestion
- B Make you feel full
- C Help reduce cholesterol
- D Give you energy

4 Which food is a person with coeliac disease unable to eat?
- A Peanuts
- B Milk
- C Wheat
- D Seafood

FOR YOU TO DO

5 Excess consumption of which food is most likely to contribute to dental caries?
- a Fats
- b Sugar
- c Salt
- d Alcohol

6 Of the following vitamins, which are water-soluble?
- a Vitamin A
- b Vitamin B complex
- c Vitamin C
- d Folate
- e Vitamin K
- f Vitamin E

7 Which of the following disorders may be associated with over-nutrition?
- a Type 2 diabetes
- b Coeliac disease
- c Anaemia
- d Osteoporosis
- e Coronary heart disease
- f Hypertension
- g Colon cancer

TAKING IT FURTHER

8 Debate the following topic: 'Carbs are bad for you.'

9 Find two recipes rich in fibre; one wholegrain based and one vegetable based.

10 Produce a mind map on the benefits of fibre in the diet.

11 Assess the nutritional value of two snack foods, one sweet and one savoury, enjoyed by teenagers. Use the nutrition information panels on their packaging to compare and contrast the nutritional value of the snack foods in relation to the needs of teenagers.

12 Interview a person who has lived in Australia for over 50 years on how they believe food in Australia has changed in the last 50 or more years. Prepare a list of questions before the interview, such as 'What did you eat for breakfast when you were a child? What did you take to school? How often did you eat out? What foods did you have when you went out?' Write up the results of your interview in a report.

9780170383417

Fattoush salad

Salads are an easy and tasty way of eating vegetables.

 **Preparation time** 15 minutes  **Cooking time** 8 minutes  **Serves** 2

INGREDIENTS

6 cherry tomatoes, halved
½ Lebanese cucumber, sliced
1 stick celery, julienned
⅓ Spanish onion, diced or sliced
2 tablespoons parsley, chopped
2 tablespoons mint, chopped
1 pitta bread

DRESSING

Juice of ½ lemon
Grated rind of ½ lemon
1 garlic glove, crushed
2 teaspoons olive oil

METHOD

1. Pre-heat oven to 180°C.
2. Cut the pitta bread into bite-size triangular shapes and toast for around 8 minutes until golden and crisp. Cool.
3. Combine the chopped vegetables.
4. Mix the dressing and pour over salad.
5. Toss in the pitta chips and serve immediately.

QUESTIONS

1. Name some cutting techniques that can be used on the vegetables.
2. From where in the world do you think this recipe originates?
3. Why do you serve the salad immediately after adding the dressing?
4. Which ingredients in this recipe would be rich in complex carbohydrates and fibre?
5. What ingredient in the recipe would be a rich source of fat?

Mark Fergus

Mark Fergus

Chow mein

This Chinese dish is popular worldwide. You can fill some lettuce cups with the filling to make San choy bau.

 **Preparation time** 10 minutes  **Cooking time** 10 minutes  **Serves** 2

INGREDIENTS

- 100 grams lean mince (beef, pork or chicken)
- 1 tablespoon oil
- ¼ onion
- ¼ carrot
- 1 stick celery
- 2 cabbage leaves
- 1 tablespoon soy sauce
- ½ teaspoon curry powder
- 1 packet of chicken instant noodles or 100 grams fresh noodles

METHOD

1. Dice the onion, carrot and celery, and shred the cabbage.
2. Heat the oil in frying pan, sauté the onion.
3. Add the mince and curry powder and cook until the meat is no longer pink.
4. Stir-fry the carrot, celery and cabbage for 3 minutes.
5. Soak the noodles in hot water.
6. Add the drained noodles and soy sauce.
7. Serve.

QUESTIONS

1. What shape is a dice?
2. Why should the meat not be left pink?
3. What is meant by the term 'lean' in reference to meat?
4. What are the main nutrients in minced meat?
5. What ingredient in the recipe would be high in sodium?

9780170383417

RECIPES

Yoghurt and muesli parfaits

A simple, healthy and great looking breakfast meal.

 **Preparation time** 5 minutes  **Cooking time** Nil  **Serves** 2

INGREDIENTS

1 cup toasted muesli
1 cup vanilla yoghurt
⅔ cup frozen mixed berries, thawed
2 teaspoons honey

Tip: You can recycle by using glass jars as serving glasses.

METHOD

1. Divide half the muesli between four 1 cup-capacity glasses. Spoon half the yoghurt over muesli, then add half the berries.
2. Repeat the layers. Drizzle each one with honey and serve.

QUESTIONS

1. How is muesli made?
2. How is yoghurt made?
3. What other fruits could be substituted for the berries?

Mark Fergus

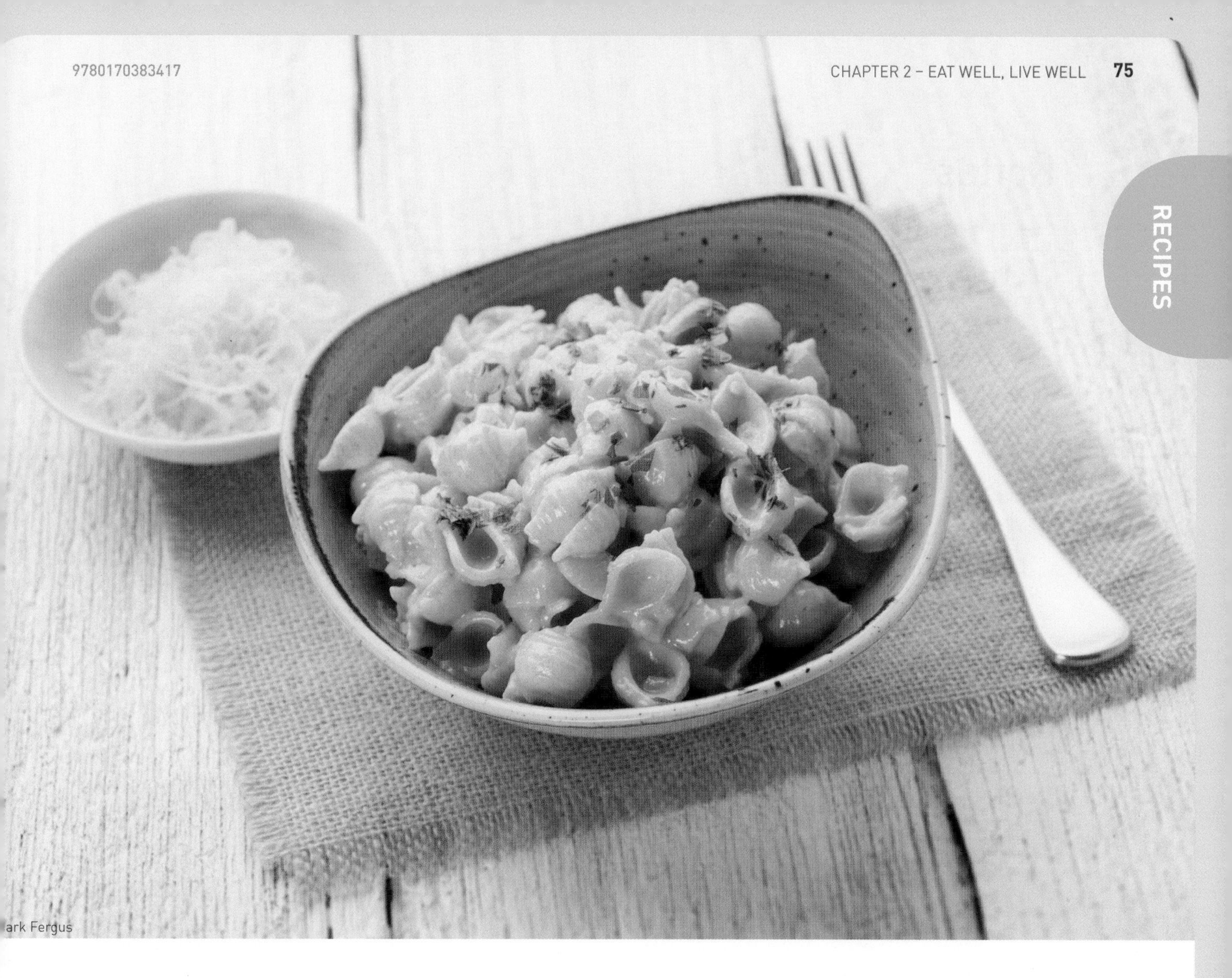
ark Fergus

Quick creamy pasta

Lipids such as butter and cream add flavour to food. These are alternatives to reduce your lipid intake.

 **Preparation time** 10 minutes  **Cooking time** 15 minutes  **Serves** 2

INGREDIENTS

15 grams butter
2 teaspoon tomato paste
¼ onion, diced
200 grams pasta shells or elbows
150 mL cream or evaporated milk

METHOD

1. Boil a saucepan ¾ full of salted water.
2. Add the pasta and cook around 10 minutes until soft. Drain.
3. Melt the butter in saucepan and cook the onion until clear.
4. Add the tomato paste and cook for 1 minute.
5. Add the cream and heat slowly until it reaches boiling point.
6. Add pasta and toss through cream until pasta is coated. Serve.

QUESTIONS

1. How do you know when pasta is cooked?
2. How is evaporated milk different to cream, fresh milk and condensed milk?

RECIPES

Koftas

Add some salad and they make a healthy meal.

 **Preparation time** 10 minutes  **Cooking time** 10 minutes  **Serves** 2

INGREDIENTS

250 grams lamb mince
¼ cup breadcrumbs
1 teaspoon tomato paste
¼ small onion, chopped
1 garlic clove, crushed
1 pinch dried chilli flakes
1 teaspoon fresh flat-leaf parsley leaves, chopped
1 teaspoon chopped fresh mint
1 tablespoon of beaten egg
½ teaspoon ground cumin
¼ teaspoon ground coriander
Salt and pepper for seasoning
Oil for cooking
4 pre-soaked bamboo skewers

METHOD

1 Place the mince, breadcrumbs, tomato paste, onion, garlic, chilli, mint, parsley, cumin, coriander, egg, salt and pepper in a bowl. Mix to combine.
2 Shape ¼ cup of the lamb mixture at a time into four 8 cm long sausages and thread a sausage onto each skewer.
3 Heat an oiled barbecued plate or fry pan on high.
4 Add the koftas. Reduce heat to low. Cook, turning occasionally, for 10 minutes or until cooked through.
5 Serve with yoghurt and flat bread.

QUESTIONS

1 What herbs have been used in the recipe?
2 What spices have been used in the recipe?
3 Why are the skewers soaked?

Hint: Do not turn meat until it is well browned. It is also best to refrigerate koftas before cooking for 20 minutes.

Mark Fergus

RECIPES

ark Fergus

Cauliflower fried rice

A healthy twist on a classic family recipe.

 **Preparation time** 10 minutes  **Cooking time** 10 minutes  **Serves** 2

INGREDIENTS

- **2 cups finely processed cauliflower flowerets**
- **1 egg**
- **2 teaspoons vegetable oil**
- **2 bacon rashers, chopped**
- **1 cup frozen mixed vegetables**
- **2 shallots, trimmed, finely sliced**
- **2 tablespoons soy sauce**

METHOD

1. Using a whisk, lightly beat the egg in a small bowl.
2. Heat oil in non-stick wok or large frying pan over medium heat.
3. Add the egg. Swirl over base to form an omelette. Cook 2 minutes. Turn over. Cook until set. Transfer to a chopping board. Set aside to cool slightly. Cut into short strips.
4. Add the bacon and shallots to pan. Cook until the bacon is lightly golden.
5. Add the cauliflower and stir-fry for 3–4 minutes.
6. Add the mixed vegetables. Stir-fry 2 minutes.
7. Add the egg and soy sauce. Stir until heated through.
8. Serve immediately.

QUESTIONS

1. What are alternative ingredients that could be used in this recipe?
2. How would you multiply the ingredients in this recipe for the number of people in your household?
3. What ingredients in the recipe are rich in the following nutrients?
   - Carbohydrates
   - Lipids
   - Protein
   - Water.

CHAPTER 3

Mark Fergus

Bush tucker to contemporary cuisine

FOCUS AREA

FOCUS AREA: FOOD IN AUSTRALIA

Defining Australian cuisine is not easy. Is it meat and three vegetables, or bush tucker such as kangaroo and macadamia nuts, or foods from other cuisines with a local spin? This chapter looks at the effects migration has had on the food Australians eat and examines the history of Australian food, including the traditional bush foods prepared by Indigenous Australians and the influence of early European settlers.

CHAPTER OUTCOMES

In this chapter you will learn about:

- native Australian foods
- early European influences on food habits
- multicultural influences on food habits and methods of cooking
- Australian cuisine
- influences on food selection
- consumption patterns and related effects
- development of food production

In this chapter you will learn to:

- investigate traditional and contemporary use of foods native to Australia
- discuss the impact of early European influences on food habits
- identify the multicultural influences on the diets of Australians today
- discuss Australian food and prepare meals that reflects its evolution
- examine why food selections and eating habits change
- relate changes in consumption patterns to their various effects
- investigate the technologies that have contributed to the development of the Australian food industry.

RECIPES

- Lemon myrtle chocolate crackles, p.109
- Wattleseed jam cockles, p.110
- Quandong jam, p.111
- Traditional damper, p.112
- Bush damper, p.113
- Contemporary Aussie damper, p.114
- Nem ran (North Vietnamese spring rolls), p. 116
- Chicken, minted pea and ricotta meatballs, p. 117

FOOD FACTS

Before 1770, there were more than 750 000 Aboriginal people living in 500–600 tribal groups, each consisting of between 100 and 1000 people. Each tribe had its own territory and language.

Men and women serving in the Australian army are now trained in how to survive on bush foods in case of combat in Australia.

A pound is approximately equal to 450 grams (the size of a packet of pasta); an ounce is about 30 grams (the size of a small bag of chips); and a pint is nearly 600 mL (the size of a bottle of coke).

Food manufacturers have responded to the growing trend towards eating out with a range of chilled, frozen and microwavable 'quick fix' meals, which are promoted as the restaurant dishes you don't have to go out for.

Recent developments in food packaging and storage include flexible and lightweight plastic packaging materials, frozen-food cabinets, refrigerated dairy cases and dual-oven heating trays. Food quality control has also had an impact on food packaging.

FOOD WORDS

coolamon elongated wooden dish used for gathering plant and animal foods

food habit pattern of eating; the way in which food is prepared, served and eaten

indigenous food (bush tucker) food native to Australia that was present before European colonisation

multiculturalism existence of different cultures within the one society

ration fixed allowance of food

satiety state of being fed or satisfied

scurvy disease caused by a lack of Vitamin C

3.1 Use of foods native to Australia

Some bush foods grow wild in the backyards of city homes, often unrecognised by the inhabitants. Other bush foods hide in remote parts of the desert and are very difficult to find. Australian **indigenous foods** are commonly known as '**bush tucker**'. Bush tucker comprises a wide variety of herbs, spices, mushrooms, fruits, flowers, vegetables, animals, birds, reptiles and insects that are native to Australia.

Indigenous Australians have been eating bush tucker for at least 50 000 years. In colonial times, the European settlers who learned about local foods from Aborigines survived much better than others who did not. To many non-Aboriginal people, bush plants are still a mystery, grubs look unappetising and Aboriginal cooking methods are not properly understood. However, many non-Indigenous Australians are developing an interest in foods unique to their environment. There is also a growing interest from world-class chefs seeking 'new' tastes and innovative food combinations.

In some restaurants there is a chance that you will find examples of bush tucker on the menu. For example, wild Australian fruits make excellent jams, sauces and desserts. Nuts are used in pies, breads and sweets. New flavours from the bush are making their way into ice-creams, beverages and spices.

Figure 3.1 Traditional Indigenous foods (bush tucker), when available, provide the range of nutrients necessary for growth and survival.

CLAYTON DONOVAN

NATIVE FOODS ELDER

At 42, Clayton Donovan can lay claim to being one of the chief elders of the native foods trend.

He started foraging when he was four, when his aunt would take him along the coastline and around the bushland of NSW's mid-north to show him all the different wild berries. Back home, his mother taught him how to cook it all.

As a young chef in the 1990s, Donovan was ahead of the times in integrating ingredients such as wattleseed, rosella and myrtle into Asian and European cuisines when he joined head chef Kenneth Leung at the Watermark at Balmoral Beach.

Now, with the trend spreading like bushfire through top restaurants around the land, the indigenous chef and star of ABC TV's *Wild Kitchen* can hardly contain his boyish excitement at how far we've come.

'It's going to be a great ride the next couple of years,' says Donovan, whose recipes incorporating indigenous ingredients also feature in the expansive new book *The Great Australian Cookbook*. 'It's so exciting to see native foods being taken seriously. The trend has so much momentum now, but we've only touched the tip of the iceberg.'

The chef – who now runs his former Nambucca Heads restaurant Jaaning Tree as a sort of roving pop-up – points to ongoing work by the CSIRO to formally identify new native ingredients. Another three types of wattleseed had been found in Broome just in the past month, for example.

'I reckon bush foods are very much like the Australian personality,' he says. 'Strong, in-your-face, bold, extroverted. They're a scary thing for a lot of chefs because of that; you have to learn to use just a pinch of ingredients like lemon myrtle and pepperberry, otherwise they will overwhelm.'

So why has it taken the white man so long to appreciate what's growing right under our noses? Sydney journalist and author John Newton has just completed a doctoral thesis that asked, in part, that question.

'The short answer is food racism,' says Newton. 'When we came here over 200 years ago, we didn't see a people who had lived here for over 50 000 years and done a pretty good job of caring for the country. We only saw wandering savages.'

Attempts to change our attitudes by pioneering chefs such as Jean-Paul Bruneteau, Andrew Fielke and Raymond Kersh in the 70s and 80s failed to gain traction, and it wasn't until 2010 that the tide started turning. That was when visiting Danish chef Rene Redzepi, of Noma, asked why we weren't eating all these wonderful foods.

'A new generation of younger chefs who weren't burdened with the racism of their older colleagues agreed,' says Newton.

But there's a fair way to go yet. Newton recalls one of the best native foods dishes he has eaten, a 'sensational' roast magpie goose cooked by Bruneteau at the Sydney restaurant Riberries during the early 90s.

'Sadly, you can't buy magpie goose, only go up north and hunt it in the season.'

Source: 'Clayton Donovan, of Wild Kitchen, is a native foods elder', by Necia Wilden, Food Editor, *The Australian*, 19 September 2015, Photo: AAP Image/Jamie Williams, City of Sydney

ACTIVITIES

1 Research a piece of legislation with which a chef such as Clayton Donovan would need to be familiar.
2 What evidence of Clayton Donovan's passion for Australian native foods do you see in this profile?
3 Research the availability of Indigenous foods in your local area. List and then discuss your findings with the rest of the class.
4 Research, compare and contrast one European vegetable and one Indigenous vegetable. Why do you think they might need to be prepared differently?
5 Research the meaning of the word 'sustainability'. Do you think that Indigenous foods are more sustainable than other food sources?
6 Investigate the sustainability of farming beef, lamb, pork and kangaroo, and rank them according to their environmental impact.
7 Visit the CSIRO website to investigate the role they play in formally identifying new native ingredients.

CSIRO

Traditional use of bush foods

Before European settlement in 1788, Aboriginal Australians lived successfully off the land. They travelled great distances in search of the available food supply. When food supplies become limited, an Aboriginal family would move to a new area where the supply was plentiful. These areas were more commonly on the coast, near rivers or where there was high rainfall, as animal and plant food sources were more accessible.

Aboriginal Australians were very successful at providing for their needs through hunting and gathering activities. The men were the hunters and went out each day to hunt, but often returned with very little. What they did bring back – larger animals or fish – was then shared among their tribe. It was the women and children who supplied the more substantial part of the food requirements. They gathered and prepared plant foods, caught small animals and collected delicacies such as witchetty grubs, ants, bogong moths and emu eggs. However, the men's catch was always valued more highly than the food the women gathered because it was much more difficult to come by. Aboriginal Australians killed only enough food for their immediate needs.

The diet of Aboriginal Australians was rich and nutritionally well balanced. Most diets were high in protein, which supplied up to half of their energy needs. This was due to the consumption of fish and shellfish by those tribes living along the tropical coast or near rivers, and goanna and kangaroo for those living in or near open scrub, mountains or deserts. Diets were also high in fibre and vitamins, which was due to the variety of fresh fruits and vegetables available. There was a small amount of sugar from fruit, nectar and honey gathered from wild bees or extracted from honey ants. The intake of fat was low, as it largely came from game or fish.

The influence of geography

The geographical location of the tribe and the particular season of the year primarily governed the choice of diet and the way in which food was prepared and cooked.

Animal foods were generally cooked, either over an open fire or steamed in pits. Kangaroo, for example, was usually laid on a fire and seared for a short period, so that the interior flesh remained practically raw; at other times the kangaroo was placed in a large hole, surrounded by hot coals and sealed from the air.

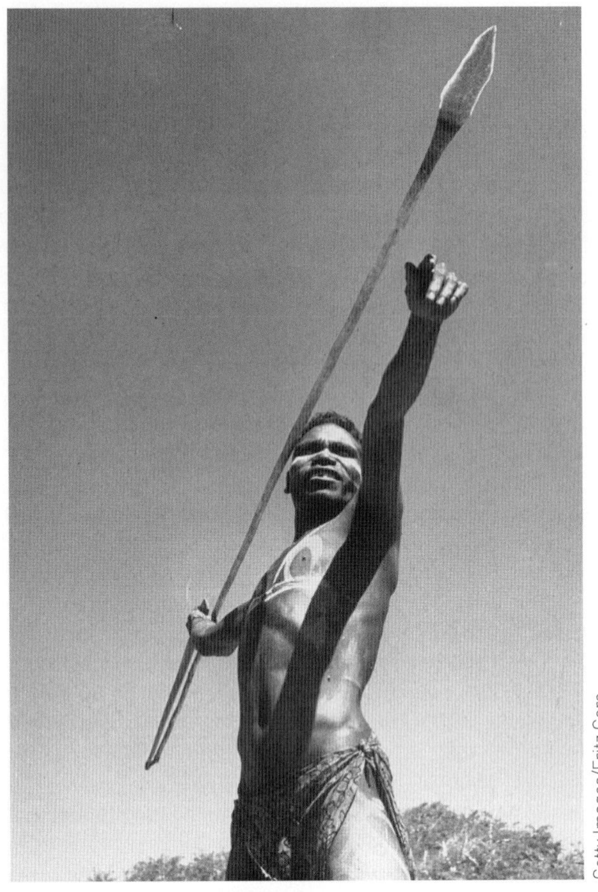
Getty Images/Fritz Goro

Figure 3.2 An Aboriginal Australian man dressed traditionally for hunting food

Sometimes food was wrapped in paperbark or leaf matter to protect the flesh from the open flame. For example, flying fox was wrapped in the leaf of the Alexandra palm for cooking. When it was cooked, the leaves were unwrapped, pulling off the skin and fur at the same time.

Plant foods required more careful preparation since many of them are difficult to digest and even poisonous. Aboriginal women spent many hours washing, grinding, pounding, straining, grating, boiling and cooking plant foods. The water used in these preparations and cooking methods was boiled in bark troughs or in large seashells.

Living off the land worked well for the Aboriginal people because their population and needs were relatively small and they moved around seasonally in search of food supplies. This prevented the overuse of any specific food source.

Contemporary use of bush foods

Bush tucker is presently being seen as an important part of our ever-evolving national cuisine. The contemporary bush tucker industry uses native animals and plants in different ways from those originally used by Aboriginal Australians. For example, plants are used primarily for flavour rather than nutrition. Consequently, many animal and plant food sources are now known as gourmet items and are becoming increasingly available in specialty shops and local delicatessens. Consider the gourmet items below that have foods native to Australia as their main ingredient:

- kangaroo meatballs
- bush tomato salsa
- lemon myrtle cheese
- wattleseed pasta
- buffalo steaks smoked over banksia cones
- witchetty and bunya soup
- rosella jam
- lilli pilli chutney.

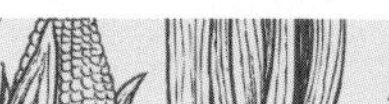

FOOD IN FOCUS

LEARNING TO HUNT AND GATHER

Much learning and experience is required to become an expert hunter and gatherer. Children generally started the learning process at an early age.

Traditionally, it was customary for children to go out with adults on hunting and gathering trips so that knowledge and skills could be passed from generation to generation. Children also played games in which they acquired hunting and gathering skills, learned how to work with others and practised using the appropriate equipment. Boys played with toy spears, spear throwers and boomerangs; girls played with toy digging sticks and **coolamons**.

In areas where watercraft was important, toy canoes were used. Children also practised tracking, an important skill. Children used their fingers or special wooden objects to make imitation animal tracks on the ground.

Because of the learning experience required, a man might not become an expert hunter until his mid-twenties. However, as hunting was such a physically demanding activity, he might retire from active hunting in his late thirties.

ACTIVITIES

1. Why do you think male and female children were encouraged to play with different hunting and gathering equipment?
2. Consider the equipment needed to hunt and gather. Outline what each would be used for and provide an example of a bush food that could be collected using this equipment.
3. The expert hunter has a limited career. Think about the expert gatherer. How long do you think female gatherers continued to work for? Explain your answer.
4. Investigate the variety of skills hunters and gatherers needed to learn before they were known as experts.

Alamy Stock Photo/Bill Bachman

Figure 3.3 Equipment used by Aboriginal people to hunt and gather food

RAYLEEN BROWN
OWNER OF KUNGKAS CAN COOK

ALICE SPRINGS BUSINESS OWNER IS THE FACE OF BUSH TUCKER BOOM WITH GROWING OVERSEAS INTEREST

ALICE Springs business owner and bush tucker extraordinaire Rayleen Brown says the Central Australian native food industry is booming, with an increase in international interest the past three years.

Owner of Kungkas Can Cook, a catering business that uses native ingredients, and an Australian Native Foods Industries Limited board director, she said the industry has grown significantly from 10 years ago.

'In the beginning when I first started talking with scientists and we first started working with the women who were doing the wild harvesting, when they first saw the bush tomato farm they were shocked – it's growing in rows,' Ms Brown said.

'There still will be some wild harvesting but as the market grows you need more sustainable crops.

'The last three or four years, interest picked up nationally and even more internationally.'

She said producers are based from Ti Tree and Utopia, stretching down to the APY Lands, with bush tomatoes the main developed crop and wattleseed starting to take off commercially.

For example, bush tomatoes picked in Rainbow Valley are sent to Melbourne-based company Outback Spirit and feature in sausages, sauces and chutneys sold nationally.

'There's a great interest at the moment for central wattles, the different varieties, they're doing some testing,' Ms Brown said.

'I think there is the potential to say 'this wattle has more oil and would be good for this' – there's a lot of work that needs to be done by food scientists.

'Then it can be ready to go into mainstream industry, chefs, and chefs need to be educated about native foods.'

Ms Brown said the next step would be to carry out more processing on country, such as drying the produce.

'There are still people doing what they used to do for thousands of years of harvesting,' she said.

'Hopefully a lot more indigenous businesses will pop up now and use the research and information.'

Department of Primary Industries' Cameron McConchie said they acted as brokers for communities and ladies' groups, providing labs and semi-commercial facilities for product development.

'If they need help with food technology or formulations or testing, we're ready to help,' he said.

Source: 'Alice Springs business owner is the face of bush tucker boom with growing overseas interest', Toyah Shakespeare, *NT News*, 14 August 2015.
Photo: Newspix/Rex Nicholson

ACTIVITIES

1 Develop a promotional video clip or television advertisement that could be used to promote the advantages of using Australian native foods to national and international chefs. The promotional material should include:
- the nutrient value of the Australian native foods
- why Australian native foods make versatile, healthy and tasty ingredients
- recipe ideas.

PRACTITIONER FOCUS

Table 3.1 Traditional Aboriginal food sources (animal)

ANIMAL FOOD SOURCES	PLACE	SEASON
Crayfish	Fresh rivers, creeks, lagoons	Winter
Emu	Mallee	All year
Fish	Rivers, lakes	All year except breeding time
Frog	Waterways, swamps	Summer
Galah, pigeon, other small birds	Bush, scrub	All year except breeding time
Honey ant	Mulga scrub, land and plains	After rain in dry areas
Kangaroo	Mallee, river plains	All year
Lizard	Sandhills, scrub, bush	Summer
Possum	Scrub, bush	All year
Snake	Sandhills, scrub, bush waterways	Summer
Water rat	Banks of water holes, rivers	Winter
Witchetty grub	Red gums	Spring to summer
Wombat	Burrows in higher ground	Spring to autumn

Table 3.2 Traditional Aboriginal food sources (plant)

PLANT FOOD SOURCES	PLACE	SEASON
Acacia wattle	Scrub	Autumn
Berries	Mallee, bushes	Autumn
Bracken (roots)	Anywhere	Autumn
Bulb and stalk	Billabongs	Spring to autumn
Fungus	Underground	Autumn
Grass seed	Mallee, scrubland	Summer
Honey	Hollow trees	Spring to summer
Nectar	Flowers	All year
Salt	Dry water course	All year
Shoots	Marshes	Summer
Wild tomato	Mallee	Summer
Wild onion	Dry areas	Autumn
Water	Rivers, lakes, tree roots, dew on grass and leaves	All year

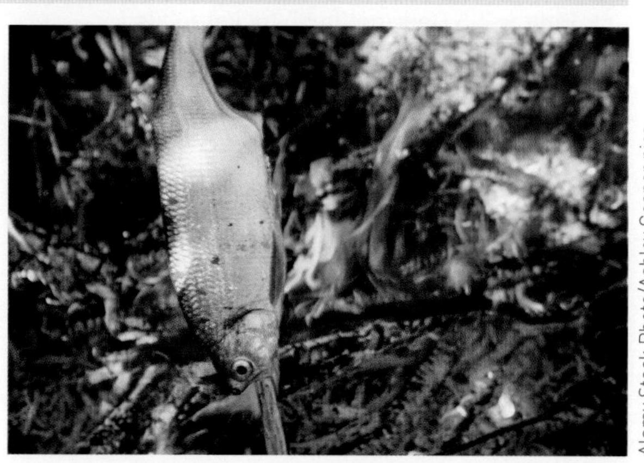

Alamy Stock Photo/Ashley Cooper pics

Figure 3.4 Aboriginal people traditionally cook animal foods over an open fire.

9780170383417

FOOD IN FOCUS

OUTBACK FACTS: BUSH TUCKER

The seemingly stark Australian landscape is like one giant supermarket for those in the know. It has sustained Indigenous communities for thousands of years. Many Aboriginal Australians continue to eat traditional foods, well aware of their health-giving properties.

Bush tucker varies according to particular regions and seasons. Traditionally, Indigenous groups living along Australia's coastline existed on marine animals, roots, fruit, small game and reptiles. Depending on the area, freshwater crustaceans may also have been available. In the Torres Strait Islands, communities also ate dugong, while in the tropical north stingrays were eaten. Some wildflowers, such as grevillea and banksia, were sucked for their sweetness, and the tips of the Western Australian Christmas trees were chewed as gum.

Hunting and gathering was more difficult in the harsh desert climate, where hunters frequently had to travel great distances to find kangaroos or emus. Other game included snakes, small lizards and goannas.

In the desert regions, game was thrown on a campfire and cooked whole. Vegetables were either cooked in coals or eaten raw. If the rivers and creeks had permanent water holes, fish such as mullet and bream were added to the menu.

Communities in the tropical parts of northern Australia enjoy a totally different lifestyle from that of the desert peoples and communities further south. The Kimberley, the Northern Territory and Far North Queensland experience lush growth and tropical downpours. The rivers and lagoons supply yabbies, catfish, barramundi, turtles, crocodiles and wild ducks. Edible water reeds grow in lakes and swamps, while the land offers a plentiful supply of meat, including wild turkey, flying fox and snake.

In the inland non-desert areas of south and south-western Western Australia, there are six distinct climatic changes. Traditionally, inland communities foraged for insects, birds, reptiles and mammals, including kangaroos and possums. They also ate a variety of fruits, flowers, berries and nuts. Various seeds were collected, ground up and mixed with water and either eaten as a paste or baked in coals to make damper.

Figure 3.5 Gourmet bush tucker foods

ACTIVITIES

1. List the types of environments in which Aboriginal Australians live.
2. Brainstorm the bush foods available in each environment.
3. The environment in which people lived mainly determined the diet of Aboriginal Australians. Design a menu that tells a story of native foods in one environment of Australia.
4. Select one bush food listed above and find a suitable recipe that could be adapted to incorporate this food. Make your adapted recipe and present it in a traditional way. Analyse and evaluate your final product using the following criteria:
   - appearance: What does it look like?
   - aroma: What does it smell like?
   - flavour: What does it taste like?
   - texture: What does it feel like?

UNIT REVIEW

LOOKING BACK

1 Define the term 'bush foods'.
2 The traditional diets of Aboriginal people varied across the country. Why was this so?
3 Explain how the traditional food supply of the Aborigines met their nutritional needs. Use the Australian dietary guidelines outlined in Chapter 2 to assist you.
4 Outline what influenced the types of food eaten by different Aboriginal groups.
5 Explain why the men's food supply was valued more highly than the women's.
6 Why did women supply the tribe with more food than men did?

FOR YOU TO DO

7 Using resources in your library, undertake the following research tasks.
   a Compare the nutritive value of the traditional diet of Aboriginal Australians to the diet of Australians today.
   b Research the health problems encountered by Aboriginal Australians today.
   c If you were stranded in the Australian bush, what foods could you eat to survive?
   d Write a newspaper article on the main reasons that Australians should include bush tucker in their diet.

TAKING IT FURTHER

8 Select a traditional Indigenous food from Tables 3.1 and 3.2 and use the internet to investigate:
   a where and when you would locate it
   b how you would capture or collect it
   c how you would prepare it for eating
   d an example of a recipe that uses the food.
9 Arrange for a local Aboriginal Australian student or family member to talk to the class about Aboriginal Australian eating patterns and how traditional bush foods are prepared and cooked. As an alternative, you could make contact with your local Aboriginal Australian organisation and invite a representative to come to your class.
10 Design and make a traditional cooking tool. Use the tool to prepare a meal using Australian bush foods.
11 Investigate the significance of food in traditional Aboriginal Australian culture. Prepare an oral or computer presentation about the technologies used by traditional Aboriginal Australians for food collection and preparation.
12 Use the website link to visit Barbushco and Outback Pride, two companies with a commercial interest in Australian bush foods. Use the information you find, along with additional resources, to complete the following questions.
   a Outline the nutritional benefits of using bush foods.
   b Investigate the benefits of marketing bush foods.
   c As a class, discuss who actually benefits from the marketing of bush foods.
   d Explain why companies such as Barbushco and Outback Pride are becoming increasingly successful.
   e Predict some of the challenges facing producers of bush foods.

Outback Pride

Barbushco

3.2 Early European influences

The ships of the First Fleet, sent from England with convicts to found a penal colony in NSW, landed at Sydney Cove on 26 January 1788. This date is celebrated annually now as Australia Day.

The first settlers came ashore to a country different from England. Their immediate needs were exactly the same as those of the Australian Aborigines – food and shelter – but they went about providing these in a very different manner.

Food brought on or with the First Fleet voyage consisted of flour, rice, salted meat, sugar, salt, alcohol, vinegar, seeds and vine cuttings. Livestock were brought as a source of fresh meat. The food supply was **rationed** and consumed within two years. During this time, the seeds and vine cuttings were used to grow crops that seemed to fail early on, as the new settlers knew little about the Australian land

and climate. They expected to be able to live off the land using the same techniques they had used in England.

Early settlers succeeded in growing corn, wheat and barley. These crops did not significantly alter the variety of the food eaten.

State Library of NSW [A1474008r]

Figure 3.6 Early settlement in Australia

HANDS-ON

RATION YOUR FOOD

PURPOSE

To plan and prepare a meal using the weekly rations provided for early settlers in Australia.

STEPS

1. Work in pairs and plan a day's menu using only the ingredients listed on page 65. Try to make the menu as interesting as possible.
2. Prepare one of the meals from the menu.
3. Rate the meal for appearance, flavour and texture.
4. List the problems people would have experienced trying to vary the menu over many weeks and months.
5. Use the five food groups to rate the diet of early settlers in Australia. As a class, discuss your findings.

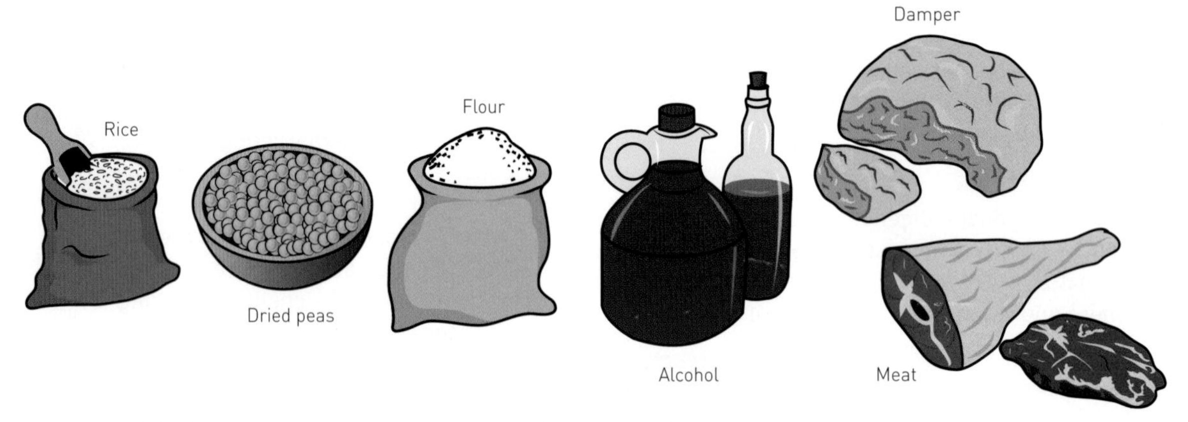

Figure 3.7 Weekly ration for sailors and convicts in 1778 on the First Fleet

Diet of early Europeans

Women received two-thirds of the rations that men received, and children were eligible for one-third. As you can imagine, people on these rations were often very hungry and the meals were rather uninteresting because of the limited ingredients and quantities.

The meat rations were usually made into stews and pies and the dried peas were boiled to a thick type of porridge. The flour was used to make damper or sometimes added to the cooking of any available green leaves or vegetables. This made a meal very similar to vegetable porridge. These rations were generally served hot and were high in fat, protein and refined carbohydrates, and particularly low in dietary fibre. These rations remained the basic diet for most Australians for the first fifty years of settlement. Eventually this diet was supplemented with fruit and vegetables as the crops began to grow successfully.

Some people were able to catch fish to supplement their restricted diet and a few ate some of the wild plants and fruits, such as wild currants and native spinach. However, for the general population, a regular lack of Vitamin C meant that **scurvy** was a constant problem.

The convicts had little or no knowledge of Australian native foods. Some proved poisonous or had a terrible taste, which did not encourage experimentation. The main beverages were water, tea (which was brewed black and strong) and rum, which was the most common alcoholic drink. Both tea and rum had to be imported to Australia from England.

Traditional damper versus the contemporary Aussie alternative

The word 'damper' was first used in England – meaning to dampen the appetite – and describes hard, unleavened, crusted bread. Damper was very important to the first settlers in Australia when flour, salt and water were the only ingredients for making bread. Originally, the bread was cooked in the ashes of a bush fire for about 10 minutes. Then the damper was covered with ashes and cooked for another period of about twenty to thirty minutes until it sounded hollow when it was lightly tapped. Today it is usually cooked in a camp oven (iron pot) that is buried in the hot coals.

Introduction of new foods to Australia

During this time, Aboriginal people started to gravitate towards the settlements and helped with the cleaning, maintenance of crops and day-to-day chores. In return for their services they were paid with rations of food. Over time, they became dependent on the rations for their food supply, and their traditional diet of bush tucker changed to one of white flour, sugar, tea, meat, salt and alcohol. This diet lacked the fresh foods of the bush, sea and rivers and contributed to nutritional disasters such as weight gain, diabetes and alcoholism, which are evident in some Aboriginal communities to this day.

Early Australian **food habits** were linked to those of England. Even today, many of the well-established eating patterns from earlier times have remained dominant in Australians' food habits. Vegetables other than potatoes and peas were not popular. Sugar and sweet foods became popular and the habit of eating large quantities of meat has persisted. This has contributed to our current high intake of protein and fat. The high consumption of alcohol also goes back to the time of the first settlement of the colony; Australians today consume more alcohol than any other English-speaking nation.

Mark Fergus

Figure 3.8 Making traditional damper

Mark Fergus

Figure 3.9 Making bush damper

UNIT REVIEW

LOOKING BACK

1 List the different ways that European settlement influenced the food habits of many Aboriginal Australians.
2 Outline how the early settlers' diet compared with that of the Aboriginal Australians.
3 Explain why the early settlers didn't include more bush tucker in their diets.
4 Why was rationing of foods necessary during Australia's early settlement?
5 Why do you think the meat was salted, instead of left fresh?
6 Explain how European settlement has influenced the food supply in Australia today. Use as many examples as possible.

FOR YOU TO DO

7 a Convert the rations shown below into metric measurements (1 lb = 450 grams; 1 pint = 600 mL; 1 oz = 30 grams).
- 7 lb flour
- 7 lb beef
- 4 lb pork
- 3 oz dried peas
- 6 oz butter
- ½ pint vinegar
- 8 oz rice.

b Multiply these rations by two-thirds to calculate the women's rations and by one-third to calculate the children's rations.

TAKING IT FURTHER

8 Research the causes and symptoms of scurvy.
9 Research the health problems of Aboriginal Australians today. Compile a class report on your individual findings.
10 Describe the problems faced by Aboriginal Australians adapting to non-Aboriginal food habits.
11 Explain some of the problems early settlers experienced in ensuring an adequate food supply for their families.

3.3 Multicultural influences

3.3a Multicultural influences

Effects of immigration on lifestyle and food habits

Since the early years of European settlement, Australia has developed as a diverse **multicultural** society. For example, in the 1840s many Germans settled in the Barossa Valley in South Australia, and established vineyards for wine making. The first Chinese arrived in 1848 and many more followed to work as farmhands on the waterfronts. By 1880, there were 20 000 Chinese in Australia. Mass immigration was introduced after the Second World War, in an attempt to bring thousands of people to Australia very quickly, which was necessary for the expansion of the country.

Today, the proportion of Australians who were born overseas has hit its highest point in 120 years, with 28 per cent of Australia's population – 6.6 million people – born in another country. All these people have brought their food habits to Australia, which has resulted in the expansion of the diet of all Australians.

Alamy Stock Photo/Steve Cavalier

Figure 3.10 Roast dinner is a traditional English meal.

Types of foods and flavourings

As people migrated here, they brought their traditional food habits, preparation techniques and cooking methods with them. They then had to adapt these to the foods available within Australia. Many foods with different ingredients and flavours have been introduced to Australia. Cuisines borrow traditional ingredients from other cultures and incorporate them into dishes. For example, a chef may use some Thai ingredients, such as lemongrass, in a meal that is of Italian origin.

Food habits differ from person to person and culture to culture. Your present food habits have developed throughout your life and will possibly change as you find yourself in new and interesting environments. Our food habits usually develop around the cultural group to which we belong, the lifestyle choices that we make and the personal needs that we have at different times in our life.

Food habits are affected by:

- religious taboos and requirements
- cultural customs
- knowledge and understanding of food and nutrition
- availability of food
- availability of cooking utensils and appliances
- food distribution methods
- social and peer pressures
- advertising
- travel
- climate and geography
- availability of technology
- food preparation techniques and cooking methods.

Preparation techniques and cooking methods

Migrants from every country have brought different styles of preparation, cooking and eating. Over time, other groups have integrated these styles into their patterns of meal preparation and consumption. Until the 1950s, most European Australians ate a traditional British-style diet, which rarely varied from the following pattern.

- Breakfast: porridge; bacon and eggs or grilled chops; toast and jam served with tea.
- Lunch: a hot cooked meal of meat and vegetables; soup and bread if you were home on the farm. If you were at work or school you had sandwiches, fruit and home-baked cake.
- Dinner: soup; meat (beef, mutton, lamb, rabbit, fish) with vegetables (potatoes, peas, cabbage, beans, carrots); sweets would usually consist of pudding and stewed fruit.

Now we are able to eat a variety of food products that can be prepared to suit our individual needs and lifestyles. We are able to purchase pre-prepared foods from all over the globe at local delicatessens and grocery stores, or dine out at restaurants and cafes that prepare international cuisine. We can prepare our own meals using traditional methods with guidance

iStock.com/laflor

Figure 3.11 Does this family represent multicultural Australia?

from cookbooks, magazines and television lifestyle programs. Cooking classes are also available to cater for every nationality as people continue to travel and bring their food experiences home with them.

Our diet has changed from traditional and rather uninteresting to one of the most varied in the world. We now readily combine ingredients and cooking styles from different cultures in the one meal. This is known as fusion cooking. We owe this mix to the various cultural groups who now reside in Australia.

Table 3.3 Top 15 birthplaces of the Australian population, excluding Australia, 2011 census

COUNTRY	PERSONS
England	911,592
New Zealand	483,396
China	318,969
India	295,363
Italy	185,401
Vietnam	185,039
Philippines	171,233
South Africa	145,683
Scotland	133,432
Malaysia	116,196
Germany	108,003
Greece	99,938
Sri Lanka	86,413
USA	77,010
Lebanon	76,451

Source: Australian Bureau of Statistics, 2011 Census (3412.0) Creative Commons Attribution 2.5 Australia licence

Figure 3.12 A food stall in the Northern Beaches on Avalon Market Day, where quesadillas and gözleme are on the menu.

UNIT REVIEW

LOOKING BACK

1 List the foods or dishes that you would regard as traditional Australian foods. Explain why.
2 Identify the popular foods that are available today because of migration. Where have these foods come from?
3 Outline how the different cultures in Australia have influenced the types of food we eat.
4 Outline the positive effects of multiculturalism on our Australian lifestyle and food habits.

FOR YOU TO DO

5 a Collect a variety of food magazines and evaluate the way in which some of the dishes are advertised and presented.
  b Design and prepare your own multicultural recipe that could be published in a food magazine.
  c Use a digital camera to take photos of the finished dish. Save them onto your computer and prepare a one-page magazine spread of your finished food product. You may like to include your recipe and any helpful hints.

TAKING IT FURTHER

6 List the different nationalities represented in your school. Compare this data with that represented above and identify any similarities and/or differences.
7 Make a list of the foods your family eats that are typical of your cultural background, and those that are from other cultures. Evaluate how this may represent multicultural Australia.
8 Keep a diary of the foods you eat over three days. Highlight the country of origin of each food on a world map.

3.3b Multicultural influences

Shutterstock.com/Stopped_clock

Figure 3.13 A variety of food products reflecting multicultural Australia

3.4 Australian cuisine

Australian cuisine was traditionally based on what would grow in Australia and what was available. The migration of many cultures to Australia over the past 60 years has resulted in a diverse mix of food. The size of Australia means that when the season for a particular fruit finishes in one state, the same fruit is starting to ripen in another part of the country. With quick and efficient transportation methods, food is easily delivered to other parts of the country.

A love of the outdoors is typically Australian. The warm, sunny conditions for much of the year mean that Australians spend more time outside and meals play an important part in our lifestyle, from barbecues with family and friends, fish and chips at the beach, meat pies while watching weekend sport or picnics in the bush. Australians are now eating in a way that truly reflects our own relaxed casual lifestyle and climate. We now have a diversity of foods and cooking methods.

FOOD IN FOCUS

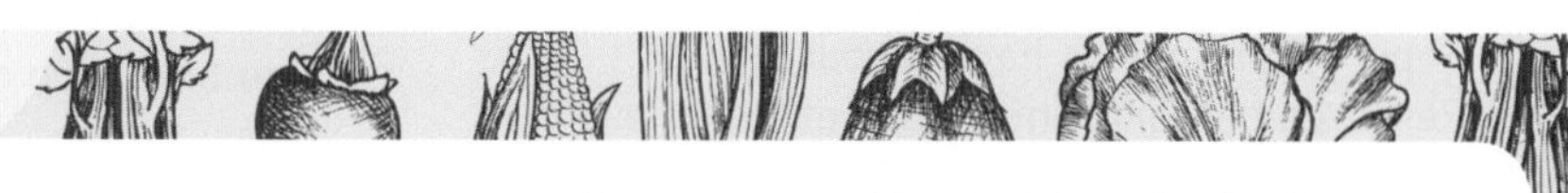

TUKKA ADVANCED AUSTRALIAN FARE

Tukka's vision is to take the flavours of the bush to the tables of the world. Tukka restaurant offers a world-class dining experience where quality ingredients from Australia's bountiful countryside, oceans and bays are combined with native herbs, spices and berries to produce a true taste of Australia.

Tukka aims to lead the development of Advanced Australian Fare, a modern and innovative international cuisine celebrating the flavours of Australia's native foods.

Source: Adapted from information on the Tukka website

>

ACTIVITIES

Using the weblink, visit the Tukka restaurant website and answer the following questions.

Tukka restaurant

1 Brainstorm words that you would use to describe current Australian cuisine.
2 Read the 'Discover Australian Fare' section to learn more about the evolution of Australian cuisine and where it is heading. Do you think this information accurately reflects Australia's current cuisine?
3 Outline why Australian cuisine is able to evolve constantly.
4 Identify three foods mentioned that you have not eaten before. Using library and internet resources, research the sensory aspects of each food and its use in food preparation.

Tourism Australia/Oliver Strewe (http://australia.com)

Figure 3.14 Australian fruits and berries are a source of ingredients for Tukka.

UNIT REVIEW

LOOKING BACK

1 Discuss the factors that have influenced the development of an Australian cuisine.
2 List the foods you eat that were not originally part of your cultural background.
3 Make a list of your 10 favourite foods. Outline why you like each and where the food originates.
4 Outline the foods that could be included in modern Australian cuisine.

FOR YOU TO DO

5 Visit the Taste.com.au weblink and read the article 'How Australian food has evolved' by Nicole Senior. Form groups of four and map, on butchers paper, the development of our Australian cuisine from 1788 to present.
6 Visit the Australian Food History Timeline. Select one milestone that has been significant for the way we eat currently in Australia. Report your findings back to your class.
7 Research how and where each of the following 'Australian' dishes originated: lamingtons, pavlova, meat pies, Anzac biscuits.
8 Debate the following: 'Australian food doesn't have its own identity. Australian food is a product of its past.'
9 Predict what the Australian diet will consist of in 2110.

Australian Food History Timeline

Taste.com.au: How Australian food has evolved

TAKING IT FURTHER

10 a Collect a range of recipes from around the time of Federation. The Country Women's Association in your local area should be able to help you locate a variety of recipes.
   b Analyse the recipes using the following criteria:
   - preparation time
   - ingredients used
   - measurement system
   - utensils used
   - writing style and layout.

   c Compare the features of the Federation recipes to modern recipes of today. What did you find?
   d Choose one of the Federation recipes you have researched and prepare and share it with your class.
11 In small groups, prepare a visual display about one of the characteristics of Australian cuisine. Combine each group's work to present as a class display.

3.5 Influences on food selection

What factors influence people to select one food over another? Very rarely would only one factor influence our selections – most decisions are a result of many interrelated factors.

Physiological influences

Physiological factors affect the body's need and desire for food. You may select foods because you are hungry, because you have an appetite, or because they make you feel satisfied after you have eaten them. For a food to make you feel satisfied it has to have a high **satiety** value. If you have eaten and you do not feel satisfied, then the food you ate is considered to have a low satiety value.

There is a distinct difference between hunger and appetite.

- Hunger is the unpleasant feeling created when the body says that it needs food straight away.
- Appetite is the desire for certain foods even when you are not hungry. It is generally the result of personal experiences as well as your senses of sight, taste and smell.

Each person experiences hunger, appetite and satisfaction differently, as each is developed through the experiences that you are exposed to and the food habits that you have. It is very important to understand the differences between hunger and appetite, as sometimes people eat when they want to rather than when they are hungry. Over time, this pattern of food consumption can lead to diet-related disorders such as obesity.

People vary in many ways, as does the amount of food each person needs to remain healthy and continue to function. To do this, you must supply your body with an adequate amount of food that contains all of the essential nutrients. The amount required depends on factors such as growth and development, gender, body size and activity level. (Refer to Chapter 8 for special requirements throughout the life cycle.)

Psychological influences

There are many psychological factors that influence your food choices. These differ from person to person.

Values

Values are personal feelings that you have about what is important to you. These can influence the foods that you select to eat. For example, individuals who value the rights of animals may not consume any animal products, including dairy products.

Attitudes

Attitudes are opinions. They are not as strong as values. Attitudes may exist around the eating of certain animals.

Experiences

Experiences can influence your relationship with food. Foods commonly eaten when the family is together, or special friends or relatives are visiting, often remain firm favourites, such as your mother's

Figure 3.15 From left to right, top to bottom, a selection of foods from high to low satiety value

Clockwise from top left: Shutterstock.com/Peter Zijlstra; Shutterstock.com/Cloud7Days; Shutterstock.com/Serhiy Shullye; Shutterstock.com/Roman Tsubin; Shutterstock.com/funnyangel; iStock.com/ © zts; Shutterstock.com/Nattika; iStock.com/indigolotos.

baked dinner. Eating them helps you to remember those pleasant experiences. Unpleasant experiences or memories can also be associated with the foods that you select and the way that they make you feel. For example, people who have been ill after eating a particular food often have difficulty eating that food again.

Emotions

Emotions are linked to feelings and often the foods that you consume can be used to express feelings. You may eat because you are feeling happy, worried or nervous, and there may be particular foods that you choose at these times.

Habits

Habits are routines that you can perform without thinking much about them, which are usually hard to break. They can be something that you do at a particular time each day or in a certain situation. Food habits relating to food selection can be both good and bad. Good food habits may include not skipping breakfast, eating the correct daily servings of fruit and vegetables, or drinking water instead of soft drinks.

Beliefs

Beliefs are what you accept to be true and generally come from interaction with the people around you. They can be based on your experiences, traditions passed down in your family, or religion. For example, your family may always say a prayer before they consume any food, or may not eat meat on Fridays.

Geographical influences

Geographic location and climate dictate the type of foods that can be grown – for example, desert areas are not suitable for growing crops of wheat, fruit or vegetables – and influence how and where you eat them.

People who live in hot climates tend to eat foods with high water content, such as salads and fruits, and generally take on a more casual and relaxed approach to eating, while people in cooler areas tend to eat more hot meals indoors, such as baked dinners and casseroles. People living in rural or coastal areas often prefer to eat fresh local produce, as it is generally cheaper than foods that need to be transported from city areas. This also allows them to keep their money within the local area. City people regularly eat out , because it is seen as a very social activity, and consume many take-away and convenience foods.

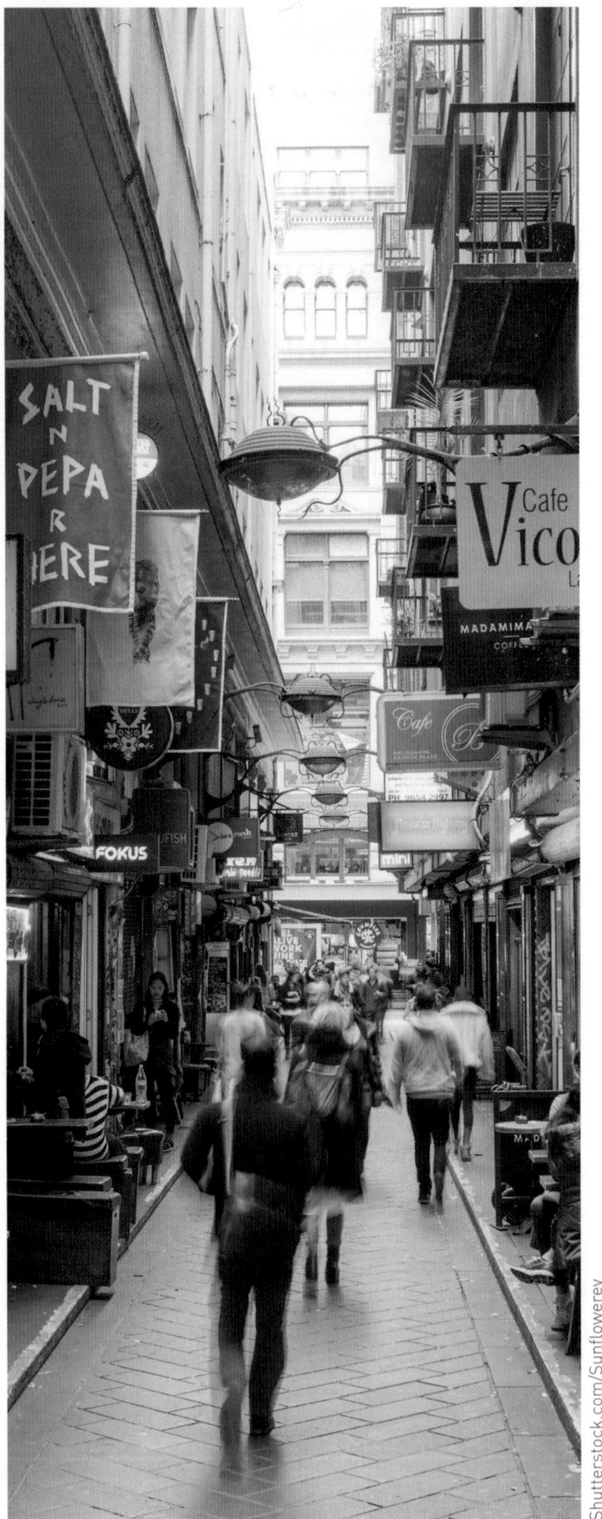

Shutterstock.com/Sunflowerey

Figure 3.16 Many people in cities regularly eat out

Social influences

Eating is a social activity and the sharing of food is often central. Any time people come in contact with others there is a social interaction and sharing of ideas, customs and lifestyles. Family and peer groups are certainly the most important socialising groups and both have a significant influence on the foods that you select when you are with them.

When families entertain, special foods and traditions are shared. It is on these occasions you learn about new foods, food presentation and how foods are used to create a desired impression. What birthday party would be complete without a birthday cake? In Australia, barbecues are often the focus of many social events. There is very little more typically Aussie than a barbecue, whether it is the latest gas-fired piece of hi-tech equipment or an old grill propped on a couple of rocks. A bit of steak and the odd sausage is as much a part of the tradition as the fire itself. As a result of influences from other cultures, Australians are cooking a greater variety of foods on the barbecue.

Economic influences

In any community the foods that people select to eat depend to some extent on price and the money that is available to spend. The cost of food can vary depending on the following factors.

- The amount of food purchased: If individuals have enough money to purchase foods in bulk and sufficient storage then it is cheaper to buy in bulk than to run to the corner store daily for the same product.
- The time of purchase: When foods are in season and locally produced they are often a lot cheaper than foods that are out of season and need to be transported from elsewhere. The costs associated with transportation add a considerable amount to the final price of the food product.
- The place of purchase: Where you purchase a food product has an impact on the price that you will pay for it. For example, if you buy your fruit and vegetables from a roadside stall, it is usually less expensive than going to your corner store for the same product.
- Whether the food is processed or unprocessed: Processed food products are more expensive than unprocessed. Frozen pizza that has undergone a range of processes before you find it at your local supermarket will cost more than if you were to make a pizza of the same quality at home.
- The age of the food: Foods that are past their use-by dates are usually sold at a reduced price.

HANDS-ON

DEVELOP STRATEGIES TO IMPROVE EATING PATTERNS

PURPOSE

To determine how individuals and families can improve their eating patterns.

STEPS

1. With a partner, consider the following general statements.
   - When income rises, the consumption of more refined, processed and convenience foods also rises.
   - When income rises, the proportion of income spent on food decreases although the actual amount spent on food increases.
   - When income rises, the proportion of protein and fat-rich food consumed increases and the proportion of carbohydrate-rich foods decreases.
2. Discuss these statements with your partner and try to work out why these patterns occur.
3. As a class, share your answers.
4. In a paragraph, describe how individuals with high incomes could improve their eating habits.
5. Predict the outcome if there was a decrease in income rather than an increase.

Religious influences

Within cultures, people select foods to conform to particular styles of eating. They also may have religious beliefs about particular foods and how they should be prepared and served.

Religious food practices vary widely. Prohibitions and restrictions even within a particular faith may change between one suburb and the next. National variations are also common. Further, individual adherence to a religious diet is often based on how strongly the person believes in the religion to which they belong. The following table provides a brief overview of some religious food practices.

Table 3.4 Common religious food practices

COMMON RELIGIOUS FOOD PRACTICES								
	Seventh-Day Adventist	Buddhist	Eastern Orthodox	Hindu	Jewish	Mormon	Muslim	Roman Catholic
Beef		■		●				
Pork	●	■		■	●		●	
All meat	■	■	▲	■	▲		▲	▲
Eggs/dairy	■	⬟	▲	▲	▲			
Fish	■	■	▲	▲	▲			
Shellfish	●	■	⬟	▲	●			
Alcohol	●			■		●	●	
Coffee/tea	●					●		
Meat and dairy at the same meal					●			
Leavened foods					▲			
Ritual slaughter of meats					◆		◆	
Moderation	◆	◆						
Fasting*		◆	◆	◆	◆	◆	◆	◆

● Prohibited or strongly discouraged
■ Avoided by most devout
▲ Some restrictions regarding types of foods or when foods are eaten
⬟ Permitted but may be avoided in some instances
◆ Practised
* Fasting varies from partial (abstention from certain foods or meals) to complete (no food or drink)

Technological influences

Technological developments within the food industry not only influence the type of foods that you eat, they also influence the way in which you prepare, store and serve foods. As technology constantly develops and improves, consumers are seeing an increase in the variety of food products available. A decrease in the amount of food wastage has also occurred, as technology has provided many more ways to successfully preserve and store food products that in the past would have been thrown away due to food spoilage. Some technological factors that affect your food selection include developments in:

- cooking methods
- food production
- food storage
- food packaging
- preservation methods
- genetic engineering
- appliances.

HANDS-ON

RESEARCHING TECHNOLOGICAL DEVELOPMENTS

PURPOSE

To determine the effect of technological developments on the food you eat.

STEPS

1 Divide the class into seven groups according to class numbers.
2 Each group is required to research one of the technological developments listed above using a variety of resources, such as books, magazines, the internet or other people.
3 Compile a poster presentation that can be displayed in your classroom. This presentation should show the changes that have occurred and the implications of these changes for food selection.
4 Prepare a one-page information sheet summarising the information you have collected. Distribute this sheet to other class members.

Media and advertising influences

The media and advertising can influence your lifestyle, what you choose to wear, the foods you eat and your perceptions of body image. The media promote images of thinness and fitness, often portraying unrealistic male and female body images. To obtain the ultra-thin look that is considered fashionable, males and females often select foods that do not provide them with adequate dietary requirements. Many women are dissatisfied with their bodies, and constant dieting in the pursuit of thinness has become a normal behaviour among women in Western societies. Thinness has also come to represent success, self-control and higher socio-economic status. Eating disorders such as anorexia nervosa and bulimia can also develop as people take drastic measures to look like the people they see in the media.

The media also target children, as they are a major influence on household food selection and are highly influenced by advertising. They are an important force in the marketplace, not only on their own account but because of the influence they have on their parents' food purchases. Many products advertised on television are also displayed in the supermarket at children's eye level. Many television advertisements promote food that is high in fat, salt and/or sugars. Advertisements directed at children screen before and after school. However, Australia does not allow advertisements during television programs aimed at the preschool age group.

Sophisticated and aggressive techniques are used in marketing. Techniques used by advertisers include prizes, giveaways, animation, special effects and jingles. Popular personalities are often used to endorse products. The predominant messages are related to having fun, being cool, looking attractive and enjoying tasty food.

To determine the extent of junk food advertising that Australian children are exposed to on television, the *Sydney Morning Herald*, in 2008, analysed free-to-air television between 6 a.m. and 9 p.m. They found that 15 commercials an hour were broadcast for fast food outlets, chocolate and other unhealthy products during the afternoon and early evening – prime viewing slots for children. Many of the food advertisements broadcast during children's television programming hours were for food high in fat, sugar and/or salt with little nutritional value.

FOOD IN FOCUS

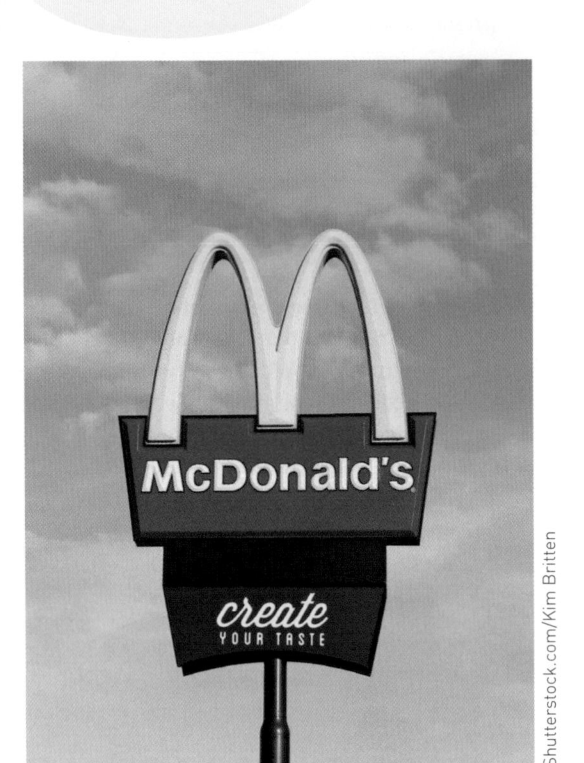

Shutterstock.com/Kim Britten

PARENTS' JURY BITES BACK OVER MCDONALD'S CASH REBATES FOR SCHOOL ORDERS

'Cash rebates' are being offered to schools that place lunch orders at McDonald's, in a move slammed by a parent group as a cynical marketing ploy.

In a letter to Mildura schools this month, a local McDonald's representative said school lunches could be used as a fundraiser, end of term treat, or delivered to school functions.

The letter promises that if the whole school participates, McDonald's will 'give your school a 9 per cent cash rebate on your order to do with as you wish'.

Free cordial, cups and ice and delivery are also on offer to schools that place lunch orders.

The letter includes an order form for burgers, fries and desserts, and notes that 'McDonald's offers' are also available for students receiving encouragement or achievement awards.

The Parents' Jury spokeswoman Dimity Gannon, whose group represents 7000 Australian parents, said she was 'quite horrified' by the promotion.

'From a parent's perspective, they feel that children at school should be protected from this kind of aggressive marketing,' Ms Gannon said.

'With one-quarter of Australian children currently overweight or obese . . . this completely undermines the healthy eating messages that we are trying to get across to children.'

Ms Gannon said the offer of a 'cash rebate' if the whole school ordered lunch put parents under pressure to participate, 'despite perhaps being against their normal practices'.

'This is a marketing strategy for McDonald's, it's not about helping schools raise money,' she said.

'We strongly encourage schools to steer away from unhealthy fundraisers; there are lots of healthy alternatives out there.'

Obesity Policy Coalition executive manager Jane Martin said McDonald's was putting profit ahead of children's health by promoting unhealthy food in schools.

The letter to Mildura schools says the promotion was not created at a national level, 'but is an initiative produced locally at McDonald's Mildura to help support our schooling community'.

A McDonald's spokesman said the chain's restaurants were sometimes approached by schools to provide assistance for fundraisers and other initiatives.

'Provided the school has parental permission, we do not see an issue with them hosting a McDonald's lunch as a fundraiser or end of term treat,' he said.

'McDonald's does not engage in any communications in Australian schools unless requested by, or agreed with, the school administration.'

Source: 'Parents' Jury bites back over McDonald's cash rebates for school orders', by Kate Hagan, *The Age*, 17 February 2015, This work has been licensed by Copyright Agency Limited (CAL). Except as permitted by the Copyright Act, you must not re-use this work without the permission of the copyright owner or CAL

ACTIVITIES

1 Evaluate the marketing techniques used by McDonald's to target school-aged children.

2 With a partner, discuss your thoughts on unhealthy food advertising – do you think it needs to be monitored more strictly or banned altogether? Share your results with the class.

UNIT REVIEW

LOOKING BACK

1. List the foods you often crave. When are you most likely to crave these foods? How nutritious are they?
2. List the foods that fill you up quickly but leave you hungry in an hour.
3. Why is food such an important aspect of social gatherings? Describe how your family entertains. What foods are offered to guests and how are they presented? Do your classmates' families do anything different? How?
4. Identify some of your good and bad food habits and suggest reasons why they occur.
5. List and explain the different physiological factors that influence food choices. Use examples to justify each point.
6. Using examples, outline how your geographic location may influence what you eat.

FOR YOU TO DO

7. Conduct your own survey. Survey the advertisements shown during a children's show on television. Note the program watched, the length of the program and the number and type of food advertisements shown.
   - **a** Describe the advertising techniques used, including jingles, colour, famous people, humour or special gimmicks.
   - **b** Are the foods advertised nutritious, or are they high in fat, sugar and salt content?
8. In small groups, collect four magazine or newspaper advertisements for food. Evaluate each advertisement, highlighting the target audience and the techniques used to attract the buyer.
9. Which food practice is the most common among the religions listed in Table 3.4 on page 98? Can you explain why?

TAKING IT FURTHER

10. Working in pairs, complete the following tasks.
    - **a** Design a low-cost main meal.
    - **b** Write up a work plan, including time, sequences of steps, processes and equipment used, health and safety considerations.
    - **c** Visit the supermarket to cost the items for your meal and determine the total cost.
    - **d** Prepare your meal and undertake a sensory and nutritional evaluation of the finished product.
11. Design a survey to be completed by all students in your class as a means of analysing what each person eats. Survey an older year group at your school. Compare results from each year group and draw conclusions.

3.6 Consumption patterns and related factors

The food we consume in Australia has significantly changed over the past 200 years. It reflects changes in our society, lifestyle and Australian cuisine.

Factors affecting the food consumption patterns of Australians can be divided into the following categories:

- social
- economic
- nutritional
- environmental.

Social factors

What we eat, when we eat, where we eat and with whom we eat have all noticeably changed in the last 20 years. The structure of Australian families has also changed, with more adult-only households and more single-parent households with dependent children. Nearly 70 per cent of women now work and approximately 60 per cent of households with two parents and dependent children have both parents working. With more single-parent families and women working full-time, the time available for food selection and preparation has decreased.

Australians are cooking less frequently, and spending less time preparing the meals that are cooked – there are homes currently being built in Sydney without proper kitchens! Men are taking a greater role in food selection and food preparation compared with previous generations. However, younger men tend to purchase meals away from home rather than cook for themselves. Children are more responsible for choosing their own snacks and meals, with many preparing meals for themselves without adult supervision.

These changes have increased our reliance on and interest in pre-prepared foods and foods with an extended shelf life. Australian consumers are also experimenting with a diverse range of new, exotic and culturally different foods, including raw animal food products such as sushi.

Socialising

Food is playing a far greater role in Australian entertaining and socialising, with many people eating out more regularly. Supermarkets, fast food outlets and restaurants have all extended their opening times to cater for consumers' lifestyles and work arrangements. The foods consumed are often picked up on the way home from work, ready for the family to eat, or ordered by phone and delivered to the door. When Australians do eat at home, they are more likely to eat alone, or in the company of the television.

Economic factors

Economic factors such as income and the cost of living affect the foods we select. The cost of food often takes priority over considerations such as nutritional value, social desirability and taste.

Income

Income is one of the most important determinants of food consumption. Families living on pensions or social welfare benefits, or other very low incomes, have difficulty paying for the necessities of life – food, clothing, heating and housing. With most of their income tied up in rent or regular payments to power utilities, the only part of the budget in which low-income families can exercise any discretion is that part set aside for food. People with low incomes spend a greater proportion of their family income on food compared with families on higher incomes. Low-income families may suffer from poor health if they cannot afford an adequate food intake. This may result in dietary deficiencies such as a lack of iron or Vitamin C. High-income families often make poor food choices, as they may consume too much food. This can contribute to diet-related disorders such as coronary heart disease and diabetes.

Groups at risk

There are a number of groups within the Australian population that have been identified as being at risk of under-nutrition: the unemployed, low-wage earners, single-parent families, pensioners and Aboriginal Australians. These groups may have major problems maintaining a healthy and balanced diet. Sometimes they have to go without food until they receive their pension or benefit.

Individuals who live in low socio-economic and remote areas of Australia are further disadvantaged by a lack of large chain food retailers in their region, and by the lack of accessible transport to such shops. They are forced to depend on smaller corner stores where the variety is limited and the cost is higher. These same people may not have access to appropriate food storage and cooking equipment.

Figure 3.17 The low cost and convenience of pre-prepared meals often takes priority over other considerations, such as nutritional value, social desirability and taste.

Nutritional factors

The role of diet in promoting health and preventing lifestyle-related health disorders has been increasingly promoted to Australians over the past 20 years. This nutrition education has focused on the dietary guidelines and improved lifestyle choices in order to encourage a healthy lifestyle.

Concern about the food consumption patterns of Australians led to the development, in 2003, of a revised set of dietary guidelines for Australian adults and for children and adolescents. These guidelines provide information for health professionals and the general population about healthy food choices. They encourage healthy lifestyles that minimise the risk of the development of diet-related disorders within the Australian population. The guidelines highlight the groups of foods and lifestyle patterns that promote good nutrition and health.

While these guidelines cannot guarantee good nutrition, as many other factors such as heredity and lifestyle also play a role, an intake from a wide variety of food will help improve and maintain health. No single food contains all of the nutrients the body needs in the amounts necessary for optimum growth and health. For that reason, it is important to remember to consume a wide variety of foods in the correct amounts and in the context of a total, balanced diet.

The guidelines are linked to the Australian Guide to Healthy Eating (see Chapter 2).

Environmental factors

Food has to be grown, processed, packaged, distributed and stored. All these processes have environmental impacts as they all use energy and natural resources. While some of these resources are renewable, many are not. For example, the growing of rice has meant that irrigation has to be used to supply sufficient water. Extensive irrigation can have significant effects on surrounding land. Land also has to be cleared to grow crops and this can cause erosion.

Packaging

Technological developments have meant that many food products are now available in different forms of packaging. Also, packages are lighter and thinner, and fewer raw materials are used in their manufacture. This is desirable as it means that fewer of our natural resources are used to produce the packaging.

Much of the landfill in Australia consists of food packaging. It is important for the food industry to find ways of using less packaging while maintaining high food quality. However, with the trend towards consumption of convenience foods, 'heat and serve' meals and single-serve food items, most developed countries are producing more packaging than ever before. Consumers also demand packaging that is convenient and will protect the food, keeping it safe and free from contamination.

Creating a balance

Australians have to maintain a balance between producing food for consumers and maintaining the environment and resources. Consumers also have to take responsibility for the environment by using resources wisely, minimising food wastage, minimising packaging materials and disposing of packaging materials in an environmentally aware manner.

Consumer demand for different food products has changed in important ways over the last thirty years, driven by increasing per capita incomes, demographic shifts and lifestyle changes.

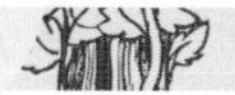

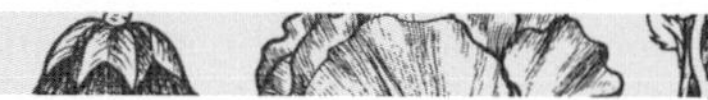

FOOD IN FOCUS

AUSTRALIA: WHAT WE EAT

Australians are some of the largest consumers of meat in the world and the quality is extremely high. Our meals tend to be lighter with less heavy sauces and gravy. However, there is an increased use of marinades and lighter sauces, such as soy sauce and teriyaki sauce. Classic European dishes are still popular but the current food fad is Asian in origin – with Japanese and Thai joining Chinese, which has long been a favourite. Meat is increasingly being used in stir-fries, kebabs, rolled roasts and noisettes. Interestingly,

>

Australians still enjoy a traditional roast of beef, lamb or pork and visitors will find these on many restaurant menus.

There is a seemingly endless supply of ethnic restaurants catering to the many nationalities that now live here. The combination of high-quality local foods with ethnic cooking methods has produced some of the best meals in the world.

Barbecues are very popular with Australians, who like a casual lifestyle and eating outdoors. Most homes would now have outdoor barbecue facilities or portable barbecues. Many parks and beaches have designated barbecue areas. You will hear the steady sizzle of meats and seafood over coals around the country on summer evenings and during weekends. Many small hotels and clubs also have outdoor eating or barbecue facilities where customers can choose and cook their own steak.

Milk-fed lamb and veal are served in top restaurants. They usually cost a little more but are valued for their fine, delicate flavour. Top restaurateurs have their own breeders who are contracted to supply them with suckling lamb (5–6 weeks old). Cattle is predominantly pasture-fed and produces lean beef by world standards; a 100 gram-portion of cooked lean rump steak contains about 6.7 grams of fat and 80 milligrams of cholesterol. Grain-fed beef would have approximately twice this fat content.

Australians are not big pork eaters but there has been an active campaign by breeders to produce a leaner meat and new cuts. Chicken is the most commonly used bird for eating. Duck, goose, turkey, quail and guinea fowl are also bred, and Australian pigeon is rated as among the best in the world.

Kangaroo and water buffalo are also slaughtered for consumption in some states. At present, kangaroo can legally be served in restaurants in South Australia, Tasmania, the Northern Territory, Victoria and the Australian Capital Territory. The taste is similar to venison but less gamey. The cuts are not large, so kangaroo is usually served in small medallions.

Migrants have had a significant influence on the introduction of many varieties of smallgoods and a growing awareness of their uses over the past 10 years. Most of the varieties available originated overseas but they are increasingly being made here by experts from those countries. Salamis include Danish, Polish, Italian, Milano, Pepperone and Hungarian. Speck, coppa, bratwurst, coteghini, smoked beef, bastourma, berliner, pate, bloodwurst, cabanossi, strasburg and csabai can be bought at most good delicatessens. Ham and bacon are popular, bacon being an integral part of the traditional Australian breakfast.

Sausages are eaten by 70 per cent of Australian households at least once a week. Australians have always regarded the sausage, or snag, as a cheap form of meat to fry, grill or barbecue. But this is changing. Immigrants from countries where the sausage has gourmet status are introducing new varieties and more interesting ways of cooking and serving.

Source: Adapted from information on the Global Gourmet website

ACTIVITIES

1 List the foods that you consume over a one-week period. Does the information outlined above represent the foods on your list? Discuss your findings.
2 Outline why Australian food consumption patterns continue to change.
3 Select one product named above that originated overseas. Use the internet to research its origin and describe how it is currently consumed in Australia.
4 Research two interesting ways of cooking and serving traditional Aussie sausages.

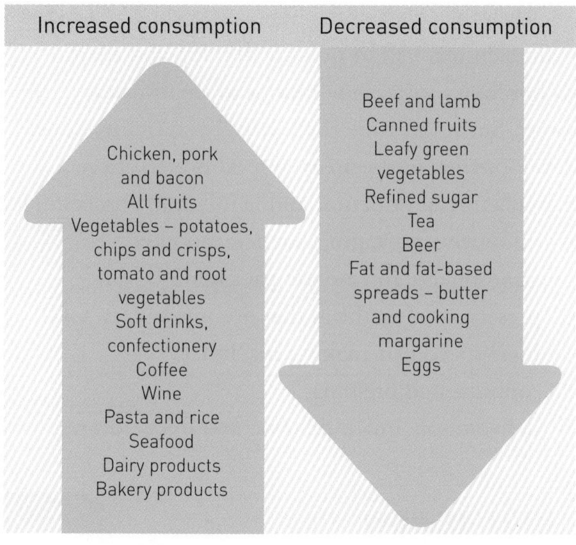

Figure 3.18 Summary of changes in consumption of food over the past two decades

UNIT REVIEW

LOOKING BACK

1 Identify one social factor affecting food consumption patterns of Australians and provide an example.
2 How may the structure of a family and the life cycle stages of family members affect the type of food eaten?
3 Outline how income may affect the food purchases of families.
4 Explain how family income may be linked to diet-related disorders.
5 Examine Figure 3.18 and discuss why people are now eating more chicken than beef or lamb.

FOR YOU TO DO

6 Visit your local supermarket or use the internet to research new convenience foods that are available to purchase.
7 In class, brainstorm the ways in which you as a food consumer can have a positive effect on the environment.

TAKING IT FURTHER

8 Conduct a survey to determine whether Australians are concerned about the environmental impact of the foods they consume. What conclusions can you draw from your results?

3.7 Additional content: production and processing

The food industry did not really start in Australia until the middle of the 19th century. Significant developments in the 19th and 20th centuries are summarised below.

- Sugar growing and refining started in the early 19th century.
- Wine making began in the middle of the 19th century, although the quality was not very good.
- Refrigeration began in Australia in 1851, although it was not really successful until much later.
- In 1855, William Arnott, a baker and pastry cook, began to make biscuits in Newcastle. His became a dominant name in Australian biscuit making.
- Advances in the later part of the 19th century included roller milling of wheat, advanced sugar milling, margarine manufacture, developments in

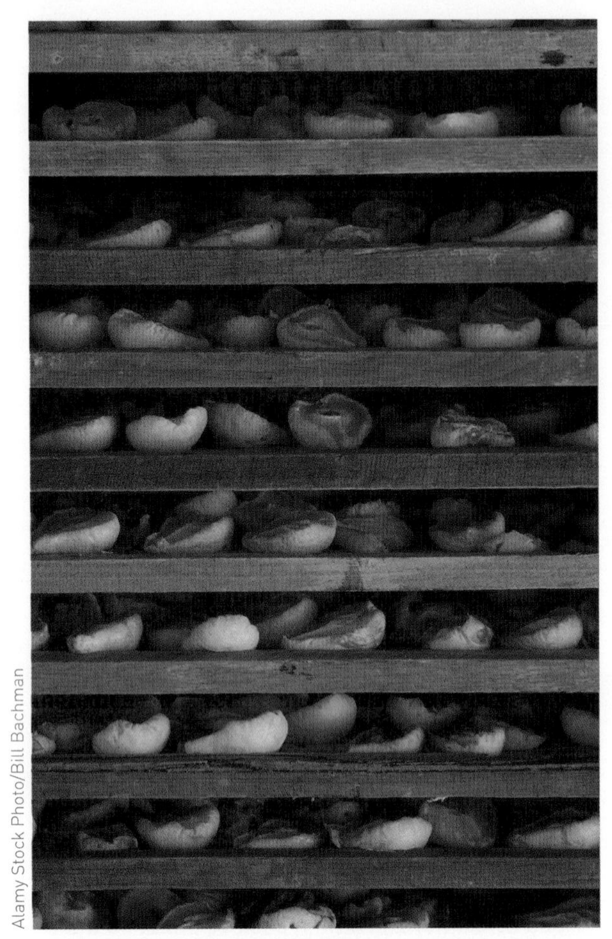

Alamy Stock Photo/Bill Bachman

Figure 3.19 Drying fruit at Angus Park, Loxton, Murray Riverland, South Australia

milk separators and the sun-drying of fruits in the Sunraysia district. Research into the growing of wheat and the farming of sheep also assisted the industry.

- In 1865, the Australian Meat Company was set up and the canning of meat, meat extract and fat began in Australia. Other meat canneries were established around the country, although most failed due to lack of money and experience. Larger canneries survived and exported meat to Britain. The canning of jam began in Hobart in 1861 and later extended to mainland Australia.
- In 1906, R. G. Edgell started growing vegetable crops at Bathurst and by the end of the 1930s he was canning a range of vegetables.
- The dairy industry developed slowly, although cheddar cheese was really the only cheese available until the 1950s. Processed cheese made by the Kraft Walker Cheese company was available in 1926.

With the onset of the Second World War, food manufacturers had to quickly develop and adapt technologies to provide rations to the overseas forces. Since this time there have been tremendous developments in the areas of food preservation, manufacture, production and legislation. For example:

- innovative packaging, such as modified atmospheric and aseptic packaging
- preservation methods that do not damage food
- high-speed can-making machinery
- sophisticated brewing
- dehydration, quick freezing and freeze-drying.

UNIT REVIEW

LOOKING BACK

1 Why did Australia lag behind Britain in producing and processing food in factories?

2 At what stage in Australia's history did commercial food processing and production really begin?

FOR YOU TO DO

3 Select one of the following sectors of the Australian food industry and research its development:

- meat industry
- sugar industry
- dried fruit industry
- milk industry
- dairy industry
- canned vegetable industry
- biscuit industry
- snack food industry.

TAKING IT FURTHER

4 Over the past 50 years there have been many technological developments in the preservation of food. Select one of the following developments and research how it has affected the food available in Australia:

- modified atmospheric packaging
- freeze-drying
- aseptic packaging
- microwavable meals
- dual-oven heating trays.

Chapter review

LOOKING BACK

1 Identify and describe two native foods. Investigate how each is grown, harvested and prepared for use in Australia today.
2 Explain why Aboriginal Australians were healthier prior to European settlement.
3 List the foods that the early settlers consumed.
4 Outline why meat was preserved and the process used to preserve it.
5 Explain why European settlers struggled to cultivate a healthy lifestyle.
6 Discuss how migration and travel have affected the range of foods now available in Australia.
7 List the ethnic origins of different restaurants in your suburb or closest town. Which are more popular? Why do you think this is the case?
8 Explain, using examples, how the mass media influence food consumption patterns.
9 Provide one example of how a person's economic position can influence their food choices.
10 Provide examples of the impact of migration on Australian cuisine.
11 List some cooking methods introduced to Australia through immigration.
12 How do social practices influence what is consumed today? Think about your own social practices and outline how they affect your food choices.

FOR YOU TO DO

13 Using the internet, research live native foods eaten by Arrernte people in their region, including a honey-like food that could be used as a sweetener. Organise for a representative of the Arrernte people to visit your classroom. Ask them to translate the names of the native foods for you in Arrernte. Prepare a range of questions that you can ask your visitor so that you gain a better understanding of this chapter.
14 Research a famous Australian restaurant and investigate the foods that are included on their menu. Try to identify what influences the choice of foods they present to their customers.
15 If you were the owner of a new restaurant that served modern Australian cuisine, what would you name it and why?
16 Visit the Parents' Voice weblink and investigate the regulations about Australian TV advertising. After reading, reflect on what you have learned in this chapter and explain what future action needs to be taken by the regulators.

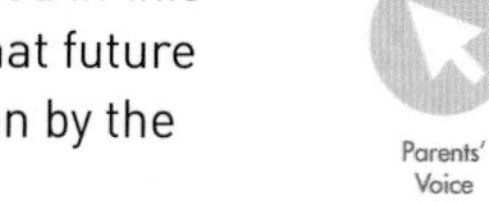

17 Brainstorm how to reduce a child's exposure to food marketing.
18 Research a recent technological advancement and investigate how it has affected food and its development. Present your findings in a newspaper article.
19 Explain the impact of the following factors on current food consumption patterns:
- social factors
- economic factors
- nutrition factors
- environmental factors.

20 Outline the social changes that have led to Australians eating more meals away from home and more convenience meals.
21 Research the food rituals, taboos and laws of the Islamic religion.

TAKING IT FURTHER

22 Conduct an accompanied excursion to a national park or local bush food trail to view and taste-test Australian bush foods. As a class, identify the foods that can be grown and included in your current diets.
23 Investigate the significance of food in traditional Aboriginal culture. Prepare an oral or PowerPoint presentation about the technologies used by Aboriginal Australians for collection and preparation of their food.
24 Discuss why many Aboriginal Australians continue to eat traditional bush foods.
25 In small groups, plan a day's menu for an adult male Aboriginal Australian who lives on the coastline of Australia. Use the internet to check your answer.
26 Predict the aspects of Australian life that would be different if Australia had been colonised by a country other than Great Britain. In groups of three, brainstorm examples. Design a poster highlighting what life in Australia might have looked like today.

27 Visit your local shopping centres and list all of the restaurants, take-away and fast-food establishments. Consider how you would describe Australian cuisine in your own town or city. Compare your description with those of your classmates.

28 Select one food product that you regularly consume and determine how far it has travelled to get to you. Do you think it is an environmentally sound choice? Will you continue to include this food in your diet?

29 Select one cultural group in Australia.

- **a** Research the types of traditional foods served.
- **b** Explain how this culture has influenced food preparation techniques and the foods eaten in Australia.
- **c** Outline the availability of food from this culture in Australian supermarkets.
- **d** List the foods from this cultural group that have become take-away and fast foods.
- **e** With a partner, select and prepare one of these foods.
- **f** Using a nutritional database, analyse the nutritional value of the food you have prepared.

30 Using the information on the Australian Geographic weblink, design a food blog summarising Australia's historical and current cuisine culture.

Australian Geographic

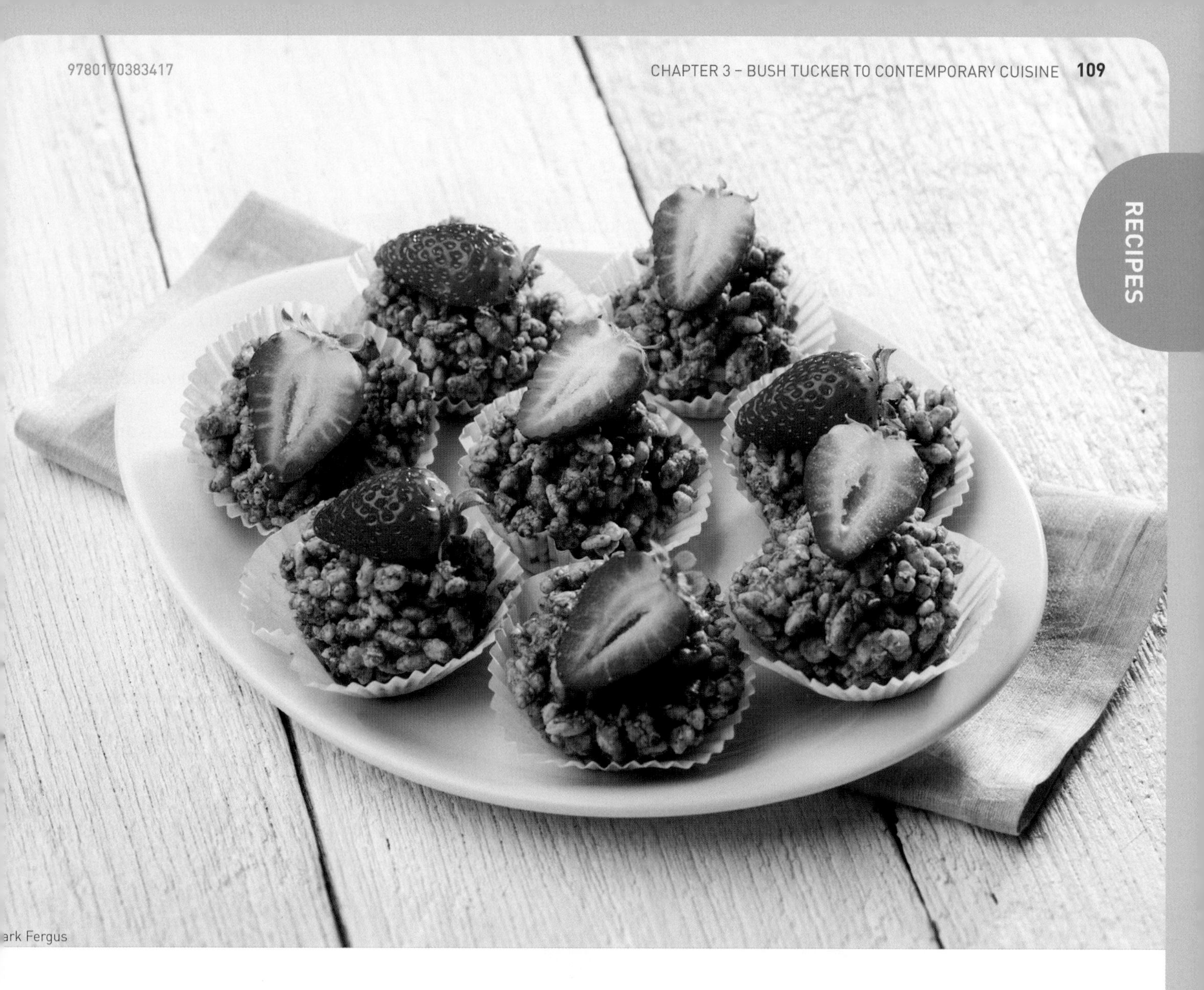
ark Fergus

Lemon myrtle chocolate crackles

 **Preparation time** 15 minutes and refrigeration  **Cooking time** 5–10 minutes  **Serves** 12

INGREDIENTS

2 cups rice bubbles or rice crispies
½ cup icing sugar
½ cup desiccated coconut
125 grams copha
1½ tablespoons cocoa powder
1 tablespoon lemon myrtle
12 strawberries

METHOD

1. In a large bowl, mix the rice bubbles, icing sugar, coconut and cocoa.
2. Melt the copha in a saucepan over low heat. Stir through the lemon myrtle. Add to the rice bubble mixture and stir to combine.
3. Spoon the mixture into patty cases, top with a strawberry and refrigerate until firm.

QUESTIONS

1. Research copha. What is it made from?
2. Suggest two safety tips when working with the melted copha and adding it to the dry ingredients.
3. After you have made and tasted the lemon myrtle chocolate crackles, evaluate your success. Did they turn out the way you hoped? Did they taste the way you expected?

RECIPES

Wattleseed jam cockles

 **Preparation time** 25 minutes  **Cooking time** 10 minutes  **Serves** 12

INGREDIENTS

120 grams butter
120 grams sugar
2 eggs
120 grams plain flour
120 grams cornflour
1 teaspoon baking powder
1 tablespoon wattleseed extract
quandong jam (see next recipe)

METHOD

1. Whip the butter with sugar until light and fluffy.
2. Add the eggs one at a time.
3. Add the plain flour, cornflour, baking powder and wattleseed extract and mix.
4. Use a dessert spoon to roll balls of the mix and place on a greased oven tray, flattening each slightly with the back of a fork.
5. Cook for 10 minutes in an oven preheated to 200°C.
6. When cold, sandwich two biscuits with quandong jam.

QUESTIONS

1. How do you know when the butter is sufficiently whipped?
2. Suggest why the recipe may be more accurately made measuring by weight (grams) rather than volume (cups).
3. If wattleseed extract is not available, what ingredient may be a suitable alternative?

Mark Fergus

Mark Fergus

Quandong jam

 **Preparation time** 8 hours (or overnight)  **Cooking time** 60–70 minutes  **Makes** Several jars

INGREDIENTS

- 100 grams dried quandong
- 2 Granny Smith apples
- 600 grams caster sugar
- 1 lemon or lime, juiced

METHOD

1. Put the quandong in a bowl, cover with water and leave to soak for about 8 hours or overnight. (Yields about 500 grams of fruit)
2. Peel, core and roughly chop the Granny Smith apples and then place the apple and the quandong in a pan with the quandong soaking water and lemon juice.
3. Simmer for about 30 minutes until soft, stirring from time to time.
4. Remove the pan from the heat and add the sugar, stir until sugar has dissolved. Bring to the boil and boil rapidly for 20–25 minutes, stirring frequently to prevent sticking. Remove from the heat, blend if required, but be sure to leave chunky pieces of apples and quandong in the jam. Pot and cover as usual.

QUESTIONS

1. Working with hot jam can be very dangerous. List two safety tips for jam making.
2. If quandongs are not available, what fruit could be successfully substituted?
3. Pectin is often used in jam making to help the jam to set. Research which ingredient in this recipe helps it to set.

RECIPES

Traditional damper

This recipe is made from the basic products that were available at the time of early European settlement – plain flour, water and salt. There is no quantity given for these; you just make it up as you go!

 **Preparation time** 15 minutes  **Cooking time** 30 minutes  **Makes** 1

INGREDIENTS

plain flour
water
salt

METHOD 1

1. Preheat oven to 220°C.
2. Take a quantity of flour (try 250–500 grams) and a pinch of salt.
3. Add enough water to the flour to make dough. The dough should be rather firm and not sticky, so add the water slowly.
4. Take a large sheet of foil and place the dough in the centre. Wrap the dough in the foil and place in the oven.
5. Take the damper out occasionally to do the 'tap test'. If the bottom sounds hollow when tapped, it is done.

METHOD 2

You could try cooking damper using the traditional method – under the coals of a fire. This method should only be used with your teacher or another adult.

1. Dig a hole about 60 cm wide and about 30–40 cm deep.
2. Arrange a fire on top of this hole.
3. When the fire has burnt down to small red-hot coals, place the damper wrapped in foil into the hole. Coals should be positioned both under the damper and on top of it.
4. Take the damper out occasionally to do the 'tap test'. If the bottom sounds hollow when tapped, it is done.

Mark Fergus

ark Fergus

Bush damper

 **Preparation time** 15 minutes  **Cooking time** 45 minutes  **Makes** 1

INGREDIENTS

2 cups self-raising flour
1 teaspoon salt
1 tablespoon wattleseeds
1 teaspoon lemon myrtle
1 cup buttermilk
1 tablespoon olive oil
milk, to brush
golden syrup to serve

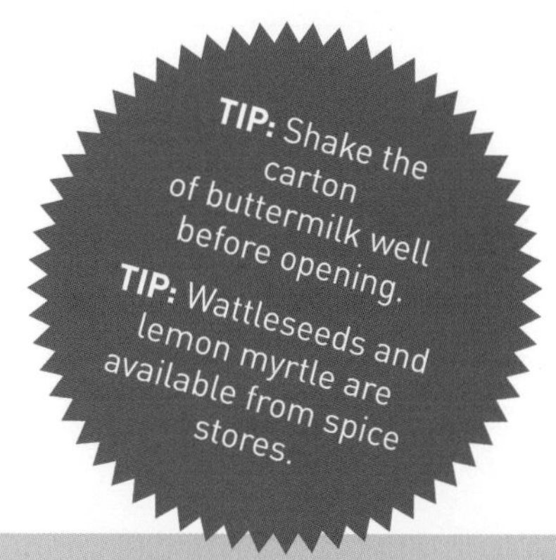

METHOD

1 Preheat oven to 180°C. Grease a baking tray.
2 Sift the flour and salt into a large bowl. Add the wattleseeds and lemon myrtle and stir to combine.
3 Make a well in the centre of the dry ingredients, add the buttermilk and oil and use a fork to mix into a soft dough.
4 Place on a lightly floured board and knead to a smooth dough.
5 Shape into a round cob, place on a greased baking tray and brush with milk.
6 Bake for 40–50 minutes, or when the 'tap test' (the bottom sounds hollow when tapped) indicates that it is done.
7 Slice and serve warm, drizzled with golden syrup.

QUESTIONS

1 If you have made both the traditional damper and the bush damper, what differences in the results can you identify?
2 How would you know if you have been too vigorous or over-kneaded the dough?
3 From your tasting, name some complimentary flavours for wattleseed and lemon myrtle to spread on your damper.

9780170383417

Contemporary Aussie damper

 **Preparation time** 15 minutes  **Cooking time** 45 minutes  **Makes** 1

INGREDIENTS

2 cups self-raising flour
1 teaspoon salt
1 teaspoon sugar
1 teaspoon butter
1 cup milk

METHOD

1 Preheat oven to 220°C.
2 Mix the dry ingredients and the butter together.
3 Add the milk and mix well.
4 Knead for approximately 5 minutes.
5 Shape the dough into a flat ball, and place the dough on a greased and floured baking sheet or in a greased and floured round cake tin.
6 Bake for 30 minutes or until the 'tap test' indicates that it is done.
7 Remove and wrap in a clean tea towel and cool on a wire rack.
8 Serve immediately in moderately thick slices.

QUESTIONS

1 Draw up a table in your notebook similar to the one below and use it to evaluate the original, bush and contemporary dampers.

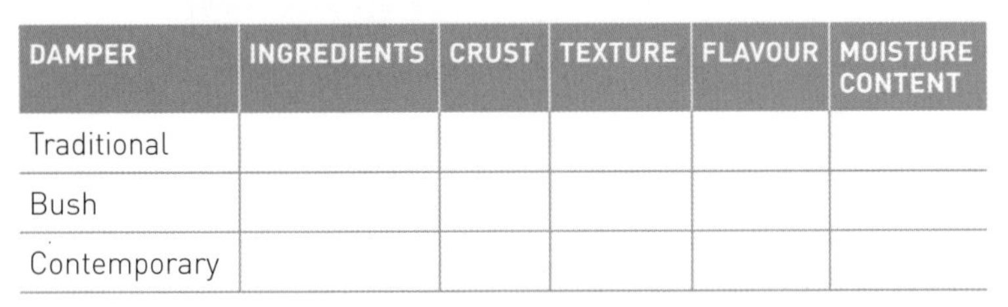

DAMPER	INGREDIENTS	CRUST	TEXTURE	FLAVOUR	MOISTURE CONTENT
Traditional					
Bush					
Contemporary					

2 Evaluate the nutritional properties of each damper by plotting the ingredients on the Healthy Eating Pyramid (see Chapter 2, page 66).
3 Suggest alterations that could be made to the ingredients to improve the nutrient value of any of the dampers.
4 Why do you think damper was a popular form of bread for the first settlers?

Mark Fergus

Nem ran (North Vietnamese spring rolls)

 **Preparation time** 20 minutes  **Cooking time** 30 minutes  **Makes** 5

INGREDIENTS

30 grams (½ cup) wood-ear fungus

700 grams pork mince

300 grams green banana prawns, peeled, de-veined, finely chopped

10 shallots, ends trimmed, finely chopped

1 × 50 gram packet cellophane (green bean thread) noodles, cut into 2 cm lengths

2 medium yam beans, peeled, finely chopped

2 eggs, lightly whisked

2 teaspoons freshly ground black pepper

1 teaspoon salt

1 teaspoon chicken stock powder

1 teaspoon sugar

1 × 375 gram packet 18 cm-diameter round rice paper sheets (bánh tráng)

Vegetable oil, to deep fry

with Permission from Pham Bao Khanh Linh

METHOD

1. Place the wood-ear fungus in a heatproof bowl. Cover with warm water and set aside for 30 minutes to soak. Drain and finely chop. Combine the chopped fungus, pork, prawns, shallots, noodles, yam beans, eggs, pepper, salt, chicken stock powder and sugar in a large bowl.
2. Fill a bowl with warm water. Dip a sheet of rice paper in the water for 10–15 seconds or until it begins to soften. Drain and top with 1 tablespoon of filling. Fold in the sides and roll up firmly to enclose the filling. Repeat with the remaining rice paper sheets and filling.
3. Add enough oil to a large saucepan or deep frypan to reach a depth of 6 centimetres and heat over medium heat. To test when oil is ready, a cube of bread turns golden brown in 20 seconds.
4. Add one-third of the spring rolls and deep fry for 10 minutes or until golden brown and cooked through. Use a slotted spoon to transfer to a plate lined with paper towels. Repeat in two more batches with the remaining spring rolls, reheating oil between batches. Cut each spring roll into 3 equal pieces.

QUESTIONS

1. The wood-ear fungus is a dehydrated product. Explain why it is soaked in water prior to use.
2. How could you modify this recipe for a person who was vegetarian?
3. What is the purpose of draining the freshly fried spring rolls on paper towels?

Mark Fergus

ark Fergus

Chicken, minted pea and ricotta meatballs

 **Preparation time** 30 minutes  **Cooking time** 20 minutes  **Serves** 4

INGREDIENTS

- ⅓ cup frozen peas
- 1 teaspoon finely grated lemon rind
- 2 tablespoons mint, chopped
- 2 tablespoons mint leaves, extra, to serve
- 175 grams chicken mince
- 1 clove garlic, crushed
- ¼ cup dried multigrain breadcrumbs
- 1 egg yolk
- 2 tablespoons low-fat ricotta
- 1 tablespoon olive oil
- 400 grams canned Italian diced tomatoes
- 2 teaspoons balsamic vinegar
- 130 grams wholemeal spaghetti
- 2 tablespoons shaved parmesan

METHOD

1. Preheat oven to 240°C or 220°C fan-forced. Grease and line a small shallow baking dish with baking paper.
2. Place the peas in a medium heatproof bowl. Cover with boiling water. Stand for 1 minute. Drain, reserving 1 tablespoon of the water. Blend or process the peas, reserved water, lemon rind and chopped mint until just combined.
3. Combine the mince, garlic, breadcrumbs and egg yolk in a medium bowl. Stir in the pea mixture and ricotta.
4. Roll a level tablespoon of mixture into balls using wet hands. Place in the prepared dish. Drizzle with oil. Roast for 15 minutes, turning once. Add the tomato and vinegar. Cook for 5 minutes or until the meatballs are cooked through and sauce is hot.
5. Meanwhile, cook the spaghetti in a large saucepan of boiling water until al dente. Drain.
6. Serve the spaghetti with the meatballs and sauce. Top with the extra mint and parmesan.

QUESTIONS

1. Imagine you are making this recipe for a person who is gluten intolerant. Which ingredients will need to be substituted for gluten-free alternatives?
2. As a class, present your dishes individually. Take photographs of each dish and display them. What was the most challenging aspect of presenting this dish?
3. Research the flavours most commonly associated with Italian food. What would be a typical Italian dessert to follow this dish?

CHAPTER 4

Mark Fergus

Fare's fair

FOCUS AREA

FOCUS AREA: FOOD EQUITY

There is enough food produced in the world to feed everyone; the problem is that it is unevenly distributed. Why do **developing nations** continue to suffer from hunger, while developed nations suffer from diseases associated with too much food? This chapter investigates the factors that contribute to this unequal distribution of food in Australia and around the world. These factors include the balance of international trade, the distribution of food between those who 'have' and those who 'have not', and the impact of difficult geography, poor roads and unstable political situations.

CHAPTER OUTCOMES

In this chapter you will learn about:

- circumstances that cause food inequity
- groups that may experience food inequity in Australia and overseas
- the factors influencing food availability and distribution
- food production practices
- malnutrition
- providing aid
- support for people suffering from food inequity.

In this chapter you will learn to:

- explain the circumstances causing food inequity
- identify people at risk of food inequity
- recognise the relationship between food availability and distribution and food equity
- examine food production and distribution
- describe the consequences, including diseases, associated with malnutrition
- identify the role of aid agencies
- design, plan and prepare meals appropriate to specific circumstances
- investigate a specific group that suffers from food inequity.

RECIPES

FOOD FACTS

Australians use a million litres of water per person per year, while those in water-scarce places such as the Middle East have access to less than 1000 litres per person per year.

There are approximately 1 billion farmers in the world. Only 50 million of these farmers live in industrialised countries.

Women perform two-thirds of the world's work, grow half of the world's food, earn one-tenth of the world's income and own one per cent of the world's property.

Two-thirds of the world's population has access to less than one-third of the world's food.

Some of the world's most valuable agricultural crops are cotton, grapes, sugar and coffee beans.

Australia's biggest primary agricultural products have not changed in generations – our greatest income continues to be from cattle (beef) and wheat, both from local and international customers.

The fastest growing age group in Australia are those who are aged 65 and over. Tasmania has the population with the oldest average age, while New South Wales has the fastest growing population.

Statistics state that at any one time, there are currently over 105 000 homeless people living in Australia. Homelessness may be defined as having no suitable accommodation, no permanent accommodation, or accommodation that is below the accepted standard.

FOOD WORDS

developing nation country with low levels of production of goods and services and limited technology

export send goods or produce out of the country

finite having bounds or limits; many natural resources are available in fixed amounts and once used cannot be replenished

malnutrition lack of proper nutrition, caused by not having enough to eat or not eating enough of the right things

refugee person who flees for safety to another country

self-sufficient ability to support oneself; communities that are self-sufficient need very little outside assistance to produce food

staple food basic food that a population eats on a regular basis and in large quantities

subsistence farmer small farmer providing only enough produce to meet the family's food needs

tariff tax on imports so that locally produced goods become less expensive by comparison

4.1 The reasons for food inequity

Food inequity is a global issue, not just a problem for underdeveloped countries. There are a number of reasons for this inequity.

Access to continuous and safe water

Clean water is a finite resource. There is a fixed amount available that cannot be increased, but is under threat from overuse and pollution.

Human health is dependent on clean water in sufficient quantity and safe, sanitary conditions. Poor hygiene and unsafe water quickly spread diseases. Children suffer the most from these diseases. Some diseases are water-borne, caused by drinking water that has been contaminated by faeces. This often occurs where conditions are unsanitary, or after a natural disaster such as a flood where sewerage treatment works are damaged. Water-related diseases are caused by insects that feed or breed in water, such as malaria that is carried by mosquitoes. Water-based diseases are caused by parasites that spend part of their lifecycle living in water. Examples include guinea worm, which infects people drinking infested water.

The supply of water is crucial to the increase in food production. Irrigated land is twice as productive as crop land relying on rainfall alone. One-sixth of the world's crop land is irrigated but produces about one-third of the world's food. One serious global issue is the irresponsible use of fresh water and groundwater for irrigation in a manner that cannot be sustained.

Access to water is crucial. Without an adequate water supply, factories that depend on water may have to close temporarily, crop yields may decline, sick workers may be unproductive and fisheries may be destroyed. In countries that have limited water supplies, women and children spend many hours a day carting water and so have less time and energy to grow food, earn an income or gain an education.

Availability of safe and nutritious food

Eighty per cent of farmers in developing countries are small farmers growing food for their families, perhaps with a little left over to sell. Many of these subsistence farmers are women.

Getty Images/UniversalImagesGroup

Figure 4.1 Flood irrigating Australian rice fields: The production of rice is only possible in warm climates with reliable rainfall, or in areas where river irrigation can be accessed.

Shutterstock.com/Monkey Business Images

Figure 4.2 All members of the community must have equal access to appropriate nutrition to have food security.

Food security means:

- food availability; that is, sufficient food available for all people in a country
- food access; that is, households and individuals have adequate resources (such as money and transport) to obtain appropriate foods so that elderly people, children, isolated and disabled people do not miss out
- food utilisation; that is, receiving good nutrition from food – this requires knowledge about food processing, storage and preparation as well as access to safe drinking water
- building up assets; that is, assets can belong to a family or community and may include grain stores, tree nurseries or savings for difficult times.

HANDS-ON

SIMULATED FARMING ACROSS THE WORLD

PURPOSE

The aim of this activity is to experience the challenges of crop farming in a variety of climatic conditions and the effect on seed germination.

Note: you may need to wait 4–6 weeks to see results.

MATERIALS

Nine small plastic pots

Seed raising mix or potting mix for each pot

Three types of seeds, such as corn, beans and wheat

STEPS

1 Fill each pot with potting mix.
2 Plant three pots with one type of seed, repeating so you have three groups of three pots with each seed type. Label them!
3 Select one pot of each seed type and place on a sunny windowsill.
4 Place a second group of pots (one for each seed type) in the refrigerator.
5 Place the final group of seeds in their pots under a desk lamp.
6 Water each pot well.

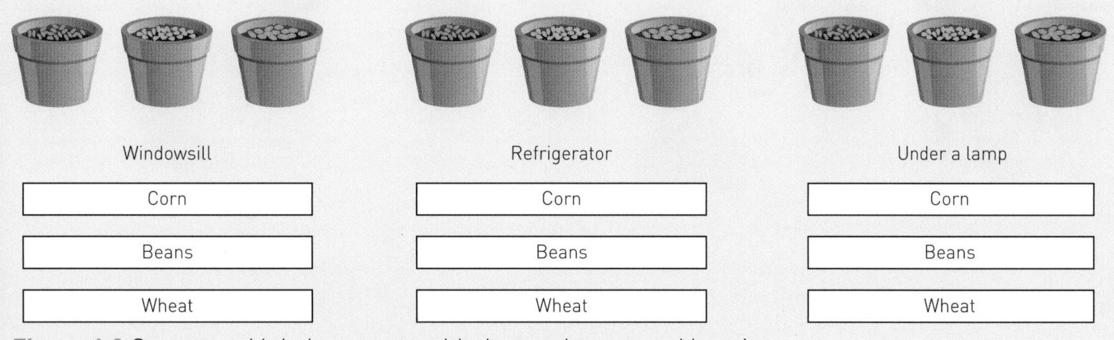

Figure 4.3 Set out and label your pots with the seed name and location.

ACTIVITIES

1 Make some hypotheses early in the experiment.
2 How long did it take to produce a result in any of the pots? How did your results match up with your hypotheses?
3 From this experiment, what can you conclude about climate, seed type and availability of water?

Meeting food needs

The wealth of a country is measured by its Gross Domestic Product (GDP). This is calculated by adding up the dollar value of all the goods and services produced by the country in a year. The gross national income of a country's citizens is calculated by dividing GDP by the size of the population.

Countries with a gross national income per person less than US$1035 are considered to be less developed. In 2014, Australia was ranked in the top 15, with a gross national income per person of US$64 680, while countries such as Cambodia, Nepal and Afghanistan record a gross national income per person of less than US$1000.

Obviously, the wealth of a country is not divided as evenly among the population. Every country has a gap between the richest and poorest.

Families' and individuals' access to financial resources determines their ability to obtain the food that they need. The ability to earn an income is dependent on:

- health
- skills
- education
- resources such as land to farm, money to buy seeds, tools and equipment for farming
- transport to get to the place of work or to go to markets to sell produce.

In the poorest countries, many of these factors rely on the availability of safe water.

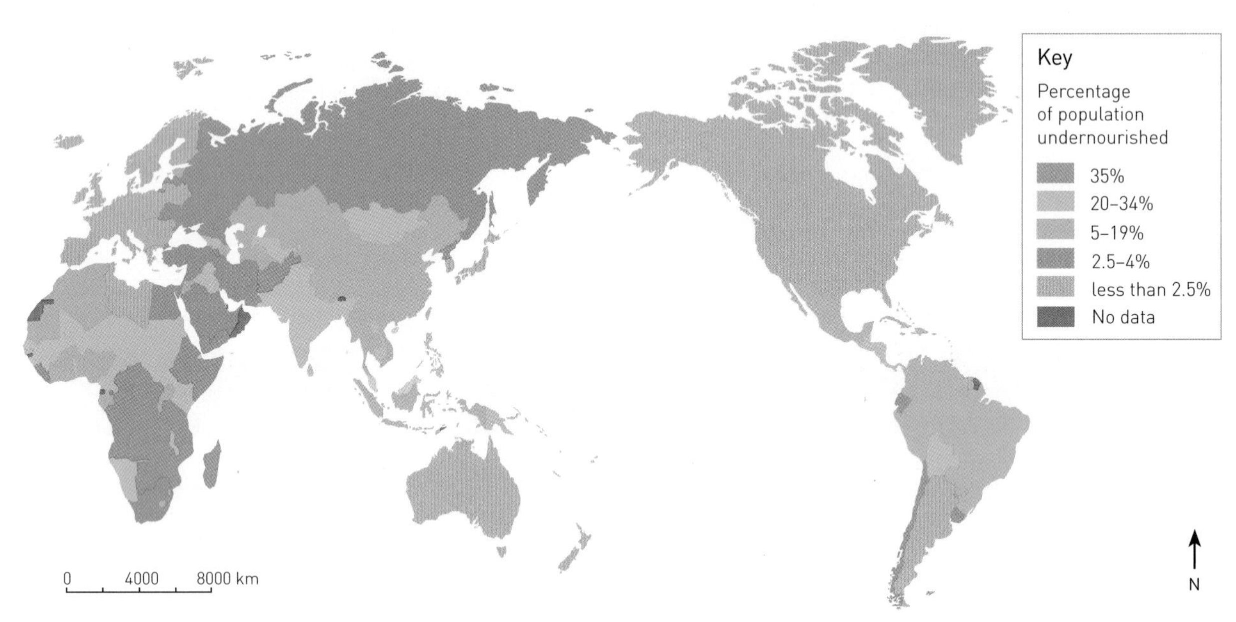

Figure 4.4 World hunger map

Selection of food

Education is highly regarded in developed countries such as Australia, where people are required to obtain a basic education before they can legally leave school to join the workforce.

Young people in other countries are not so lucky. Many of them have to work from an early age to help support their family, so their education is limited, and the result is a high rate of illiteracy among this group.

Lack of nutrition education can lead groups in the global community into making poor nutrition choices. An example of this is the marketing of infant formulas.

iStock.com/jfk_image

Figure 4.5 Knowledge of the vast benefits of breastfeeding may have decreased the influence of powerful advertisements for infant formulas in developing countries.

The makers of some such formulas devised a persuasive advertising campaign that led to mothers in developing countries rejecting the nutritional quality of breastmilk in favour of these powdered alternatives.

Families sacrificed hard-earned money to purchase formulas with labels they could not read or understand. Often the formula was mixed in incorrect amounts using unsterilised bottles and contaminated water. Sometimes, less powder was deliberately used to make a tin of expensive formula last for more feeds, thereby diluting its nutritional value. Subsequently, the numbers of infant deaths rose.

Nutrition education in Australia comes from many sources, such as the Australian Dietary Guidelines (see Chapter 2).

Your education in Food Technology gives you the ability to make good food choices, and your ability to cook gives you control over your nutrition – you know what you are eating because you prepared it.

UNIT REVIEW

LOOKING BACK

1. List three consequences of having insufficient clean water.
2. Define the terms 'water-borne disease', 'water-related disease' and 'water-based disease'.
3. Why would breastfeeding mothers in poor countries switch to formula milk? Why can this be a dangerous choice?
4. Describe the kinds of farming activities your family might engage in if they were subsistence farmers with little available income to buy food.
5. List five circumstances that lead to food inequity. Choose one and write a paragraph explaining it. Include an Australian example.

FOR YOU TO DO

6. Conduct a survey of your class to determine the average time (in minutes) spent in the shower by each person.
7. Assuming that the average shower uses 30 litres of water per minute, how many litres of water in total does your class use for showering daily?
8. Brainstorm ways of using this water more efficiently – other than having shorter showers!
9. Write a story from the point of view of a mother in a developing country trying to decide whether to purchase infant formula or continue breastfeeding.
10. Think about your local suburb or nearest town. List all the community assets that would be available to you and your family in a time of crisis; for example, emergency services and charitable organisations.

TAKING IT FURTHER

11. Investigate ways that household water may be used more than once. Consider:
    - the rinse water from the washing machine
    - the shower water that runs until the temperature is right
    - water from the bath
    - toddler's pool washing-up water.
12. Does your school canteen have a policy regarding foods sold in it? Ask your canteen director what food lines are the top sellers. How does this match with the Australian Dietary Guidelines?
13. The United Nations declared 2016 the International Year of Pulses. Access the UN website and determine the significance of pulses to sustainable farming and nutrition for less developed countries.

The United Nations

4.2 Groups that may experience food inequity

Food inequity is not the experience of poor countries alone. There are also people living in countries such as Australia who do not have equal access to food and nutrition.

Shutterstock.com/digitalreflections

Shutterstock.com/Nils Versemann

Shutterstock.com/Nils Versemann

Figure 4.6 The presence of an Aldi supermarket has increased competitive pricing among the other supermarkets in the area.

Rural and isolated

Australia is a big country and transportation of food is a major consideration in remote areas. Consider the findings of a CHOICE investigation into supermarket pricing of an identical basket of goods purchased around the country. It found that states in the east were enjoying better value than those in the Northern Territory or Western Australia. Within NSW, there was little difference (less than $2 over the cost of 30 supermarket items) between stores in large centres such as Newcastle, Sydney, Broken Hill and Dubbo. Cheaper prices were found in areas of more competition, close to or within a major city and in larger supermarkets as they could buy in bulk. Some foods were substantially more expensive in isolated areas (up to 75 per cent more expensive in very isolated towns) and the most remote areas had the least variety in grocery and fresh produce lines. In many cases, the quality and variety depends on the day consumers made their purchases, as some towns only had deliveries of fresh fruit and vegetables once a week.

People on low incomes

Low-income earners, particularly people on government benefits, may be at risk of poor food equity as their financial commitments dictate the priority of their spending. People on low incomes need to prioritise their spending. Bills such as rent are fixed amounts, and utilities such as electricity and gas are essentials, while the budget for food is flexible, so it is purchased with whatever money is left over.

Women and children

Women in many countries are the primary care-givers to children. In poor countries, women are victims of injustice and discrimination and have fewer opportunities for self-improvement. In many cases they are poorly educated and carry an enormous burden of work.

Consider the plight of women in less developed countries. In large rural communities, women are responsible for water collection. The nearest river or lake may be miles away, so the daily walk for water is exhausting. If famine and drought hits, water sources disappear and water collection takes up increasing amounts of time. Hence there is less time to cook, clean and care for sick children.

Women play an important role in the well-being of children. Female children particularly are disadvantaged by their gender in poor countries. Young children are reliant on their mothers for nutrition, particularly when being breastfed, and those women must be healthy and have access to education, training and financial resources to perform this role well.

People with disabilities

Disabilities can range in effect from mild to severe. People with disabilities can be of any age, living alone, with a family or other people, or in an institution. Disabled people have special needs according to their level of ability, mobility, literacy and general wellbeing. The situation of disabled people may be further compounded if they are aged, living in an isolated area, living in a poor country, or if they have little or limited financial resources. Disabled people may experience food inequity because of their inability to see, read or understand instructions. They may be limited in their mobility so cannot make food decisions independently, or live in institutions where others make food decisions for them.

iStock.com/Bartosz Hadyniak

Figure 4.7 Women in many developing countries are responsible for caring for children, collecting water and firewood, and working on the family farm.

HANDS-ON

MAKING A PACKET CAKE BY GUESSWORK!

PURPOSE

To gain some understanding of the difficulty faced by those who cannot read.

MATERIALS

You will be supplied with a packet cake without instructions. You must also choose any quantity of any of the following ingredients to make your cake.

- Eggs
- Butter
- Water
- Oil
- Milk

STEPS

1 Set your oven on what you think will be an appropriate temperature.
2 Select your additional ingredients. Record all your decisions including ingredients chosen, quantities and order of combining.
3 Mix the cake by hand or electric mixer. Record the time taken.
4 Select a cake tin. Record the size and shape. Pour in your mix.
5 Bake. Record the time taken until cooked.

>

HINT: The cake will be cooked when a skewer inserted into the centre comes out clean.

ACTIVITIES

1 Evaluate your result in terms of appearance and taste.
2 Describe how it felt to guess about the unknown factors such as ingredients, times and temperatures.
3 Explain in a paragraph how being unable to read would limit the opportunities for an individual to achieve food equity.

Aged people

Aged people are also at risk of food inequity. They may be less able to shop or cook for themselves, or may be dependent on others making these decisions for them, such as residents of nursing homes. Many such institutions, as well as those that cater for young people such as day-care centres, now have meal and diet policies that cater more suitably for good nutrition.

Elderly people are further at risk of poor food equity if they:

- suffer from Parkinson's disease
- suffer from Alzheimer's disease
- suffer from the aftereffects of a stroke
- are housebound after fracturing a hip or have some other disability
- have difficulty chewing and swallowing
- are socially isolated or on small, limited incomes.

Aboriginal Australians

Aboriginal Australians and Torres Strait Islanders have the poorest health statistics of any group in Australia. This can be linked to the destruction of their traditional lifestyle. They experience social isolation and financial disadvantage in the community and this leads to poor food equity.

Alamy Stock Photo/David Grossman

Figure 4.8 The elderly can have improved nutrition when they do not have to prepare meals for themselves.

Alamy Stock Photo/Davo Blair

Figure 4.9 Adoption of a Western diet and the loss of traditional foods and methods of cooking have contributed to a deterioration in Aboriginal Australian health.

FOOD IN FOCUS

THE HEALTHY WELFARE CARD

In 2014, Australian businessman Andrew Forrest was charged with the responsibility of reviewing Indigenous employment and training. One of his key recommendations was the introduction of a cashless Healthy Welfare Card to assist in income management.

His report found that almost half of all working-age Aboriginal Australians rely on welfare payments as their main source of income, compared to only 17 per cent of all Australians.

The Healthy Welfare Card introduces a new system to support welfare recipients to:

- manage their income and liabilities
- save for the occasional bigger expenses like Christmas or school camps
- encourage welfare income to be invested in a healthy life.

All individuals and families benefit from a stable financial environment where regular bills and rent are paid on time and there is food on the table, allowing them to concentrate on other concerns including returning to work and raising and educating their children.

The Federal Government needs to make the necessary changes to Australia's welfare system to empower individuals to use it as it was intended. Welfare is provided to help people build healthy lifestyles and make the best choices they can for themselves and their families – particularly their children. Forrest wrote, 'It is a social safety net of last resort and should never be a destination, or support poor choices'.

In summary, the Healthy Welfare Card would:

- allow individuals to use the banking system to manage their welfare payments
- enable the purchase of all goods and services, with the exception of alcohol, gambling products, illicit services and instruments that can be converted to cash (such as gift cards) and exclude activities discouraged by government, or illegal in some places, such as pornography
- be linked to a locked savings account that can accrue savings for major purchases, such as a deposit on a home, whitegoods, furniture or rental bond, or unexpected large costs.

The Australian government is currently looking to trial this system through selected communities throughout the country.

Source: Adapted from 'Creating Parity: The Forrest Review' by Andrew Forrest, The Australian Government, Indigenous Jobs and Training Review. Creative Commons Attribution 3.0 Australia Licence

ACTIVITIES

1. Do you think this system of income management would work? Why do you think it was recommended?
2. What might be the impact if this Healthy Welfare Card was issued to non-Indigenous welfare recipients such as veterans or elderly pensioners?
3. What may be the negatives of such a card?
4. How do people learn to manage money? What information is available to you on budgeting and spending wisely?

The traditional hunter-gatherer lifestyle and variety in plant and animal foods, lean meat, fish, low salt, little added sugar and alcohol, combined with regular physical activity, produced a good life expectancy and healthy diet. Today, common diet-related disorders in the Aboriginal Australian community include obesity, diabetes, hypertension, coronary heart disease and dental caries. These disorders are all directly related to the adoption of a more expensive 'Western' diet.

Chronically ill people

Approximately 10–20 per cent of children have a chronic or recurring illness such as asthma, chronic fatigue syndrome, diabetes or eczema.

Some children are born with or develop severe food allergies. Ingredients in certain foods can trigger these allergic reactions. The most common are peanuts and other nuts, seafood, dairy foods and gluten (contained in flour). Severe reactions cause life-threatening anaphylaxis, and many schools in particular have become 'peanut free' for the safety of these students. Chronically ill people may experience food inequity as a result of restricted choices available or a lack of energy to prepare foods.

People with dementia

Alzheimer's disease may cause dementia. Dementia is common in Parkinson's disease sufferers and it can be alcohol-related. Dementia is also related to other conditions.

In simple terms, dementia causes an individual to progressively lose the ability to do familiar tasks. Dementia sufferers will become unable to perform simple tasks such as using the kettle or toaster, or create even the simplest meals such as an omelette or toasted sandwich. Many dementia sufferers rely on Meals-on-Wheels or residential respite care for their meals.

Alcohol and drug abusers

Alcohol abuse is a serious health problem in Australia. Alcohol is a neurotoxin; this means it poisons the brain. Alcohol interferes with the absorption of Vitamin B and contributes to the poor food equity that individuals who are alcoholics frequently experience. A person who abuses alcohol to excess places a priority on having a drink over eating a meal. Poor nutrition also contributes to elevated blood alcohol levels.

Abuse of illicit drugs causes a range of diet-related reactions. Amphetamines such as ice contribute to a loss of appetite that can eventually result in **malnutrition**. Cannabis is a depressant, but causes an increase in appetite and a feeling of nausea. Ecstasy was originally developed to suppress appetite. Serious use of ecstasy can cause malnutrition or dehydration, leaving the user with excessive thirst.

Homeless people

Homeless people have no fixed address. They may be of any age or either sex, living in short-term crisis accommodation or on the street. Some obvious causes of food inequity for this group would be the lack of income and irregular or incomplete meals. Many crisis accommodation venues, such as hostels, do not have cooking facilities for residents, and frequently homeless people are evicted from these places as they suffer from alcoholism and cannot abide by the drinking bans.

Some homeless people are displaced people who find themselves involuntarily outside the boundaries of their country. In many instances this occurs because of war forcing people from their homes and over country borders. Many **refugees** arrived in Australia from Vietnam in the 1970s by boat, fleeing their war-torn homelands. Since 2011, an estimated 9 million Syrians have fled their homes since the outbreak of civil war, taking refuge in neighbouring countries or within Syria itself. According to the United Nations High Commissioner for Refugees (UNHCR), over 3 million have fled to Syria's immediate neighbours Turkey, Lebanon, Jordan and Iraq; 6.5 million are internally displaced within Syria.

Displaced people are at risk of poor food equity because of their difficult circumstances. They may be in another country with no money, very few possessions and possibly living in a crowded camp.

Figure 4.10 Overconsumption of alcohol can have serious implications for the health and nutrition of individuals.

FOOD IN FOCUS

FOOD AID IN SYRIA

SYRIAN GOVERNMENT TO ALLOW AID INTO TOWN WHERE THOUSANDS ARE STARVING, UN SAYS

Around 40 000 people, mostly civilians, live in Madaya in Damascus province, many of them displaced from the neighbouring rebel stronghold of Zabadani.

At least 10 people have died there from a lack of food and medicine, according to the Syrian Observatory for Human Rights, a British-based activist monitoring group.

Another 13 people who tried to escape in search of food were killed when they stepped on landmines laid by regime forces or were shot by snipers, according to the Observatory.

The aid deliveries were not expected to start for several days due to administrative delays, aid workers in the country said.

Outrage has been growing as images of what appear to be Madaya residents looking extremely frail after months of little food have spread on social media.

The UN said that over the past year, only 10 per cent of its requested aid deliveries to hard-to-reach and besieged areas of Syria were approved and carried out.

Source: Adapted from 'Madaya: Syrian Government to allow aid into town where thousands are starving, UN says', Reproduced by permission of the Australian Broadcasting Corporation – Library Sales A154 © 2016 ABC/Wires

ACTIVITIES

1 The United Nations has alleged that starvation in Syria is being used as a weapon of war. From your reading of the information, what does this mean?
2 Research the work of humanitarian groups and their activities 'on the ground' to alleviate malnutrition in areas of current crisis.
3 What dangers do aid workers experience when working in such environments?

UNIT REVIEW

LOOKING BACK

1 Why can food transportation in Australia be difficult?
2 Why does low income contribute to food inequity?
3 Create a mind map that illustrates the reasons for women and children, people with disabilities and aged people being at risk of food inequity.
4 Explain how the combination of factors such as being aged and living in an isolated area may compound an individual's disadvantage.
5 List three reasons why Aboriginal Australians people may experience food inequity.
6 Explain how illness, drug or alcohol abuse can cause food inequity.

FOR YOU TO DO

7 List the reasons why groceries and fresh foods may be more expensive in rural or isolated areas.
8 Design an instruction leaflet to be inserted into a packet muffin box to assist those who have difficulty reading or cannot read English.
9 In groups, design a fact sheet for each group that may experience food inequity. Use your own research to develop the information here further.

TAKING IT FURTHER

10 Visit or contact by email a local nursing home, boarding school, hospital, meals-on-wheels coordinator or day care centre and ask them to show you their meal and nutrition policy. Everyone in the class should contact a different establishment. Discuss your results in a one-page report. What conclusions can you draw about the access that people in these institutions have to appropriate food?

4.3 Influences on food availability and distribution

4.3a Influences on food availability and distribution

Geography and climate

Geographical conditions and climate dictate what crops a country can grow successfully. Geography also determines how easily or how quickly these crops can be transported to factories for processing, to markets for sale or to ports for **exporting**. The kinds of foods grown and therefore available in wealthier countries varies widely from those available in poorer countries.

Globally, **staple foods** are basic foods that a population eats regularly and in large quantities. They tend to be cheap and easily obtained, making them the basis of the diet. Many countries grow staple foods that are best suited to their climatic conditions.

Wheat

Wheat is grown throughout the world. It is the most popular cereal grain produced. Different types of wheat are grown depending on the climate, the seed used, and the end-product desired. For example, soft wheat is best for biscuits and cakes as it makes fine flour, while hard wheat is more suitable for bread and pasta. In developed countries, wheat-growing is highly technological, with the grain being harvested by machine. In developing countries, wheat is harvested by hand using tools called sickles, and processed further by hand. Wheat is not used as a grain food in seed form, but is ground into flour. It can be used to make porridge or couscous (as in North Africa); leavened or risen products such as bread; unleavened products such as tortillas or pitta bread; pasta or noodles.

Rice

Rice is primarily grown in Asia, although other countries such as Australia have an expanding rice-growing industry. Growing rice requires lots of water – it may be grown in swamps known as rice paddies, or on dry land that is flooded with water during the growing stage. Harvesting trends are similar to that of wheat – highly technological in developed areas and highly labour intensive in less developed areas. Rice may be eaten as a grain, ground into flour or made into rice noodles.

Maize

Maize is grown either as sweet corn, which we know as corn on the cob, or as field corn that is harvested to be ground into flour. This second type of maize may be processed or milled by crushing the grain using a mortar and pestle. The by-products of this milling process may be fed to animals. Maize is used to make polenta, corn tortillas and cornflakes.

Millet

Millet is a hardy crop that will tolerate extreme weather conditions such as poor rainfall and soil of an inferior quality. The grain is ground and used to make a porridge (called gruel) or breads. Millet may also be used to make straw-like brooms or brushes.

Shutterstock.com/Dreamsquare

Figure 4.11 Grains such as these can be processed to make bread-type products and the seeds can be saved to use for another crop.

Cassava

Cassava is a staple food that is not a grain but a tuber, harvested from under the soil much like a potato. It is grown in the poorest parts of Africa as it is reliable in all sorts of climatic conditions.

Sago

Sago or tapioca is a food product made from the sago palm. The spongy centre of the plant stem (known as the pith) is removed and shaped into small pellets. This plant is grown in tropical areas.

Farming in Australia

In Australia, 20 million hectares of land is devoted to produce seed and grain crops. Wheat is by far the crop that covers the largest area, with approximately 12.2 million hectares of Australian land dedicated to its production. The largest proportion of this area is in Western Australia, followed by NSW.

Canola, a plant with a yellow flower, is increasingly popular. Its seed is crushed and converted to oil and used in food manufacturing in Australia and overseas. At present, millions of hectares of land is used to grow canola, producing approximately 3.5 million tonnes of canola oil seeds annually.

Other significant crops include:

- sugar cane, grown along the coast between Far North Queensland and Grafton in NSW. There are 4000 sugar cane farms of an average size of 100 hectares
- rice, grown in Leeton and Griffith, NSW, in the Murrumbidgee irrigation area because of the need for access to a good water supply
- fruit, grown throughout Australia with different geographical areas specialising in different types.

For example, in NSW, grapes are grown primarily in the Hunter Valley, apples in cool climates such as Batlow and Orange, and cherries grown in Young in the south-west of the state.

Most farmers in Australia are engaged in beef cattle farming. A wide variety of cattle is raised in Australia, and each breed is chosen for its suitability to the geographic location and climatic conditions of the farm. For example, Brahman cattle are commonly raised in the northern part of Australia because of their ability to endure extreme temperatures that species such as the Hereford do not handle as well. Twenty-four per cent of all farms raise beef cattle. Currently in Australia there are approximately 26.6 million head of cattle and calves raised for beef and veal. Feedlot production is an increasingly popular method of fattening cattle for periods of three to twelve months by feeding them high-protein food. This method is commonly used to produce beef for the Japanese export market, as it results in meat that is marbled with fat distributed evenly throughout the muscle tissue.

iStock.com/BergmannD

Figure 4.12 Sorghum is a grain crop related to sugar cane and millet. It is the most valuable cereal crop grown in Queensland. It grows well in poor soil with unpredictable rainfall.

Shutterstock.com/David Steele

Figure 4.13 Australia has a long history of farming sheep for wool and meat. Merino sheep are suitable for both wool and meat production.

Dairying is a smaller concern, comprising 1.6 million head of cattle raised for milk production. Jersey cows are particularly well suited for this purpose. Victoria has the highest proportion of dairy farms in Australia.

Sheep are raised in a wide range of climates. There are currently approximately 75 million sheep and lambs grazing in Australia. Sheep are bred for meat, wool or both. Lambs are raised for meat in geographic areas that enjoy high rainfall and good pasture for grazing.

There are about 5 million pigs in farm production in Australia. Pig farming is an intensive, indoor production, which tends to be expensive due to the high cost of grain feed. For this reason, pig farms are often located close to major grain-growing areas.

Religious and cultural beliefs

Animal products are not generally regarded as a staple food for people of **developing nations**. Ownership of animals is usually associated with wealth or prosperity, and animals are the belongings likely to be sold first when poverty occurs. Another reason why animal foods may not be staples in some countries is because of widely held religious beliefs. For example, the eating of beef is forbidden for Hindus, while the eating of pork or blood is considered unacceptable to Muslims and Orthodox Jews.

Socio-economic status

Socio-economic disadvantage may result from poor education levels. This may be seen in poor levels of literacy or general lack of nutrition knowledge, both leading to poor food choices. Lower education levels may mean lower-paying employment with the result that food choices become limited because of a shortage of available funds.

Studies have shown that there is a strong relationship between socio-economic disadvantage and poor health. Poor nutrition is directly related to low income.

HANDS-ON

'RICH MAN, POOR MAN' FEAST

PURPOSE

This activity illustrates the difference between the 'haves' and the 'have nots' and what it feels like to be a member of one of these groups.

MATERIALS

Coloured paper in two colours

Cooked rice to feed three-quarters of the class

A favourite food, such as chocolate or pizza, enough for one quarter of the class

STEPS

1 At the beginning of the lesson, collect a coloured piece of paper from your teacher (75 per cent of the pieces of paper will be one colour, 25 per cent will be another colour).
2 One colour group (the largest group) sits on the floor with a bowl of rice to share and a cup of water for each group member.
3 The other colour group sits at a table set with a tablecloth, crockery, cutlery, glassware and their favourite food to eat and drink.

ACTIVITY

During and after the meal, discuss what you thought and felt about the experience of being in your colour group. How fair was it?

Globalisation and government policy

Globalisation is an international movement towards a global economy where national borders cease to matter. It includes the globalisation of trade, where a world market is created through freer trade practices, and the globalisation of production as large multinational corporations set up factories in many different countries.

In practical terms, globalisation means that there is a free movement of goods between countries. The advantage of globalisation is that it encourages competition, allowing consumers everywhere the opportunity to choose from a variety of products at competitive prices. Globalisation also has its disadvantages. Countries are limited in the actions they can take to protect their own industries, such as restricting imports or placing **tariffs** on them to force the imports to sell at a higher price.

Globalisation has broadened consumer tastes. For example, in developed countries, demand has shifted from mass-produced coffee to specialty beans and from processed to fresh vegetables.

4.3b Influences on food availability and distribution

Figure 4.14 Fairtrade coffee actively aims to support farmers who produce products that might be swamped in a competitive global market.

FOOD IN FOCUS

LAST AUSTRALIAN FOOD CANNERY TURNS OFF THE LIGHT

Australia's only wholly Australian-owned cannery, Windsor Farm Foods Group Ltd, has shut its doors. The closure in March 2013 resulted in 70 workers at the Cowra site being stood down with no pay.

'Losing Windsor Farms has been another nail in the coffin in the industry,' said William Churchill from AusVeg. The closure is another example of a 'death by a thousand cuts', with the tough economic climate, retail space competition, and the supermarket duopoly putting more and more pressure on both growers and food manufacturers, he added.

The cannery itself has been in use since 1943, and for much of its history was used to process vegetables, such as peas, beans, tomatoes, carrots and beetroot, grown nearby. In 1995, the factory began canning mushrooms, red kidney beans and canned bean mixes for major supermarkets. Foods were packed at the site for the 'Edgell', 'Cowra Gold' and 'Lachlan Gold' brands, as well as for supermarket labels 'Homebrand', 'No Frills', 'Savings' and I.G.A.

In 2003, facilities for processing baker's jams and fillings, and fruit straps were constructed at the Cowra site.

Australia has had a spate of food manufacturing closures in the last few years: Rosella (most well-known for its tomato sauce) ceased operations, Heinz Australia also closed its tomato sauce factory in Gigarre, Victoria, and SPC Ardmona in

>

Victoria's Goulburn Valley is currently being reorganised after a restructuring that saw many jobs being lost.

In mid 2011, Heinz's move of its Golden Circle beetroot and fruit processing facility in Queensland's Lockyer Valley was shut down as its business was partly shifted from Australia to New Zealand. This closure resulted in the immediate loss of 340 Australian jobs.

Source: Adapted from 'Last Australian food cannery turns off the light', 14 March 2013 Republished with permission of 'Australian Food News Media'

ACTIVITIES

1 What is meant by the term 'supermarket duopoly'? Why has this affected the cannery?
2 Why do you think many of the cannery operations are being moved offshore to locations such as New Zealand?
3 Research the population of Cowra. What effects might the loss of 70 jobs have on other industries in the town?

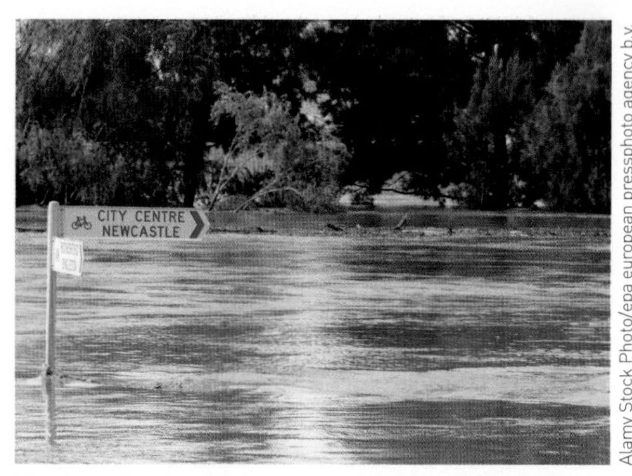

Alamy Stock Photo/epa european pressphoto agency b.v.

Figure 4.15 Flooding creates not only isolation, but also the risk of water contamination.

AAP Image/NEWZULU/JEROME CID

Figure 4.16 Refugee camps are often crowded and food availability is critical.

Natural disasters

Natural disasters such as flood or drought are impossible to plan for, but developed nations generally have resources available to cope in such a contingency. Developing countries try to plan for the unexpected by developing grain stores and saving money to purchase foods or crops destroyed by disaster.

War

War also adds to the refugee crisis – people leave their homes to escape military conflict and flee to safety.

Just as war can cause hunger, hunger can cause war. Economic and social inequality that denies people the money, land or skills they need to feed their families can lead them to take up arms.

In times of war, aid agencies endeavour to assess the needs of those affected, particularly civilians caught or forced from their homes. People who are displaced from their homes are removed from their normal activities, such as working to earn an income. Refugee camps are often set up in safe areas but getting food aid to these areas can be tricky. In some areas, such as Angola, food relief for persons displaced by war is made difficult by the remoteness of many refugee camps and the state of disrepair of the nearest airport to dispatch the food aid. Security for the aid workers is paramount, and the food must reach those for whom it was intended. Many children become malnourished as a result of war – these children are the priority for food aid.

Educational levels

Education is crucial to providing knowledge about the availability of food. In NSW there are specific pockets of metropolitan Sydney with special needs in this area.

FOOD IN FOCUS

EAT IT TO BEAT IT

Evidence suggests a link between cancer and lifestyle factors, such as poor diet and being overweight. Cancer Council NSW's *Eat It To Beat It* program, launched in the Hunter region, is aimed at developing awareness of the importance of increasing fruit and vegetable intake to help prevent cancer.

Eat It To Beat It is anticipated to become Cancer Council NSW's flagship nutrition program to reduce cancer risk. If it is successful, some of its strategies will be offered to other Cancer Council regional offices in NSW.

The program targets parents of primary school-aged children because a healthy family diet during these crucial years can have a lasting and significant influence on a child's future. In NSW, many families with primary school-aged children fall short of eating the recommended two serves of fruit and five serves of vegetables every day.

Eat It To Beat It has a very clear focus on increasing consumption of fruit and vegetables, working alongside other programs such as the *Good for Kids Good for Life* program (funded by NSW Health). This program has a wider scope with the aim to tackle childhood obesity.

Both *Eat It To Beat It* and *Good for Kids Good for Life* are underpinned by the national *Go for 2 & 5* campaign.

Eat It To Beat It strategies include free sessions and workshops for parents of primary school-aged children, helping them to understand why fruit and veg are so important.

Fruit & Veg Sense workshop runs for 90 minutes and helps parents with advice on fussy eaters, meal planning and shopping tips.

Healthy Lunch Box session targets parents who have children starting kindergarten. The importance of packing fruit and veg in the lunch box every day is emphasised, and the sessions provide easy ideas to make this possible.

ACTIVITIES

1. Why does *Eat It To Beat It* target parents of primary schoolers?
2. What is the recommended daily intake of fruits and vegetables?
3. What other children's health and nutrition campaigns link in with *Eat It To Beat It*?
4. Research *Eat It To Beat It* using the Cancer Council NSW website. What other resources are available to parents on this site?

Eat It To Beat It

Multinationals

Multinational corporations are large corporations with operations in many countries. They often control every stage of the production process from farmer to consumer. For example, Kellogg's has farmers in Australia under contract to produce corn for specific use in the production of cornflakes to be made in local factories and sold in local supermarkets. Of course, multinational companies are in business to make a profit. They frequently set up supplementary branches in poor countries as these countries supply abundant cheap labour, land and plentiful resources. In theory, this can benefit poor countries by employing local people. However, despite injecting money into the country, the investment is made for the benefit of the large corporation. Multinational corporations specialise in controlling the production of lucrative raw materials such as coffee, sugar, bananas, forest products, tea and tobacco. They have the financial resources to increase the supply of their own food products worldwide but, in the case of many fast food chains, this may not be the best choice for good health.

Technological developments

Physical barriers to food distribution can have an impact on food sources. In Mozambique, in Africa, poor roads between the north and the south of the country make it cheaper to import food from Europe than to provide a market for farmers within the country. Even when the roads and weather are at their best, the transportation costs of moving grains from farms in the north to the capital Maputo contributes one-third to one-half of the selling price.

Shutterstock.com/Egyptian Studio

Figure 4.17 Large supermarket chains are able to access goods for lower prices than the local greengrocer or butcher, causing many of these small businesses to struggle.

This means that what should be a highly competitive, locally grown food becomes uncompetitive.

Similarly, in Tennant Creek in the Northern Territory of Australia, residents are forced to pay high prices for food, particularly fresh fruit and vegetables, due to the fact that these products are transported by refrigerated truck from Katherine or Darwin. This contributes to the high rate of malnutrition and disease because of the difficulty of affording a balanced diet.

State-of-the-art transport and refrigeration will improve food availability and ease distribution considerably, but inevitably will continue to be reflected in the cost of goods.

UNIT REVIEW

LOOKING BACK

1 List some common end-products of wheat, rice and maize.
2 Define the term 'staple food'.
3 Why are pig farms often located near cereal grain farming areas?
4 Provide two reasons why livestock are not generally owned by people in developing nations.
5 Draw a mind map brainstorming the reasons why people who are disadvantaged may suffer from poor nutrition.
6 Define the terms 'globalisation', 'multinationals' and 'tariffs'.
7 Why do multinationals become established in many countries?
8 Why is the cost of transporting food significant?

FOR YOU TO DO

9 Obtain an illustration of a cereal. Use it to explain why refining rice to form white rice, or refining wheat to form white flour, reduces its nutritional value.
10 In groups, design a staple foods buffet menu.
11 a On a map of NSW, locate the Hunter Valley, Batlow, Orange, Young and Leeton. Find out why these areas are climatically suited to these crops.
b Research the grain-growing areas. Shade these areas on your map.
12 In groups, use a calendar and highlight the times of the year when foods are not permitted to be eaten by certain religious groups.
13 Research the factors that contribute to socio-economic status. Use the Australian Bureau of Statistics weblink.

Perspectives on Education and Training: Social Inclusion, ABS

14 Based on the information given, choose a partner and have a silent debate. On one piece of paper, take turns arguing 'That globalisation is the only way to a successful future'. Each partner must choose to be either affirmative or negative, write one point on the paper and pass it to their partner to be refuted or a new point added.
15 Design an innovative bumper sticker to encourage consumers to buy local produce.

TAKING IT FURTHER

16 Obtain some maize grains and use a mortar and pestle to grind these into flour. Add this flour to your tortilla recipe.
17 Use the newspaper *The Land* or go online to the website of ABC television program Landline to determine the current issues affecting farmers. Write a report on your findings.

Landline, ABC

The Land

18 Investigate the food laws of Islam and Judaism. What do their laws say about eating meat? Why?
19 Go to the website for the NSW Department of Primary Industries. Research the government support that is available for farmers and graziers in NSW.

NSW Department of Primary Industries

20 Using a map of the Northern Territory, calculate the distance between Tennant Creek and Darwin or Katherine. If a truck is travelling at 100 kilometres per hour, how long would it take to drive this distance?

4.4 Food production practices

Food production methods throughout the world fall under one of the following categories, although the reasons for a country practising one type of agriculture over another may vary.

Subsistence farming

Traditional agriculture is practiced by subsistence farmers, usually in developing nations. This method of farming depends on rainfall and uses very few resources. Crops are grown that provide multiple end uses such as food, fuel, building materials and feed for animals. There are many advantages of this simple type of farming. No chemicals are used, the environment is not polluted, little energy is expended to produce food, and a variety of crops are grown as many traditional farming communities are self-sufficient.

Green revolution agriculture

Green revolution agriculture is a more modern type of farming practised in areas that are flat, tropical and abundant in natural resources.

The method uses farm machinery, high-yielding seeds, irrigation systems, fertilisers and pest control. The green revolution disregards traditional crops and farming techniques of the local area – everything is imported from developed countries.

Industrial agriculture

Industrial agriculture is practiced in developed countries such as Australia. This type of farming uses high levels of technology, water, energy, chemicals and fertilisers. Many farms specialise in the production of vast quantities of only one or two crops. Multinational corporations, which control every aspect of food production from farm to consumer, control much of the industrial agriculture for their products. There are also many successful farming businesses that are privately owned in Australia and overseas.

Industrial agriculture is a commercial process – the food product undergoes many changes before reaching the consumer. Industrial agriculture alters the environment because the chemicals used have a range of effects; for example, they may reduce the number of pests but also alter the soil quality.

Cash cropping

Cash crops are those agricultural products that are grown for a luxury market. They are not crops that allow a family to be self-sustaining, but if they can be sold, cash crops command a high price. They generally also require smaller amounts of land to yield the same profit as traditional crops. Common cash crops include coffee, sugar cane, rubber, cotton, nuts, tea, tobacco and oil seeds. In areas such as the

Figure 4.18 Tea is a lucrative cash crop that provides little practical benefits to a struggling farmer.

Amazon River in South America, the lure of cash crop profit must be balanced against the need to provide food for the family, and the problems caused by clearing native rainforest land to plant more crops. Many cash crops are produced for **export** to other countries.

Cash cropping undoubtedly creates opportunities for wealth, but in poorer countries it can pose a risk to farmers. Large farms are required for large profit, and access to world markets for export is necessary. Freer trade between countries has meant that poor countries specialising in the production of one key product must compete with increasing quantities of competitively priced imported foods as well. This can have a devastating effect on a small farmer's income if the farmer has only one type of food product to sell.

UNIT REVIEW

LOOKING BACK

1 List the three main categories of food production.
2 What is the main feature of a sustainable system of agriculture?
3 Describe the main differences between traditional and green revolution agriculture.
4 Are cash crops harmful to the environment? Explain.

FOR YOU TO DO

5 a Design symbols for your workbook to represent each type of farming.
 b Expand these symbols to design single frame cartoons that explain the inputs and outputs of each type of farming.
6 Draw up a table and list the advantages and disadvantages of cash cropping.

TAKING IT FURTHER

7 What is organic farming? From your research, explain the reasons organically grown foods tend to be more expensive.
8 What is crop rotation? Why is it a technique that supports self-sufficiency?
9 Use the internet to research one of the cash crops mentioned. Where is it grown?

4.5 Malnutrition

Malnutrition in children is hunger at its most devastating. It can affect their mental and physical development for the rest of their lives, which in turn deepens the hunger cycle as they not only suffer ill health but also have less access to education and opportunities for work later in life. Hunger and malnutrition seriously affect adults too, impeding their productivity and creating a host of associated problems.

Some forms of malnutrition such as a deficiency in vitamins and minerals or protein occur when individuals do not receive correct amounts of these. It is the world's most common nutritional problem. The symptoms of these deficiencies, known as protein-energy malnutrition, are not immediately visible but have debilitating effects on people and societies. In poor countries, this disease causes growth retardation, chronic starvation and protein deficiency. The result is lost productivity and a great increase in health costs that affect a government's attempts to improve the quality of life.

In Australia, a high proportion of the Aboriginal Australian population suffers from diseases that were previously unknown to them, such as diabetes. This is largely due to a shift in diet from fresh meat, vegetables, fruits, legumes and nuts to the highly processed foods of the 'modern' society, which are often very nutrient-deficient.

Overconsumption is also a form of malnutrition. Consuming more kilojoules than is required is often accompanied by a deficiency in vitamins and minerals. Compounded by greater meat consumption and reduced physical activity, people become overweight and obese. In our consumer-driven society, food companies respond to demand for prepared foods by delivering attractive, expensive but often nutritionally empty products. The health care costs, missed productivity (sick days off work due to diet-related illness) and environmental costs associated with over-eating are huge.

The World Health Organization estimates that 795 million people in the world do not have enough food to lead a healthy, active life.

UNIT REVIEW

LOOKING BACK

1 Explain the difference between malnutrition and overconsumption.
2 Describe the costs to the community of poor nutrition.
3 What reason is given for the decrease in health status of Aboriginal Australians?

FOR YOU TO DO

4 Draw a cyclic diagram to explain how hunger in childhood can lead to disadvantage later in life.
5 Research the disease diabetes. What are the risk factors for contracting diabetes later in life?
6 Design a menu for a workplace canteen that promotes healthy lunches and snacks.

TAKING IT FURTHER

7 Design an information poster discussing the nutritional risks of under-consumption and overconsumption of food.
8 Research Rotary International and the work they do in health. Write a one-page report on your research.

Rotary International

4.6 Provision of aid

Aid agencies and emergency relief

Aid agencies support developing countries by providing help in the form of long-term developmental help such as education, agriculture assistance, road building and securing water supplies. Some agencies also specialise in emergency and disaster relief providing food, money, shelter and medical care in an emergency.

The World Bank is made up of 188 member countries, which appoint a Board of Governors to make decisions on their behalf. The World Bank has the goal of ending extreme poverty in a generation, and to boost shared prosperity. The Bank provides low or zero interest loans to less developed countries for projects such as roads, transport networks, communication networks, agriculture and industrial development, health or education programs.

The Australian Government allocates a foreign aid budget every year, whereby money is directed to less developed countries. Government-sponsored aid generally benefits countries that are important to the country offering the aid. The Australian government also supports some non-government aid agencies.

Foreign aid agencies that are not government controlled, such as the United Nations International Children's Emergency Fund (UNICEF), can be self-directed in the type of aid they provide, as they are not restricted by political factors. Generally, non-government foreign aid agencies strive to help poor communities to help themselves by using long-term solutions rather than short-term 'band-aids'.

Getty Images/Helen H. Richardson

Figure 4.19 The World Bank supports projects that provide clean, safe water and waste-water treatment.

Developmental aid

Developmental aid can be described as 'back to basics' aid. It encourages developing countries to place greater emphasis on long-term solutions by:

- giving priority to primary and secondary schooling, therefore increasing workforce skills
- promoting breastfeeding as the best nutritional start to life
- investing money in health care
- encouraging and supporting individuals to start their own businesses and develop agricultural skills.

Some specific examples of developmental aid include:

- research into and elimination of diseases, including smallpox, polio (eliminated in many countries), diphtheria and measles
- development of new crop varieties and increasing skills to support green revolution agriculture
- finance for projects to extend and improve energy, communication and transport infrastructure, bringing more nations into the 'modern' economy.

Getty Images/Anthony Asael

Figure 4.20 Education of children and allowing them to attain higher levels of schooling is a key element of a country's prosperity.

UNIT REVIEW

LOOKING BACK

1 What is the role of aid agencies?
2 Why are non-government aid agencies freer to operate in less developed countries?
3 Why is developmental aid described as being 'back to basics'?
4 Explain how funds can be used to:
  - improve local health care
  - help individuals in business in a developing country.

FOR YOU TO DO

5 List the advantages to a recipient country that receives a World Bank loan.
6 Devise a fundraising venture to support the aid agency of your choice.
7 Design a poster that promotes breastfeeding for display in a local area health centre.

TAKING IT FURTHER

8 Who is the Federal Minister for Foreign Affairs and Trade? Using the weblink, find out the budget allocation for foreign aid during the past financial year. What percentage of GDP does it amount to?

Australia's aid program, DFAT

9 Ask an overseas aid agency representative to talk to the class about their work.

4.7 Additional content: support networks for groups

Government

In Australia, Centrelink, on behalf of the Commonwealth Department of Human Services, provides a range of benefits for individuals and families who are struggling financially. Support is available on the basis of marital status, employment status, whether the person is a carer, in crisis, or in special circumstances such as a farmer in drought-declared areas. The assistance can come in the form of a one-off payment for those who meet the criteria, or ongoing payments until a person's circumstances change.

Voluntary

There are many voluntary agencies in Australia that provide both monetary and practical aid to individuals and families in crisis. These include church-based and non-religious organisations such as the Smith Family, Salvation Army, St Vincent de Paul, the Red Cross and the Samaritans. Assistance may can be in the form of a food hamper, a soup kitchen for the homeless or vouchers to spend at the local supermarket.

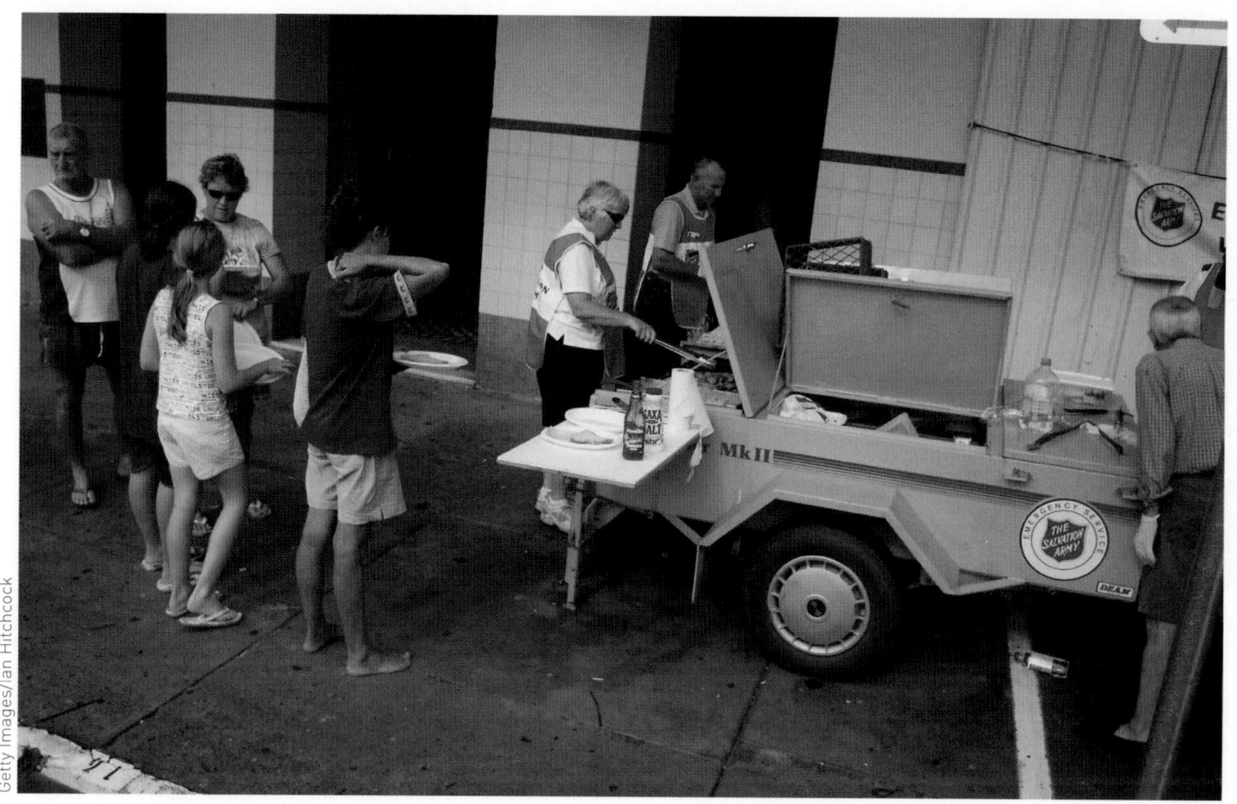

Getty Images/Ian Hitchcock

Figure 4.21 Salvation Army Emergency Services provide help and encouragement to victims of disasters such as bushfire and flood.

UNIT REVIEW

LOOKING BACK

1 Identify the government agency that provides social security benefits.

2 'Charitable agencies provide practical help to those in need.' Do you agree with this statement? Justify your answer using examples.

FOR YOU TO DO

3 Design and cost a two-course meal that meets the Exodus Foundation's (see page 142) budget of approximately $5 per person, to be served on NRL Grand Final Day.

TAKING IT FURTHER

4 Visit a local charitable organisation. Interview the person in charge and write a report that discusses:
- the services the charity provides to your community
- how the charity is funded
- how volunteers are recruited.

REVEREND BILL CREWS
EXODUS FOUNDATION

Reverend Bill Crews is a former electrical engineer turned Uniting Church minister. In the late 1960s he visited the Wayside Chapel in Kings Cross and became involved in visiting housebound and elderly residents of inner Sydney. This association led to Bill quitting his job as an engineer and working full time at the Wayside Chapel.

In 1983 he was studying Theology and was ordained a minister of Ashfield Uniting Church in Sydney's inner west. Building on his experience at the Wayside Chapel, he established the Exodus Foundation to assist homeless and abandoned youth and other people in need.

The Exodus Foundation began by providing a hot dinner for the homeless on a Monday night from the Church hall. It now encompasses the Loaves and Fishes Free Restaurant, serving up to 400 meals to the needy every day. It is open between 9 a.m. and 10.30 a.m. for breakfast and 11.30 a.m. to 1 p.m. for lunch.

The annual Exodus Christmas Day lunch in Ashfield is the centrepiece of the Exodus food program calendar, and ensures those with very little don't miss out on the joy of a Christmas Day lunch.

More than 2000 guests took part in Exodus's 2014 Christmas Day lunch, and more than 300 volunteers were involved.

In 2009 the Exodus Foundation began operating a night food van to provide hot meals to the 150–200 homeless who live in Sydney's CBD.

In 1996, the Foundation began its first Ashfield-based Literacy Tutorial Centre program for needy children. This was followed about a decade later by a similar Literacy Centre in Redfern, primarily for needy Aboriginal Australian children.

The Foundation also operates a free dental and medical clinic, and provides social workers, counselling and food parcel assistance.

Source: Adapted from billcrews.org & The Exodus Foundation. Photo: AAP Image/Paul Miller

ACTIVITIES

1 Why does a person decide to do volunteer work, such as Bill Crews did with the Wayside Chapel?
2 Why do you think the Exodus Foundation started by providing a hot meal as a way to assist the homeless?
3 Why do you think the restaurant has expanded to serve breakfast and lunch, but not dinner?
4 What is the significance of the Tutorial Centre? Why do you think it was added to the types of assistance along with medical and other counselling services?
5 Suggest foods that could be included in a food parcel given to a family in need.

PRACTITIONER FOCUS

Chapter review

LOOKING BACK

1 List four factors that influence the food security of farmers.
2 Why is income the defining factor in meeting food needs?
3 Explain the link between literacy and nutrition choices.
4 Why are women and children particularly at risk of food inequity?
5 How does having a physical or mental disability place an individual at risk of food inequity?
6 What is the definition of a chronic illness? Provide examples of illnesses that may influence food inequity in your answer.
7 Why are certain staple foods grown in particular areas of the world?
8 How does religious belief influence food availability?
9 Explain the effect of globalisation or free trade on poor nations.

FOR YOU TO DO

10 Construct a mind map to illustrate the reasons for food inequity.
11 Design a menu of lunches for a day care centre's preschool section for five days.
12 Explain the link between drug addiction and food inequity. Conduct some further research and develop a fact sheet for a particular drug.
13 Write a list of pros and cons for taking on a part-time job after school at age 14. Then write a list of pros and cons for a young person taking on full-time work in an under-developed nation at the same age. What differences are there in your lists?
14 Design a low-cost healthy meal for a family using an inexpensive meat as the base ingredient. Use an online supermarket or a letterbox catalogue to determine which cuts of meat are cheaper or on special.
15 What facilities are available in your local area for aged care? Investigate their nutrition policy.
16 Conduct research and then design a display of products that support Fairtrade.
17 Design a presentation for the class on a disease of malnutrition. Outline the causes, symptoms, treatment and any long-term effects.

TAKING IT FURTHER

18 Research Vision Australia online to find available services to assist the vision impaired. Present a one-page report on what assistance is available for shopping, cooking and other food-related tasks.

Vision Australia

19 Locate an isolated outback town on a map of NSW. Conduct research on the services and facilities available for the town's population. Identify and list any food inequity factors that you think may be present in this town.
20 Research the work of a speech therapist in assisting the elderly who have swallowing difficulties. What causes these problems, and how can they be treated?
21 Investigate the government support for communities to replace services destroyed by natural disasters such as fire or flood.
22 Research a volunteer group operating in your state. What opportunities exist to help the less fortunate in your local area?

9780170383417

RECIPES

Orange and chick pea couscous

Couscous is actually made up of tiny pasta balls of semolina flour. It is a staple of North African cooking; traditionally, women dampened the flour and rolled it, so the particles stuck together as little balls.

 **Preparation time** 15 minutes  **Cooking time** 15 minutes  **Serves** 4

INGREDIENTS

- 1½ cups couscous
- 1 vegetable or chicken stock cube
- 1½ cups boiling water
- 1 cup canned chick peas, drained and rinsed
- ¼ cup currants
- 2 oranges, peeled and segmented
- ½ cup shallots, chopped

HINT: Ask your teacher to demonstrate the method of peeling and segmenting an orange for best presentation.

METHOD

1. Prepare the oranges and shallots.
2. Place the couscous in a large heatproof bowl. Dissolve the stock cube in the jug of boiling water and pour over couscous. Cover and leave for 10 minutes, then fluff up grains with a fork.
3. Gently stir the remaining ingredients into the couscous.
4. Serve dressed with your favourite salad dressing – try a balsamic or creamy Caesar salad dressing, or make one of your own!

QUESTIONS

1. Describe the texture of this meal.
2. Can you think of meals for which this dish might be an accompaniment?
3. Suggest other herbs, spices and vegetables that could be added to this dish.

Mark Fergus

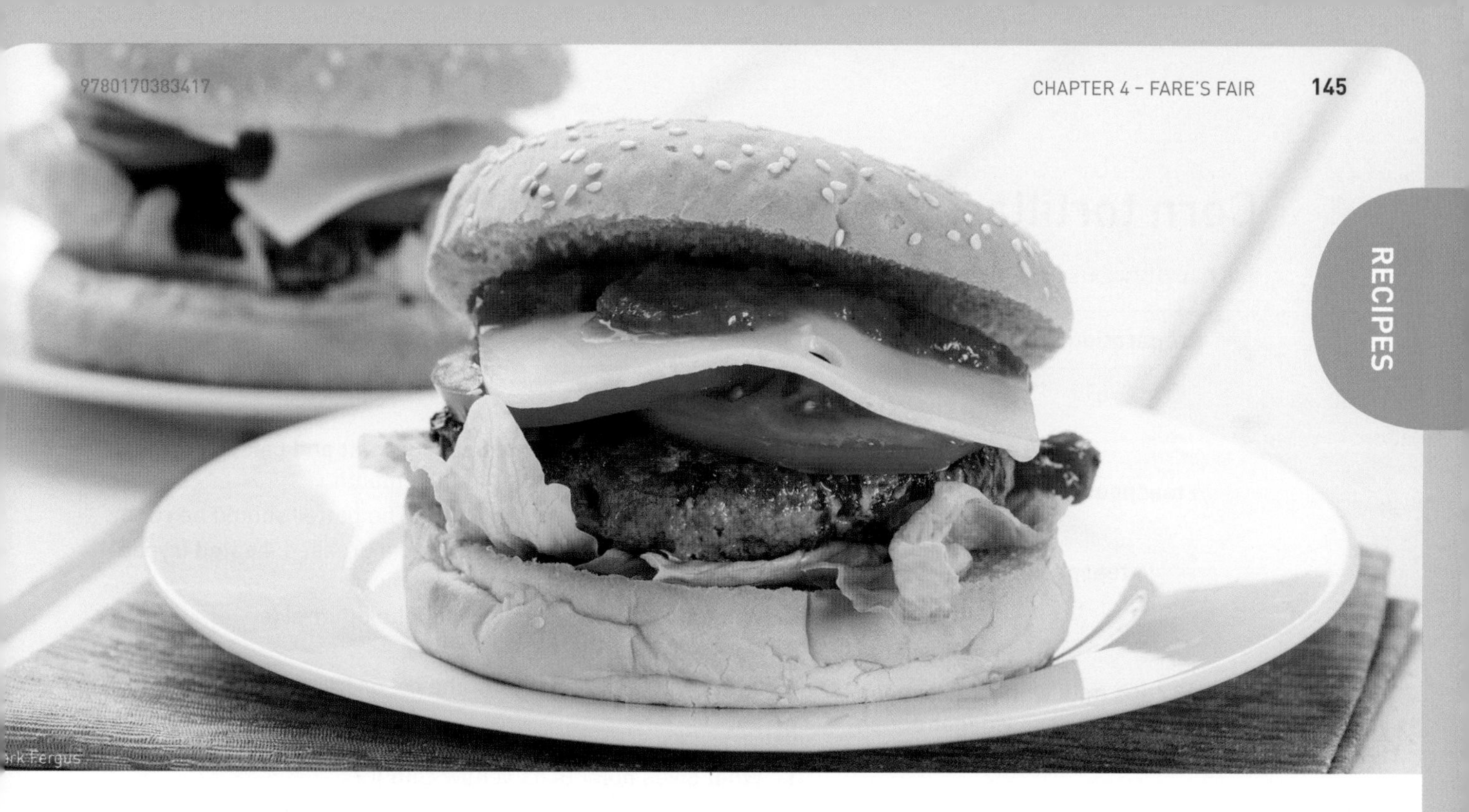

Lamb burgers

This is a low-cost take-away-style recipe suitable for families on limited incomes. It is easily varied by substituting chicken, beef or turkey mince, and the addition of fresh salad accompaniments or different breads to finish.

 **Preparation time** 10 minutes  **Cooking time** 15 minutes  **Serves** 4

INGREDIENTS

250 grams lamb mince

½ box frozen spinach (approximately 70 grams)

1 clove garlic

¼ teaspoon ground cumin

2 teaspoons fresh chopped parsley

2 tablespoons oil (for frying)

4 hamburger buns

4 large lettuce leaves

4 slices cheese

8 slices tomato

tomato or barbecue sauce for serving

METHOD

1. Defrost the spinach by draining over a bowl or the sink.
2. Prepare the parsley and garlic as stated in the ingredients list.
3. Place the mince, spinach, garlic and cumin in a bowl. Use very clean hands to mix until well combined. Divide the mixture evenly into four and shape into patties about 6 cm in diameter.
4. Heat a frypan over medium-high heat and add a small amount of oil.
5. Cook the patties, turning once only during cooking. They will take approximately 10–12 minutes to cook through.
6. Meanwhile, wash and prepare salad ingredients for the burgers. Prepare the buns and top with salad.
7. Serve hamburgers in buns with cheese, salad and sauce.

QUESTIONS

1. Calculate how much this recipe would cost to feed a family of four. Add a serving of homemade chips and milkshake. What is the total cost of the meal?
2. How much does this work out to be per person? Compare this to a 'meal deal' at a local take-away – is your version cheaper?
3. Suggest other take-away-style meals a family could prepare at home if they were on a tight budget.
4. Design a burger patty using one of the alternative mince meats suggested. Vary the additional ingredients.

 9780170383417

Corn tortillas

Corn tortillas are an unleavened bread product and a Mexican staple food.

 **Preparation time** 5 minutes  **Cooking time** 20 minutes  **Serves** 2

INGREDIENTS

½ cup plain flour
½ teaspoon bicarbonate of soda
½ teaspoon salt
½ teaspoon baking powder
1 cup polenta
1 egg, beaten
1½ cups milk with 1 tablespoon vinegar added

METHOD

1. Sift the flour, bicarbonate of soda, salt and baking powder into a medium bowl. Stir in the polenta.
2. Stir in the beaten egg and milk. The batter should be runny.
3. Pour 2 tablespoons of batter into a greased, heated frypan. Tip the pan to make the tortilla thin.
4. When done, turn over and cook the other side.
5. Remove and cool.
6. Serve with a filling of your choice, or freeze until required.

QUESTIONS

1. What gives polenta its yellow colour?
2. When vinegar is added to the milk, what happens? What is this called?
3. Suggest three savoury fillings for these tortillas.

Mark Fergus

Sago custard and peaches

This recipe combines the staple of sago (tapioca) and a popular Australian fruit crop (peaches).

 **Preparation time** 10 minutes  **Cooking time** 30 minutes  **Serves** 2

INGREDIENTS

- 2 tablespoons sago
- 1 cup water
- 2 cups milk
- 2 tablespoons sugar
- 2 eggs
- ½ teaspoon vanilla
- pinch of nutmeg
- 6 peach slices
- 2 tablespoons of thickened cream

METHOD

1. Place the sago in a sieve and rinse. Preheat oven to 160°C.
2. Bring water to the boil in a medium saucepan. Add the sago. Stir until transparent.
3. Add the milk and sugar. Mix well. Allow to cool.
4. Beat the eggs and vanilla together. Add to the sago mixture and pour into a casserole dish. Sprinkle on nutmeg.
5. Place the casserole dish in a baking dish filled with sufficient water to reach halfway up the side of the casserole dish.
6. Place in oven. Bake for 30 minutes or until set.
7. Remove from water bath (bain marie) and serve with peaches and cream.

QUESTIONS

1. Ask your teacher for the ingredient purchase dockets for this recipe and provide a costing.
2. Why is a bain marie cooking method necessary for this dessert? What would happen if it was not used?

9780170383417

RECIPES

Coconut and passionfruit rice

This recipe uses rice – grown in Leeton and Griffith, NSW.

 **Preparation time** 5 minutes  **Cooking time** 25 minutes  **Serves** 2

INGREDIENTS

- 400 mL light coconut milk
- ¾ cup milk
- ¼ cup caster sugar
- ¼ teaspoon salt
- ¾ cup short grain white rice
- 3 passionfruit

METHOD

1. Place the coconut milk, milk, sugar and salt in a medium saucepan.
2. Bring to the boil over a medium heat.
3. Add the rice and stir well. Reduce heat to low and cover with lid. Cook for 25 minutes or until most of the liquid has been absorbed.
4. Stir through almost all of the passionfruit pulp.
5. Serve in individual bowls garnished with remaining passionfruit pulp.

QUESTIONS

1. What is the difference between the rapid boil and absorption method of cooking rice?
2. Find out what type of rice is best suited to making sushi and risotto. What qualities make these rice types suitable for these recipes?

Mark Fergus

Mark Fergus

Survival biscuits (also known as hardtack)

This recipe replicates biscuits given to soldiers as part of their rations when on active duty. They have also been adapted for use as food aid in famine-ravaged countries.

 **Preparation time** 25 minutes  **Cooking time** 40 minutes  **Serves** 4

INGREDIENTS

- 2 teaspoons salt
- ½ cup coconut oil
- 2 tablespoons bacon chips, crushed
- ½ cup powdered protein
- 2½ tablespoons bicarb soda
- 2½ cups buttermilk
- 2 cups rolled oats
- 3½ cups wholemeal flour
- ½ ground flax seed (optional)

METHOD

1. Preheat oven to 160°C. Line a biscuit tray with baking paper.
2. In a large bowl, mix the dry ingredients with a wooden spoon.
3. Slowly add the coconut oil and mix with clean hands until the mixture is crumbly.
4. Add the buttermilk and knead until a solid ball of dough is formed.
5. Using a rolling pin, roll the dough out on a clean, lightly floured bench top. Roll until approximately 1 cm thick.
6. Cut into squares or shapes approximately 5 cm in diameter. Prick each biscuit with a fork.
7. Bake in the oven for 20 minutes. Flip the biscuits over and bake for another 20 minutes.
8. Remove and cool on a rack until completely free of moisture, then store in an airtight bag or container.

QUESTIONS

1. How do they taste? Can you suggest any improvements to the recipe?
2. Research what else is included in a ration pack. How might the survival biscuit be combined with other items included?

9780170383417

RECIPES

Zucchini and corn fritters with salsa

Does your school have a Kitchen Garden? This recipe could be made almost entirely of school-grown produce!

 **Preparation time** 20 minutes  **Cooking time** 20 minutes  **Serves** 4

INGREDIENTS

- 1 cup self-raising flour
- 1 egg
- 1¼ cup milk
- ½ zucchini
- ½ corn cob
- ½ cup grated cheese
- 1 carrot
- 1 teaspoon oil
- 2 shallots
- Large handful of fresh basil or parsley leaves
- 2 tomatoes

METHOD

1. Peel the corn and microwave with a tablespoon of water for 4 minutes.
2. In a large bowl, sift the flour. Make a well in the centre of the flour, add the egg and milk and whisk together to make a batter.
3. Grate the carrot and zucchini. Finely slice the shallots. Cut the cooked corn from the cob.
4. Add the carrot, zucchini, shallots, corn and cheese to the batter. Stir until combined.
5. Heat the oil in a frypan.
6. Add a spoonful of mixture to the pan; turn the fritter over when bubbles appear. Place the cooked fritter on a clean plate and repeat with the remaining mixture.
7. While the fritters are cooking, chop the tomatoes and tear the herbs into small pieces. Mix together.
8. Serve the warm fritters with salsa on top.

QUESTIONS

1. Using an online supermarket website, or the purchase dockets from your teacher, cost the recipe for the class.
2. Divide the total cost by the number of people on the class to get a cost for the serving per person.
3. Now subtract the cost of all the fresh vegetables and herbs. How much money would be saved by having a Kitchen Garden? Can you use this as an argument to establish one in your school?

Mark Fergus

RECIPES

rk Fergus

Weet-Bix and banana choc loaf

Does your school have a Breakfast Club? This is an easy recipe that would be an attractive and quick breakfast alternative.

 **Preparation time** 20 minutes  **Cooking time** 30 minutes  **Serves** 2

INGREDIENTS

- 3 Sanitarium Weet-Bix, crushed
- 1½ cup white flour
- 2 teaspoons baking powder
- ½ teaspoon salt
- ¼ cup dark chocolate chips
- 60 grams margarine
- ¼ cup sugar
- 3 ripe bananas, mashed
- 1 egg, lightly beaten
- ¼ cup Sanitarium So Good Lite or lite milk

METHOD

1. Pre-heat oven to 180°C. Grease and line a loaf tin.
2. In a large bowl combine the Weet-Bix, flour, baking powder, salt and chocolate chips.
3. Using electric beaters, cream the margarine and sugar together, then add the bananas, egg and So Good Lite.
4. Add to the dry ingredients, blending well.
5. Pour into the loaf tin. Bake for 30–40 minutes, or until a skewer inserted in middle of loaf comes out clean.
6. Cool. Slice and serve on its own, or toast and serve with butter or yoghurt.

QUESTIONS

1. How could this recipe be modified for students with special dietary needs; for example, gluten-free or dairy-free?
2. Devise a Breakfast Club menu that could be served in your school canteen. Suggest four menu items that are inexpensive, filling and easy to prepare and serve.

New food!

FOCUS AREA: FOOD PRODUCT DEVELOPMENT

FOCUS AREA

Innovations are new ideas. Food product innovations are new ideas in food. Is there such a thing as a 'new' product? The food industry demonstrates the importance of the word 'new' in marketing foods. Innovations may take the form of a new product, a new method of providing the food to consumers (how it is dispensed), a new variety of an existing food, or simply a competitor for a successful existing product. It costs a great deal of money to develop a food product innovation. The aim of the development of a new food product is to increase company profits. Improvement and innovation are crucial to the continuing success of food products.

Mark Fergus

CHAPTER OUTCOMES

In this chapter you will learn about:

- the reasons new products are developed
- impacts of food product innovations on society in the past and present
- steps taken in developing a new food product
- the role of market research
- the promotion of new food products
- emerging technologies used in new food products.

In this chapter you will learn to:

- explore the purpose, characteristics and diversity of new food products
- make links between new food products and their effect on society
- outline the design process for developing new food products and apply this in the design, production and evaluation of a new food product
- describe the purpose of market research in food product design and development
- identify the elements of the marketing mix
- analyse the success of a range of marketing and promotional techniques and apply these to a new product
- investigate the application of a new technology in food product development
- design an innovative food product.

RECIPES

FOOD FACTS

Food trade shows provide the opportunity to showcase new products. Amongst the new products launched in 2015 were a rose petal and elderflower cordial syrup and a range of products aligning with the paleo diet.

The Pink Lady apple was developed in Western Australia and is a cross between the Lady Williams and the Golden Delicious varieties.

One of the biggest new product flops of all time was the US Gerber company's attempt in 1974 to market single-serve pureed baby food in a jar to the university student. Called 'Singles', the product came in flavours such as Beef Burgundy, Chicken and Mushrooms and Blueberry Delight.

The 'selfie' trend on social media extends to packaging. Expect to see a move in the coming years towards packages that are self-opening, self-closing, self-sealing, self-dosing, self-regulating and self-heating!

Dark colours on packaging can inhibit sales growth because they can make a product hard to find on supermarket shelves.

Molecular gastronomy blends chemistry and physics in the style of cooking made famous by Heston Blumenthal. Heston discovered the ability of fat to hold flavour and created a dish that had three flavours – basil, olive and onion – with each taste being perceived in sequence.

Food scientists and technology experts are working on applications in the food industry for 3D printing. Some possibilities include pasta, sugar-based cake decorations, and developing foods for the elderly who have difficulty chewing and swallowing.

Sydney's Royal Easter Show is famous for its 'carnival' food fare, beginning with the Dagwood Dog (or Pluto Pup) – a battered frankfurter on a stick dipped in tomato sauce, which made its Show debut in the 1940s.

FOOD WORDS

condiment a substance, such as a sauce, that gives a special flavour to a food

constraint restriction

consumer a person who uses a product or service

criteria standards to judge against

evaluate to determine the value of something

modified changed or altered

morbidity illness

mortality death

sensory relating to the senses or sensation

tamper meddle with for the purpose of altering

5.1 Reasons for developing food products

Consumers tend to become bored with products and are easily distracted by new products. For a company that continues to market 'the same old thing', this distraction could mean a loss of profits because consumers leave to spend their money elsewhere.

Companies devote large budgets to developing new products in order to stay competitive and to satisfy consumer demand.

Changes in the law can force food companies to alter their products. Sometimes the required changes may be relatively minor, such as a labelling requirement that declares that the product may contain traces of nuts. However, in some circumstances food companies may be required to alter the content of a product; for example, to remove certain additives from certain foods.

Market concerns

It could be said that consumers today are better informed about their health and nutrition than ever before. With this clearly in mind, food companies develop food products that target this awareness. Companies promote foods that are **modified** versions of their regular product lines; for example, foods for which the recipe has been altered to make them low in fat, salt-reduced or higher in fibre.

Mark Fergus

Figure 5.1 A vast selection of lunchbox treats are available in Australian supermarkets. Statistics show that up to eight out of 10 families use lunchbox fillers such as these to save time packing lunches.

Examples of these products can be found in supermarket lines such as peanut butter, baked beans and fresh bread.

Shutterstock.com/Elena Shashkina

Figure 5.2 Kale chips are commercially available and have gained popularity because kale is reported to be a 'superfood'.

Technological developments

Technological developments affect food product development. For example, the high level of microwave ownership resulted in the development of specifically microwavable products. As microwave technology developed, so did the products. For example, as microwave ovens have become more powerful, products, packaging and cooking times have changed accordingly. The resurgence of slow cookers and popularity of espresso pod coffee makers for the home have created a market for manufacturers of specialty recipe bases and boutique-style coffee pods. Developments in technology can also influence packaging. In 2015, Coca-Cola launched a summer marketing campaign with bottle labels that changed colour when they were chilled.

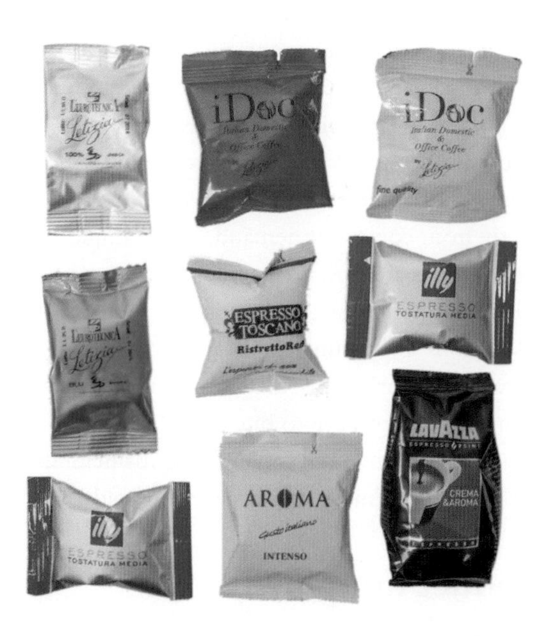

Dreamstime.com/Alarico

Figure 5.3 Coffee machines that use pods are not all alike and consumers are locked into buying pods that are the correct shape and size for their machine.

Increasing company success

A great deal of money is invested in product development. Companies can be hesitant to invest huge amounts of money in developing a new product because of the risk of financial loss if it is not successful.

There are three main types of food product innovations:

- innovative products
- line enhancements
- copy-cat products.

Innovative products

Innovative products are the result of completely new ideas. These may be a new concept in packaging, such as the functional lid that holds the tablet until opening in the Berocca Performance drink, or a product directed to a new target market, such as breakfast replacement drinks (Up & Go) or foods for consumers with a gluten intolerance.

This is a volatile form of product in the market – it is very expensive to develop and launch a completely new product, and follow up with advertising and promotion. Companies want consumers to try a new product but also to continue buying it, which can be difficult to achieve. In Australia, many of the 'new' products may have in fact already been established as successful products elsewhere in the world.

Line enhancements

As new product development is expensive, many food manufacturers opt for a line enhancement as a way of retaining their customers and improving their product.

Line enhancements might include the development of a new flavour, or a recipe revision that changes the recipe to alter its nutritive value, to make it low in salt or high in fibre. While line

Newspix/Ross Swanborough

Figure 5.4 There are a vast number of milk varieties available in the supermarket. What influences a consumer's selection of milk?

enhancements are a safer, lower-cost way of attracting customers, they produce less profit than a brand new innovation.

Some examples might include banana flavoured M&Ms or peanut butter Up & Go. In 2015, McDonalds reinvigorated their sales by introducing a new concept called 'Create Your Taste' whereby customers can 'build' their own burger at a digital kiosk and it is cooked to order.

Copy-cat products

Copy-cat products are aimed at attracting customers for an already established and successful product by creating a similar product. Often companies will copy both the competitor's product and the packaging style as well – however, the profits and market share of the copy are rarely threatening to the original. There is no substitute for being first with a good idea!

Consumer demand

Consumer satisfaction is very important to food companies, because it encourages shoppers to stay loyal to their brands. Consumers are loyal to brands they are happy with and some consumers will communicate their wishes using consumer product hotlines. Some line extension ideas have come directly from consumer demand. For example, many types of foods are now packaged in different ways for better handling or convenient use, such as 3-litre milk cartons so consumers can avoid purchasing three 1-litre cartons, and 12 × 100 gram tubs of yoghurt, so consumers do not have to buy a single 1.2 kg tub of yoghurt. Variety in availability helps to maintain the loyalty of large families and single households.

Product tampering

Product tampering occurs when a customer tampers with a product; for example, opens a packet of biscuits to try one without intending to buy.

Figure 5.5 Big supermarkets aim to trick consumers into purchasing private-label products by designing packages that are similar in style and colour to the brand-name product.

Threats of serious product contamination in the food industry over the last 10 years have led to consumer demand for greater protection of their health in the consumption of purchased products. This tampering may occur at any stage in the production and packaging process. There have been some widely publicised examples in the past, such as the threat of jars of baby food containing fragments of glass, and packets of biscuits and headache tablets being injected with deadly poison. These threats usually result in the affected products being withdrawn from sale Australia-wide, which can be potentially ruinous for a company.

Tamper-evidence devices are methods of packaging that immediately tell the consumer if the product has been opened in some way. There are many examples:

- plastic shrink seal around the lid of products such as baby foods and sauces
- screw-top jar with a 'pop top' that is released when the seal is broken for the first time
- plastic screw top with a collar below the lid, such as on milk and juice bottles
- foil seal under the lid, as on juice bottles, peanut butter and coffee
- paper strips that wrap from each side of a jar and over the lid, which tear when the seal is broken, as on some types of jam
- ring-pulls under reusable plastic lids on cans of baby formula or Milo
- cellophane wrapping around products such as boxes of chocolates or tea bags
- shrink wrap-type plastic that completely seals bottles such as herbs and spices
- packaging within packages, such as for muesli bars and breakfast cereals
- sticky-tape type seals on resealable boxes, such as for ice-cream cones or paracetamol tablets.

CAROLYN CRESSWELL
MUESLI ENTREPRENEUR

Carolyn Cresswell was 18 years old and in her first year of university with a part-time job working for a small muesli-maker who supplied their muesli to cafes in Melbourne, when the owners decided to sell the business and told Carolyn she may lose her job.

She and a friend offered the owners $2000 to buy the business. The offer was accepted and, for the next five years, Carolyn worked at developing her product and expanding the business.

The big breakthrough finally came in the late 1990s when Coles Supermarket agreed to stock Carman's muesli in 20 supermarkets around Melbourne. Carolyn personally delivered the stock to the supermarkets and presented it on the shelves!

Within six months, Coles was displaying her product throughout Victoria, followed by NSW. Carman's is now a $50 million business supplying to Coles, Woolworths and other independent retailers.

Carman's began as a premium-label breakfast cereal and has benefitted from positioning in this niche market. Sales have not been affected by discounted 'house' brands.

The brand has now diversified into muesli bars and muesli biscuits, which account for more than 50 per cent of sales. Carolyn is now exploring export opportunities and extending the product line to include a bread line.

Source: Adapted from 'Hard work paid off for muesli queen' by Miriam Steffens, *Sydney Morning Herald*, 30 July 2012, Photo: Newspix/Norm Oolrloff

ACTIVITIES

1 Why are Carman's muesli sales not affected by price cutting of other breakfast cereals?
2 What kind of consumer would purchase Carman's products? Why?
3 Explain the advantage of extending the Carman's product range to muesli bars and biscuits.
4 Speculate on the potential success of a Carman's bread range. Describe the advantages and disadvantages of entering the bread market.

Special applications

Foods developed for special purposes are frequently adapted from tried-and-true dishes, such as a pasta and sauce. This may mean developing a freeze-dried or dehydrated version of the product.

The advantage of modifying the product in this way is that it is light and space-efficient (ideal for bushwalking and camping), but with the added challenge of not compromising the flavour and texture qualities of a traditionally made sample. Special food applications include:

- defence force ration packs
- camping supplies
- space foods
- airline foods
- medical foods.

5.1a Reasons for developing food products

Defence force ration packs

Combat ration packs are developed according to Australian Defence Force specifications. The issue of weight is important, although soldiers usually only carry rations for three days at a time, as part of their average 40-kilogram pack weight.

Rations are designed to be eaten either hot or cold. Typically, a ration pack would include:

- freeze-dried meals (to be reconstituted with water)
- fruit
- Vegemite
- energy drink powder
- tea and coffee
- toilet paper
- matches
- can opener.

WS 5.2a Reasons for developing food products

Some armies use rations packed in special 'flameless heater' pouches that heat the contents when water is added. The pack is designed to meet the energy needs for all soldiers, regardless of weight, height and energy expenditure. Ration packs of today can be described as energy- and nutrient-dense.

Camping supplies

Foods for camping have come a long way from small cans of baked beans and damper.

Visit any large camping supplier and you will see an extensive range of camping-friendly foods. Experienced campers sing the praises of flat pitta breads (already flat, cannot be squashed further in a pack), dried pasta and sauce (just add long-life milk and water), dehydrated vegetables (just add water and heat) and single-portion packages of breakfast cereal for which the box becomes the bowl.

Two main issues plague the camper/trekker:

- how to minimise the weight and bulk factor of food required to carry in a pack, or packed in a camper or caravan
- how to minimise the rubbish created by pre-packaged food, which has to be carried home.

Campers also need to consider whether a campfire is needed and whether this is allowable in the camping ground.

Space foods

Considerations for selecting food for space travel is not unlike planning a menu for camping. The duration, perishability, waste created and cooking equipment required would all be considered.

New foods and methods of packaging were designed for astronauts as they explored space. Early astronauts on the Mercury capsule (1962) ate bite-sized freeze-dried foods and puréed foods from tubes, which looked like toothpaste. The biggest problems were with food crumbling and floating around the cabin, and the rather unappetising look and taste.

By the time of the Apollo missions (1969), dehydrated foods could be rehydrated by saliva in the mouth of the astronaut. Freeze-drying technology had advanced so the food looked and tasted more like it should.

More recently, foods supplied to the International Space Station reflect the preferences of the American and Russian astronauts, while also adapting to the unusual conditions. Space station crew members have a menu cycle of eight days, which means that the menu repeats every eight days. Food is mostly pre-prepared and packed in packages that form the bowl or cup the meal is heated in, and from which it is eaten. Velcro is used extensively to prevent food packages from floating away, and dry salt and pepper are not used at all as any floating elements create problems when the crumbs lodge in essential instruments! Foods are individually packaged and stowed for easy handling in microgravity. All food is pre-cooked or processed so it requires no refrigeration and is either ready to eat or can be prepared by simply adding water or heating. Fresh fruit and vegetables must be eaten within a few days because they spoil more quickly.

Foods are packed and labelled in the order in which they should be eaten. A computer on board calculates an individual diet for each astronaut.

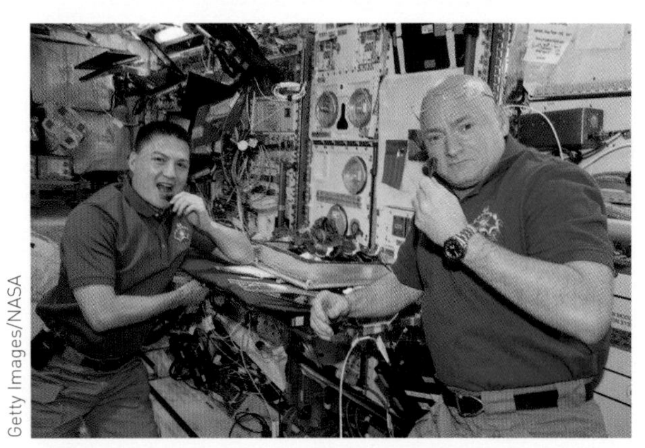
Getty Images/NASA

Figure 5.6 The International Space Station is restocked approximately every eight weeks by an unmanned freighter from Russian and US Space Shuttles.

Airline foods

The first on-board airline meal was served in 1919 on a flight between London and Paris. Sensibly, they chose to serve sandwiches and fruit. Preparation of food for flight is tricky. Food that is pre-prepared and stored prior to reheating is a potential food-poisoning hazard. Airline food must be prepared, chilled to less than 5°C then heated during the flight.

Other challenges are presented by the weight restrictions for the aeroplane, available space on each tray module, ease of movement (for example, soups can be difficult and dangerous), budget and hygiene to avoid the risk of food poisoning.

Alamy Stock Photo/RosaIreneBetancourt 9

Figure 5.7 This is a Qantas economy meal provided on an international flight. Airline meals have improved a great deal in the last decade.

Presentation of the food in an attractive way is perhaps the most difficult challenge given the confined tray space and the fact that many foods lose colour, flavour or texture on reheating.

Medical foods

Foods are also developed for special medical conditions. These are officially called Foods for Special Medical Purposes. They are developed for people who have ongoing medical conditions that mean they cannot eat, chew or swallow. These individuals are often fed for a short time using a nasogastric tube, which goes through the nose down into the stomach. This may be utilised after surgery to the mouth, or for the long term, such as in the case of an ongoing illness. These food products represent a growing area, due to the ageing population and a pattern of early release from hospitals after treatments.

Other Foods for Special Medical Purposes are specialised dietary supplements, such as protein supplements for body builders or low-energy diet formulas available under medical supervision for weight loss. These products tend to be very expensive, but have gained popularity in recent years.

Target market changes

Today's consumers have less time for cooking and yet a greater interest in gourmet kitchen appliances, television cooking shows and cookbooks. The popularity of bullet-style cyclonic blenders that create a meal in a glass is an example, as are the multi-functional thermal cooking appliances. Families show a greater willingness to purchase foods and products that reinforce the idea of home-cooked meals.

Ageing

The Australian population is ageing at a greater rate than it is reproducing. This is due to increased life expectancy and a lower birth rate. In addition, the number of single-person households is increasing in number in Australia. The availability of 'single serve' meal solutions will increase, as will smaller meal serves, given the trend to consume smaller meals rather than three large meals per day.

Alamy Stock Photo/Kumar Sriskandan

Figure 5.8 Elderly Australians living on their own will increasingly be looking for food products in single or double serves.

Reduced size of households

The 'average' household in Australia typically starts small, gets bigger with the arrival of new members, and shrinks again over time. Food companies respond to these demographics by creating family-sized packs that cater for four rather than six. There is an abundance of convenience meals aimed at the busy family whose schedules never coincide for them to eat together.

Multicultural

Australian come from a great many cultural backgrounds. This diversity is reflected on the supermarket shelves and in specialty shops providing ingredients for different cuisines. The popularity of multicultural foods has grown since the 'exotic' spaghetti bolognese and sweet-and-sour pork of the 1960s and 1970s to include naan bread, falafel, baba ghanoush and rogan josh ingredients available in supermarkets today.

Many metropolitan supermarkets reflect the cultural backgrounds of the local population; for example, Asian grocery stores are popular for specialty ingredients in some areas.

Alamy Stock Photo/Patti McConville

Figure 5.9 Asian grocery stores flourish in cities and regional towns. They supply unusual or hard-to-find ingredients that may be unavailable in mainstream supermarkets.

UNIT REVIEW

LOOKING BACK

1 List and describe the three main types of food product innovations.
2 Of the three types, which is the most expensive and why?
3 How is the success of a new product measured by food companies?
4 Define the term 'tamper-evidence device'.
5 List six examples of products that use different tamper-evidence devices.
6 Explain the need for high-energy foods in a soldier's ration pack.
7 Outline the advantages of dehydrated and long-life foods for ration packs, camping, air travel and space travel.
8 List the benefits for campers and International Space Station inhabitants of food for which the package becomes the cooking and/or serving dish.
9 What were some of the problems with early space foods?
10 What constraints limit the scope of meals served on airlines?

FOR YOU TO DO

11 Sometimes packaging, rather than product, needs a revamp to boost sales. Design a new package for an existing successful product; for example, paper-wrapped butter, tea bags or bread.

>

12 Develop a tamper-evidence device for a container to store plum jam.
13 The Packaging Council sees easier opening as a feature of packaging for the elderly. Invent a tamper-evidence device that is user-friendly for the elderly.
14 In pairs, design a menu for a four-day, three-night camping and walking trek. You must include all meals and snacks, making sure they are high in energy, lightweight and easy to carry.
15 Using the weblink, check the Department of Defence's activity for designing a ration pack.

Australian Defence Force – Defence 2020

TAKING IT FURTHER

16 Look at the websites of some well-known food companies. Are their products copy-cat products, line extensions or completely new products?
17 New food products are often introduced to Australia after being launched in the United States. Write a report on two interesting products recently released in the United States.
18 True or false: Shoppers who open products to inspect quality, or try products such as grapes in the fruit and vegetable section of a supermarket, are breaking the law. Justify your answer.
19 Why do products such as bread not use tamper-evidence devices?
20 Discuss the following topic: 'Tamper-evidence devices just increase the non-recyclable component of packaging'. Do you agree? Justify your answer with examples.
21 Purchase beef jerky, dehydrated peas and quick-boil rice. Prepare and present these as a meal for camping. Evaluate the appearance and taste.
22 Using the internet, research meals served on international flights. How do the meals differ between economy, business class and first class? (How do the fares compare?)
23 View videos from NASA about eating in space. Write a report about the challenges faced.

NASA – Eating in Space

5.2 Impact of food product innovations

Food product innovations do not go unnoticed, nor are their effects limited to our tastebuds. Innovations impact on all areas of our lives.

Social and cultural impacts

In the first half of the 20th century, innovations in food came in the form of labour-saving devices such as the refrigerator and the electric mixer, and the concept of the supermarket. These innovations changed the lifestyle of Australian home-makers. Products needing refrigeration could be bought in greater quantities, home baking became less laborious and grocery choices could be made by consumers, who could select products from the shelves at their leisure.

In the 21st century, products are developed to match our lives rather than change our lifestyles. More than ever, households with children contain two income earners, and this extra income comes at the price of reduced time spent for food preparation. Consequently, food companies develop products that reflect this changed lifestyle. Examples include traditional breakfast substitutes (breakfast biscuits), meals for one, meals for families that do not eat together as a result of clashing timetables because of sporting and other commitments, and the ever-expanding number of sauces and **condiments** that can be added to meat and chopped vegetables (the ever-popular Chicken Tonight was a pioneer in this area). The Australian acceptance of foods from other cultures has led to an expansion of products based on Mexican, Thai, Malaysian, Mediterranean, Chinese and Indian cuisines in supermarkets.

The cafe culture has also driven the development and promotion of cafe-style products and the growth of breakfast as a service period for customers.

Economic impact

Australia's increasingly busy population continues to be willing to try new products. As the population of working adults increases, so does their disposable income. This increased income means people are more willing and able to spend money to try new products, particularly products that make their lives easier. For example, lunchbox fillers such as muesli bars, pre-packaged dips and biscuits, fruit sticks and dried fruit boxes provide an appealing, time-saving, portioned alternative to home baking.

Of course, this move away from traditional home-baked foods comes at a cost. These products cost much more to make than the homemade equivalent.

Environmental impact

Food innovations affect the environment. Waste packaging generated by new processed foods causes an obvious impact, but sustainability is also a major concern.

Sustainable food production is now a priority in Australia in order to address the issues of land degradation, soil salinity and deforestation.

Environmentally friendly crop development using organic farming methods creates produce that is enthusiastically received by the purchasing public. Supermarkets now stock many more vegetarian options, from vegetarian soy sausages to chick pea patty mixtures. Vegetarianism is increasing in the community, and food producers and retailers are endeavouring to meet the needs of this group by providing a variety of new products, from fresh to frozen foods.

Developments in the area of genetically modified cereal crops have been met with caution.

Nutritional impact

Consumers are much more aware and much more demanding of food products to meet their requirements. Statistics show that the impact of diet on **morbidity** and **mortality** is of great concern to Australian scientific and dietetic professionals.

Food consumption trends have motivated public health organisations to develop strategies to reduce the impact of chronic disease linked to the foods we consume. Diet contributes in some way to the incidence of:

- heart disease
- stomach cancer
- bowel cancer
- stroke
- diabetes
- obesity.

As a result, many food products and label symbols have been developed to draw attention to ways of preventing or reducing the incidence of these nutritional disorders.

HANDS-ON

MAKING A PACKET CAKE MIX

PURPOSE

To develop a popular convenience product – a packet cake or muffin mix of your own.

MATERIALS

½ cup sugar
1 cup self-raising flour
1 tablespoon milk powder
Flavouring of your choice such as cocoa, orange rind and poppy seeds, coconut, dried apple and cinnamon or chopped apricots
75 grams butter
2 drops vanilla
1 zip lock plastic bag
Computer program to create labels
Labels for computer printing

STEPS

1. Grind the sugar to a powder using a food processor.
2. Add the other dry ingredients and blend on low speed.
3. Add the butter and vanilla. Blend.
4. Store in a zip lock plastic bag.
5. Using a computer program that creates labels, label the bag with the name of the product, instructions for mixing and cooking, and the date.

[To make the cake, add one egg and 180 mL of water. Blend. Pour into a prepared 20-centimetre round tin and bake in a moderate oven (180°C) until cooked.]

ACTIVITIES

1. Determine the cost of this mix and compare to the cost of a commercially available packet cake mix.
2. Store your mix in the pantry for one week and then make up according to your instructions. Evaluate in terms of taste and shelf life. Do you need to add a 'use by' date to your label?

UNIT REVIEW

LOOKING BACK

1 Describe the impact on lifestyle of the availability of refrigerators, electric mixers and microwave ovens.
2 Explain the popularity of products such as Chicken Tonight and risotto or pasta sauces.

FOR YOU TO DO

1 Design an advertisement for a bottled pasta sauce – aim your sales pitch at working parents.

TAKING IT FURTHER

1 Visit the Australian Bureau of Statistics weblink to find the latest statistics on households with two or more income earners. In how many families are both parents working? Use this statistic to explain why convenient foods are an expanding market.

Australian Bureau of Statistics

5.3 Steps in new product development design

An idea for a new product may come from:

- the manufacturer's need to cut costs and find a more cost-effective way to produce the product
- a drop in the market share for the product
- a competitor's successful product
- market research indicating that a consumer need is not being met
- environmental concerns.

Most big companies today employ teams of people in the research and development department. Their job is to keep up to date with sales figures of product lines and to stay aware of competitors in order to pinpoint areas for product development (for example, a revamp for a product with declining sales, or development of a new product to compete with another in an area related to the company's area of expertise).

Developing a new product is a step-by-step process. At each step, the project is **evaluated** to determine whether it should continue.

Product design is a cyclic process, as shown in Figure 5.10.

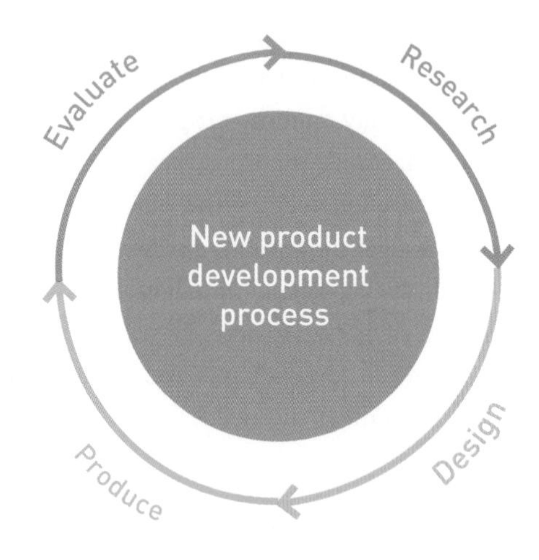

- Evaluate
- Develop a design brief
- Research
- Create ideas
- Select the most appropriate ideas
- Produce a trial product
- Launch the product

Figure 5.10 Designing a new product is a continuous process.

Creating a brief

A brief is a statement of the aims of the project and **criteria** to measure the success of the product. It may be broad or very specific and includes limitations such as budget and timing.

Often detailed briefs specify raw ingredients or a method of processing to be used and the market the product is targeting (for example, snack food for children, low-fat product for health-conscious people, packaged prepared vegetables for working parents).

Research: identify the target market

The manufacturer must investigate the consumer market to clarify the wants and needs of consumers, as well as provide a description of the proposed product and target market. The results will determine whether the original idea needs to be modified, or even dropped completely if the research shows there is no need for it.

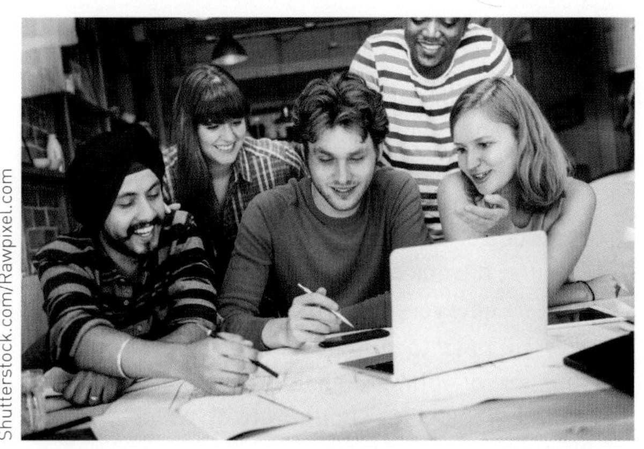
Shutterstock.com/Rawpixel.com

Figure 5.11 The research and development team meets regularly to discuss ideas and directions for product development.

Alamy Stock Photo/brt PHOTO

Figure 5.12 Test kitchens are used to develop recipes, which are then evaluated against the criteria and tasted.

Creating and developing ideas

This is the stage at which ideas are 'dreamed up', based on what the team thinks the consumer will want to buy. These ideas then generate research – looking for available ingredients, investigating processing methods, watching competitors and consulting again with consumers to clarify strengths and weaknesses of existing related products.

Assessing options and selecting the most appropriate

The list of ideas generated is examined and each suggestion matched against the criteria given in the design brief. Ideas that fail this test are discarded.

The ideas that remain (usually about 10 or so) are then investigated more closely to determine whether they will meet budget constraints. For example, will the purchase of a new processing or packaging machine be necessary? Will the ingredients be available all year round? Does this idea meet the consumers' demands?

Producing

Developing a recipe and prototype

Laboratory experimentation is aimed at developing a prototype or model product that satisfies the design brief. Samples are produced that demonstrate changes in single aspects of the basic recipe. All steps and results are recorded carefully so that recipes can be reproduced when a decision is made.

In-house sensory testing is used to evaluate colour, flavour and/or texture.

If the product is developed to compete with another established product, sensory testing will compare the new product with the competitor. Development of a trial product or prototype is a time-consuming process. There may be extensive fine-tuning to match colour, flavour and texture to consumer requirements.

During this stage, the packaged and unpackaged product is tested for spoilage and deterioration under a variety of conditions, such as refrigeration, pantry storage, sunlight exposure and heat.

Calculating the actual cost of the product will determine whether the cost of producing the product is within the financial budget, and will also act as a starting point for calculating retail price (which includes company and retail profit). The cost of a product is determined in this way:

Ingredient costs
+
Production costs
+
Packaging costs
+
Marketing costs
=
Total cost

The cost per product is the total cost divided by the number of product units manufactured.

Pilot production

Pilot production is adaptation of the successful prototype for full-scale production. Manufacturers often simulate full-scale production in smaller pilot production plants. This enables them to refine ingredient amounts, cooking times and energy costs for large-scale production.

HANDS-ON

DESIGN A VEGETABLE 'MAKEOVER' DISH!

PURPOSE

In line with current food trends and the resurgence of vegetables and the popularity of farmers' markets, you are to give a tired or traditional vegetable recipe a makeover. Your aim is to make veggies appetising and trendy!

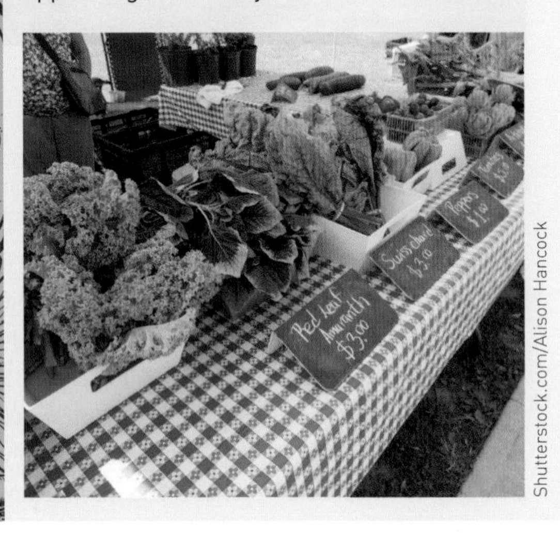

Shutterstock.com/Alison Hancock

STEPS

1 Find a traditional recipe for either a salad or warm vegetable dish.
2 Brainstorm some ideas for changes. Think about using different vegetables, cooking methods or adding other ingredients for texture, flavour or colour.
3 Write a new version of this recipe.
4 You will need to submit your recipe to your teacher for voting. Once all recipes have been collated, the class should vote on which dish they would like to make.

ACTIVITIES

1 Select the most popular veggie recipe in your class and create a standard recipe suitable for producing large quantities for sale.
2 Design a package for the individual product. Ensure the label includes a name, list of ingredients, date of baking and names of the producers.
3 Plan a production line for the large-scale manufacture of the vegetable dish. Make sure you start with collection of ingredients and finish with packaging and labelling.

Packaging

The development of a package for the new product is integral to its success and recognition in the consumer market. Package design occurs at the same time as product design and it follows a similar design process.

Getty Images/Monty Rakusen

Figure 5.13 Factory production is undertaken on a small scale first with small batches, so that production concerns can be fine-tuned before full-scale production.

Evaluating

Evaluation is a continuous process, undertaken at every stage of the product development process.

Once a product has been developed into a successful prototype and appears to be 'the one' the company would like to add to its existing line, market evaluation occurs. This means confirming that the product has met the needs identified earlier in the project.

Market evaluation: sensory assessment

Sensory analysis is carried out to determine the answers to three main questions.

- Does the target market like it?
- Does it differ from what is currently available?
- How does it differ from what is currently available?

Sensory analysis is used to evaluate and describe the new product in terms of the senses:

- taste or flavour; for example, sweet, sour, bitter, salty, strong, mild or weak
- texture or mouthfeel; for example, smooth, lumpy, juicy, moist, gritty or sticky
- aroma; for example, strong, weak, spicy, yeasty, sweet or pungent
- appearance; for example, light, dark, golden, glossy, bright or dull
- noise; for example, crunchy, crackly, fizzy or dull.

The attitudes of consumers to aspects of the new product such as the aroma are tested. Testers may have to tick on a scale where their preference lies, from 'like very much' to 'dislike very much'.

Researchers also want to know how consumers feel about a product and may ask them to select words they would use to describe the product, such as 'nourishing', 'natural', 'healthy', 'artificial' or 'enjoyable'.

Consumers often participate in taste tests, which are usually blind tests; that is, tasters do not know which product they are tasting. This is especially important if a company wants a comparison between their new product and that of a competitor. More accurate answers are gained when testers do not know which the well-known product is.

If the results from the sensory evaluation are favourable, the product is launched onto the market.

Alamy Stock Photo/Jaymi Heimbuch

Figure 5.14 In a blind taste test, the products for testing are not identified and taste-testers do not speak to each other. Can you guess why?

HANDS-ON

TASTE-TESTING NEW PRODUCTS

PURPOSE

To conduct your own market research. The success of new food products depends on the willingness of consumers to buy the product more than once, gradually building up brand loyalty.

MATERIALS (WHOLE-CLASS ACTIVITY)

6–8 new food products currently being marketed (preferably ones you have not tried) in sufficient quantities for each student to taste

METHOD

Divide each product into equal portions for each student. Devise a table to record your results using the following headings:

- Name of product
- Price
- Appearance (devise a scoring system to evaluate this and texture and flavour)
- Texture
- Flavour.

ACTIVITIES

Devise a bar chart on the computer to analyse the class results in terms of the following.

1. Which product had the most pleasing appearance?
2. Which product had the most liked texture?
3. Which product had the most favourable flavour?
4. Which product was the most expensive?
5. Take a vote among the class and determine which product is most likely to be purchased again. Why?

UNIT REVIEW

LOOKING BACK

1 Describe the characteristics of a design brief.
2 Discuss the role of research at the early stage of designing a new product.
3 Write an equation to calculate the cost of producing a product.
4 What is a pilot plant?
5 Why is package design so important?
6 Why is constant evaluation throughout the product development process crucial?
7 Explain the need for consumer opinion in the form of sensory analysis.

FOR YOU TO DO

8 Draft a design brief that could have been the driving force behind one of the new food products you have tried.
9 Select a new food product and devise a 10-question survey to gauge whether the product is well liked among your classmates.
10 Do you think products are developed with healthy eating principles in mind? Support your answer with relevant examples.
11 Why are products tested for storage under so many varied conditions?
12 Devise an experiment to test for the deterioration of a product under exposure to sunlight.

TAKING IT FURTHER

13 Brainstorm a list of ideas for new food products you would like to see. Why do you think they haven't been developed yet?
14 **a** Devise a recipe for a green juice smoothie.
b Imagine you are aiming to sell this drink to take away from a corner shop. Using a plain paper cup (milkshake size), design a logo to appear on the cup to promote your drink.
15 Obtain a recent copy of the journal *Food Australia* or visit the website. Write a report on the innovations (food, packaging, distribution, etc.) publicised in this journal.

Food Australia

5.4 Role of market research in product development

Identifying needs

Before a design brief can be developed, market research is conducted to determine whether there is a need in the consumer population that is not being met, or whether products need to be modified in order for them to remain profitable. Market researchers need to find out about consumer buying patterns, a company's competitors and business in general. The information gathered helps to identify the target market. This research is usually gathered by companies who specialise in the area of collecting consumer information.

Economic viability

If the new product will make sufficient profit for the company, it is said to be economically viable. Market research will help determine whether consumers will buy the product not just once out of curiosity, but continue to buy it. This is called product longevity. A product with a long life in the market (that is, it continues to maintain good sales for a long period of time) will reward the company with profit.

Consumer feedback

Consumer opinion is the backbone of market research. Information is collected from consumers using:

- telephone surveys
- researchers who survey shoppers in busy centres containing supermarkets
- consumer feedback lines advertised on product packaging
- formal surveys and tests such as sensory tests.

Professional research companies carry out many formal research projects.

The use of consumer feedback telephone lines enables companies to obtain direct information from their consumers. They often have Freecall numbers, which may also operate as advice lines, and allow consumers to register their likes and dislikes about a product or even to suggest changes to an existing product, such as making a low-fat or salt-reduced version of a family favourite. Some companies offer this service online as well.

Sensory assessment

As you have seen, sensory assessment is pivotal in selecting the successful product to be launched. Its specific role is to identify areas for development or adjustment, such as aspects of flavour, texture, appearance, aroma or sound. Additionally, it is important that through sensory testing, the consumer can identify a point of difference between the new product and its competitor.

iStock.com/AlbanyPictures

Figure 5.15 Have you ever been approached in a supermarket to taste a new product? Did you buy it?

HANDS-ON

USE AN EXISTING FOOD PRODUCT TO CREATE A LINE EXTENSION

PURPOSE

To create a line extension to extend the life of a successful product.

STEPS

1 Select a food as the base for your line extension; for example, ice-cream, milk, rice bubbles or bread.
2 Submit an ingredients list and numbered method.
3 Prepare and make your line extension. Take photographs of the finished product to be used in a promotional presentation to your class.

ACTIVITIES

1 What was the most difficult aspect of your product development?
2 Identify the target market for your product.
3 Make a 15-second television-style commercial for your product.

UNIT REVIEW

LOOKING BACK

1 Define the term 'product longevity'.
2 List four ways that companies receive feedback on their products. Why is interaction with the customer important to a company?

FOR YOU TO DO

3 In pairs, design a large-format flow chart to explain the design process for developing a new food product. Use a recently launched product as an example.

TAKING IT FURTHER

4 Contact a food company to request information on the development of a successful product for which they are the market leader. (Be aware that you may have difficulty getting details other than promotional material on very recently launched products or in categories in which the competition is fierce. To help, make sure you mention that you are a school student, and why you need the information.)

5.5 Promotion of new products

Products that are new to consumers must be promoted in order for the public to be aware of their existence.

Marketing mix

The marketing mix is a combination of product, price, promotion and place (sometimes known as distribution). These factors are crucial to a successful product launch.

Product

The success of a new product depends greatly on a company having the right product to promote: its product make-up (ingredients, processing methods used), the appropriateness and appeal of the packaging, the image associated with the company and its brand name (sometimes referred to as reputation).

Price

The selling price of the product will determine its 'positioning' in the market. Many consumers equate price with quality, but at the same time look to get value for money. Setting the right price is important for establishing the correct image for the product.

Promotion

Promotion involves deciding whether and how to inform and educate the consumer about the new product. This may mean media advertising, personal selling, sales promotions, sponsorships or a combination of these. The technological age means promotion takes a new direction through the use (sometimes exclusively) of social media – Facebook, Instagram, Twitter and YouTube. McDonald's used this form of promotion to great effect for their Create Your Taste burgers, creating web clips only available on their YouTube page.

Coca-Cola branding not only includes the product but also the logo. Coca-Cola are well known for changing their advertising campaigns regularly, but their logo and trademark ribbon have always remained the same.

Place

This involves deciding on the avenues to distribute the product. Some food products, described as 'boutique', may sell only at small specialty shops or delicatessens, while others will be sold widely at supermarkets in all cities and country towns in Australia; others yet may be developed solely for export. Distribution of the product must be efficient so that the product can arrive at the sale venue in good condition.

Promotional techniques

Successful promotion involves finding out what influences consumers, and what encourages them to buy your product over the competition. Researchers look for changing consumer trends and these trends, having guided the development of a new product, are then used to promote the product. For example, consumers see chocolate as an indulgence food, and market research shows that chocolate consumption is higher in the cooler months of the year. Winter promotions of people rugged up against the cold in front of an open fire, sipping a steaming cup of hot chocolate, are directed at what the research indicates about consumer behaviour. Very successful promotion campaigns identify consumer trends and needs directly.

AAP Images/Meat & Livestock Australia

Figure 5.16 The annual Australia Day lamb advertising campaign by Meat and Livestock Australia is a keenly awaited marketing strategy that has quickly gained a cult following.

Competitions, advertising campaigns, celebrity endorsements

Once the decision has been made about where the item is to be sold, promotion through advertising can be planned. The aim of advertising is to persuade the consumer to buy the product, thereby increasing the company profits.

Competitions encourage consumers to buy particular products; for example, in order to use the bar code or competition entry form on the package for the chance of winning a prize. Some competitions require more than one purchase to enter and win. Other promotions offer cash back after a proof of purchase has been sent to the manufacturer.

Advertising appears in various forms, on television, radio, billboards, magazines and newspapers, and social media as well as moving displays on buses and taxis, and more subtle promotion through sport sponsorship (for example, the Melbourne Cup or One Day Cricket).

Advertising is expensive, especially on television and radio. This cost needs to be absorbed into the final cost of the product. Greater profit is made on processed foods, particularly fast food and convenience foods, and for this reason they are the most highly advertised products.

Effective media advertising tools include jingles and slogans, and endorsements by celebrities. Other useful promotional activities occur at the point of sale, such as the supermarket, milk bar or corner store. These activities may be subtle in nature, such as positioning products at eye level on the supermarket shelf, to using flags and signs that extend from the shelves, to competitions or in-store taste testing, to large end-of-aisle displays.

UNIT REVIEW

LOOKING BACK

1 Having the right product at the right time is crucial for food companies. Why?
2 In what ways does the distribution aspect of a boutique product differ from that of a widely available supermarket product?
3 What is promotion?
4 How does advertising affect the price of a product?

FOR YOU TO DO

5 Using pictures, prices, logos and slogans from supermarket advertising brochures, explain the concepts of product, price, promotion and distribution.
6 Collect three magazine advertisements for food products that you consider to be effective. Identify the elements that make each so good.
7 Using these elements, identify the target market for your 'Bickies in a jar, (page 179) and create a simply styled image suitable for your company's Instagram page.

TAKING IT FURTHER

8 Investigate a food product that is a specialty of your local area. Gather information on it via email or interview a relevant spokesperson and present an oral report on the way this product is marketed.
9 Select a food product that is relatively new to the market. How effective do you think the marketing strategies have been for ensuring the success of this product?
10 Conduct a debate on the following topic: 'Advertising informs the consumer'.
11 Watch 30 minutes of television between 3 p.m. and 5 p.m., and then 30 minutes between 7 p.m. and 9 p.m. on any weekday. List the advertisements you see in each time slot. Are the range of products vastly different? Why?
12 Why are children's programs on commercial television stations that screen in the mid-morning on week days often free of commercials, while on weekends, commercials are quite specifically aimed at children?

Additional content: emerging technologies and products

The food industry is ever-changing. There are new developments in food and food business, such as:

- farming and cropping methods; for example, genetically modified foods
- advancements and innovations in food and packaging processes; for example, cryovac packaging
- improvements in distribution and warehousing; for example, computerised stocktaking and ordering such as Radio Frequency Identification tags
- developments in flavours and additives
- greater variety in market venues and sales opportunities; for example, the growing popularity of Aldi supermarkets and internet shopping.

Genetically modified (GM) foods are created when genes from plants, animals or micro-organisms are inserted into the genetic material of food crops and animals. Current GM crops approved for development in Australia include soybeans, canola, corn, cotton, lucerne and rice.

Any foods produced that have been genetically modified in any way must seek approval from Food Standards Australia and New Zealand before they can be sold in Australia. All foods containing genetically modified ingredients or additives must be labelled with the words 'genetically modified'.

Shutterstock.com/Fotokostic

Figure 5.17 Genetically modified Roundup Ready Canola is resistant to the weed killer Roundup. This means that this GM canola crop dusted by Roundup will not be destroyed – only the weeds will.

FOOD IN FOCUS

FARMERS GROW ALTERNATIVE CROPS TO MEET GLUTEN-FREE MARKET

Many farmers are investing in niche market crops to cash in on the growing demand for gluten-free grains. With an estimated 10 to 15% of the population unable to tolerate gluten found in wheat, rye, barley and oats, there is a substantial customer base available.

Brett Ryan grows buckwheat on his farm in Blayney. Despite the name, buckwheat is gluten-free. The seeds are sent to Parkes where the yield is dehulled, cleaned and graded. Mr Ryan sells much of his crop to Japan for manufacturing into soba noodles, but has an increasing interest from gluten-free manufacturers in Australia.

A farmer in Tasmania is growing a grain known as teff. Teff is common in Ethiopia and ground into flour to make a flat bread known as injera. It is hoped it will find popularity in the growing Ethiopian community and those looking for a gluten-free alternative.

Finally, a solution for banana growers has been developed for an end use for fruit that is rejected due to skin blemishes. A Queensland famer has discovered that when the fruit is dried. It too can be ground to form a gluten-free flour.

Source: Adapted from 'Australian farmers take up alternative crops to catch on to growing gluten-free craze' by Kerry Staight , ABC News, 5 October 2013

>

ACTIVITIES

1 Research other grains that are gluten-free.
2 Speculate the effect an increasing market for gluten-free products may have on traditional grain growers.
3 Obtain a recipe for scones. Make a batch with regular flour and a batch with a gluten-free flour. Evaluate the difference in taste and texture. How did the cost of the raw ingredients compare?

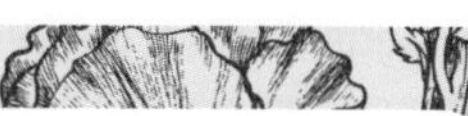

FOOD IN FOCUS

NANOTECHNOLOGY AND FOOD

Nanotechnology involves manipulating matter at the nanoscale to create new materials, structures and devices. One nanometre (nm) is a billionth of a metre and, to put this in perspective, a human hair is about 80 000–100 000 nm thick!

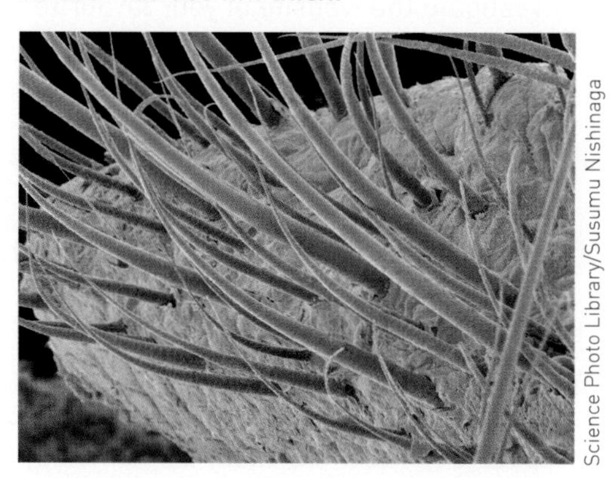

Science Photo Library/Susumu Nishinaga

Figure 5.18 A single human hair, such as each of these eyelashes, is more than 80 000 times thicker than a single nanometre!

Nanoscale particles are not new. Food is naturally composed of nano-sized particles; for example, proteins and sugars. Humans are exposed to nanoparticles of dust and smoke daily.

Possibilities for development of nanotechnology in food include:

- using nanosensors in soils for agricultural purposes to monitor soil conditions and crop growth.
- nanoencapsulated flavour enhancers
- nanoemulsions and particles for better distribution of nutrients
- antimicrobial and antifungal surface coatings in food packaging using nanoparticles of silver, magnesium and zinc.

ACTIVITIES

1 Define the term 'nanotechnology'.
2 Draw up a table of two columns and list the advantages and disadvantages of nanotechnology.
3 Brainstorm the uses for nanotechnology you would like to see.

UNIT REVIEW

LOOKING BACK

1 List three main areas of technological development in food innovation.

FOR YOU TO DO

2 Design a new-to-the-world food product using an emerging innovation.

TAKING IT FURTHER

3 Research the issue of GM food further. Read widely and present an oral report to the class on your views on GM foods – are you in favour or against? Support your views with evidence.

9780170383417

Chapter review

LOOKING BACK

1. Why do consumers lose 'brand loyalty' over time?
2. List some factors related to consumer wishes that drive food product development.
3. Suggest the criteria driving product development for products such as defence force ration packs, camping foods, food for space, airline foods and food for special medical purposes.
4. Describe the impact of an ageing population on food product development.
5. Outline the economic impacts of the increasing availability of new food products.
6. Define the term 'sustainable food production'.
7. List three diseases that have been linked with poor diet. Suggest a food product that aims to address the prevalence of each disease listed.
8. Where do ideas for new food products come from?
9. Define the term 'prototype'.
10. Define the four elements of the marketing mix.

FOR YOU TO DO

11. Identify the newest food preparation or cooking appliance in your home. What products are available that have been designed specifically for that appliance?
12. Explain why a company might develop a copy-cat product rather than an innovative product.
13. Market researchers identify consumer buying patterns and their motivation for purchasing certain brands. What do your purchasing habits say about you? Why do you stay loyal to certain brands?
14. Design a ration pack for an emergency kit (for example, to be taken in the event of a home evacuation) that will make one day's worth of dishes that can be eaten warm or cold. Consider weight, packaging, utensils and preparation required.
15. Explain the target market the following products may be aimed at:
    - breakfast biscuits
    - pre-made pizza bases
    - fresh lentil burger patties
    - heat-in-the-bowl low-fat microwave meals.
16. Debate the following topic: 'Price is the greatest deciding factor in product longevity'.
17. Watch a televised sporting event such as football (any code), basketball, motorsport or tennis. Identify any incidental promotion you see during the telecast, but not the advertisements. Where are they? How obvious are they?

TAKING IT FURTHER

18. The United States is the home of new and unusual food. Visit the weblinks and write a report on the new foods that have arrived in American supermarket shelves recently.

Supermarket Guru

19. Investigate the website of your favourite food product. What does the website cover? How could you contact the company with your ideas or complaints about the product?

The Impulsive Buy

20. Obtain a set of bathroom scales and fill a backpack to weigh 40 kg. Consider the energy required by soldiers to carry this weight while walking for three days.
21. Interview the local Meals on Wheels coordinator. What modifications are made to meals to make them easily consumed by their clients?
22. Select a food product that aims to assist in the prevention of heart disease, cholesterol or bowel cancer. Research this product and describe how this product works in this therapeutic fashion.
23. Design an innovative competition to encourage primary school children and/or their parents to purchase a food product of your choice. Will it be a game of skill or chance?
24. Conduct further research on nanotechnology or gene technology in foods. Present a short oral report on your findings.

ark Fergus

Plum jam

Jams sold at cake stalls and fetes can also use tamper-evidence devices. Make this jam and develop your own tamper-evidence device. Jars used for the jam must be sterilised.

 **Preparation time** 10 minutes  **Cooking time** 25 minutes  **Serves** 2

INGREDIENTS

500 grams plums, washed, stoned and cut into small pieces

1½ cups sugar

2 teaspoons lemon juice

METHOD

1. Place a saucer in the refrigerator to cool and use for testing.
2. Place the plums in a microwave-safe dish and cook on high for 15 minutes.
3. Add the sugar and lemon juice and stir until dissolved.
4. Return to the microwave and cook for 12 minutes.
5. Test a small amount on the saucer that has been in the refrigerator. If the mix gels or wrinkles when touched, it is ready. If not, cook for another 3 minutes and test again.
6. Pour into sterilised jars, label, date and seal.

QUESTIONS

1. What types of tamper-evidence devices are commonly used for jams?
2. This product would be regarded as a copycat as it is identical to other plum jams on the market. Suggest ways to make it a line enhancement.

9780170383417

RECIPES

First-class flathead

Creating an airline-style, complete meal in a box.

 **Preparation time** 10 minutes  **Cooking time** 30 minutes  **Serves** 2

INGREDIENTS

- 2 small foil boxes with lids
- 1 teaspoon oil
- ½ carrot, cut into thin strips
- ½ onion, finely diced
- 1 tablespoon vinegar mixed with 1 tablespoon water
- 2 white fish fillets
- 2 slices lemon plus juice of 1 lemon
- pepper
- 1 teaspoon butter
- Aluminium foil
- 2 dinner rolls

METHOD

1. Preheat oven to 190°C.
2. Brush each foil box with oil.
3. Mix the onion, vinegar and water, lemon juice and pepper to taste.
4. Place half of this mix in the base of each take-away box and put a fish fillet on top. Place the carrot and sliced lemon on top of the fish.
5. Dot with butter and cover with foil.
6. Bake for 30 minutes.
7. Serve in box with a warmed dinner roll (20 seconds in microwave on high), paper serviette and plastic cutlery.

QUESTIONS

1. Look around the class when all the meals are served. Calculate (based on the size of an individual box) the volume of space required to accommodate this meal on an aeroplane.
2. Using the ingredient purchase receipts for this dish, calculate the cost of the dish. Do you consider it to be suitable for first class, business class or economy?
3. Suggest other vegetables to add to this meal.
4. Suggest an appropriate dessert.

Mark Fergus

9780170383417

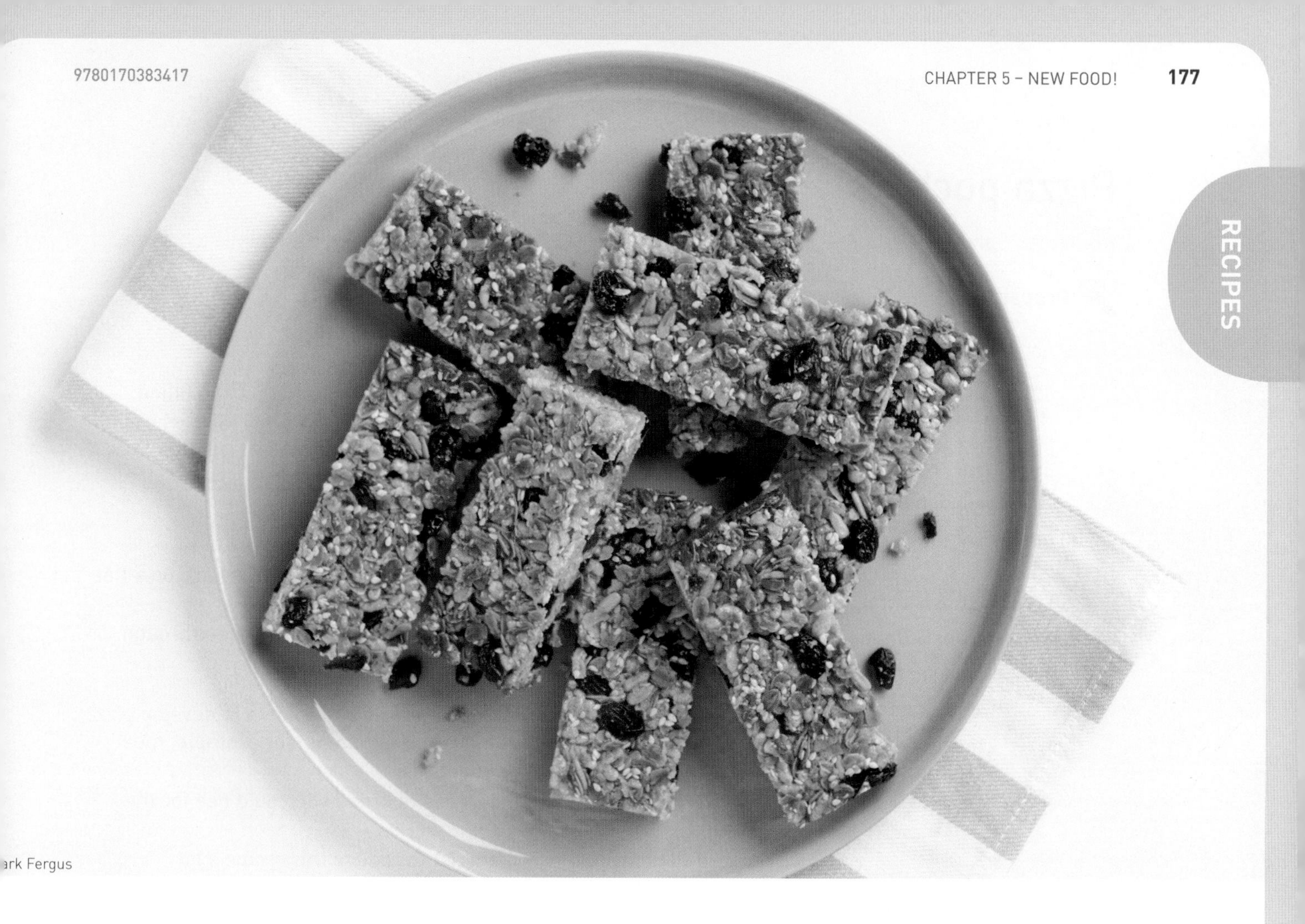
ark Fergus

School lunch muesli bars

Great snacks for school don't need to be purchased. This slice can be made, baked and ready to eat in 40 minutes!

 **Preparation time** 10 minutes  **Cooking time** 30 minutes  **Serves** 2

INGREDIENTS

- 2 cups rolled oats
- 1½ cups rice bubbles (puffed rice)
- ½ cup 'craisins' (dried cranberries)
- ½ cup sultanas
- ½ cup desiccated coconut
- ¼ cup sesame seeds
- ¼ cup sunflower seeds (or linseeds)
- 150 grams butter, diced
- ⅔ cup honey
- ⅓ cup caster sugar

METHOD

1. Preheat oven to 170°C.
2. Grease and line a lamington tin with baking paper.
3. Mix the oats, rice bubbles, craisins, sultanas, coconut, sesame seeds and sunflower seeds in a large bowl.
4. Heat the butter, sugar and honey in a saucepan, melt the butter and stir to combine all ingredients. Simmer until mixture becomes more syrupy and a honeycomb colour. Remove from the heat.
5. Stir liquid gradually into the dry ingredients and, when well combined, spoon into pan/tin. Press down well. Put into the oven for 20 minutes.
6. The mix should be golden brown on top when you take it out of the oven. Cool in the tray.

QUESTIONS

1. Cut the slice into muesli bar-sized pieces. Work out the cost per piece.
2. List five other school snacks that can easily be prepared at home.

9780170383417

Pizza pockets

Homemade pizza pockets are a healthy alternative to the processed versions commercially available.

 **Preparation time** 15 minutes  **Cooking time** 5 minutes  **Serves** 2

INGREDIENTS

- 2 round flat hamburger rolls
- 1 teaspoon butter
- 1 teaspoon pine nuts
- ½ teaspoon garlic, crushed
- 50 grams salami or low-fat ham, chopped
- 50 grams mushrooms, chopped
- ¼ green capsicum, finely diced
- ¼ teaspoon basil, finely chopped
- ¼ cup grated mozzarella
- 1 teaspoon tomato paste

METHOD

1. Split each roll carefully to create a pocket for the filling.
2. Remove some of the centre and process in a food processor to form breadcrumbs.
3. Combine the butter, pine nuts and garlic.
4. Cook in microwave on high for 30 seconds.
5. Add all other ingredients. Stir to combine.
6. Spoon the mixture carefully into pockets and place on a flat dish covered with absorbent paper.
7. Microwave on high for 1 minute or until warmed through.

QUESTIONS

1. Using the basic recipe, alter the ingredients to develop pizza pockets using traditional pizza names; for example, meat eater, Hawaiian or vegetarian.
2. Suggest three other types of bread you could use for this recipe.
3. Modify the ingredient list to reduce this recipe's fat content.

Mark Fergus

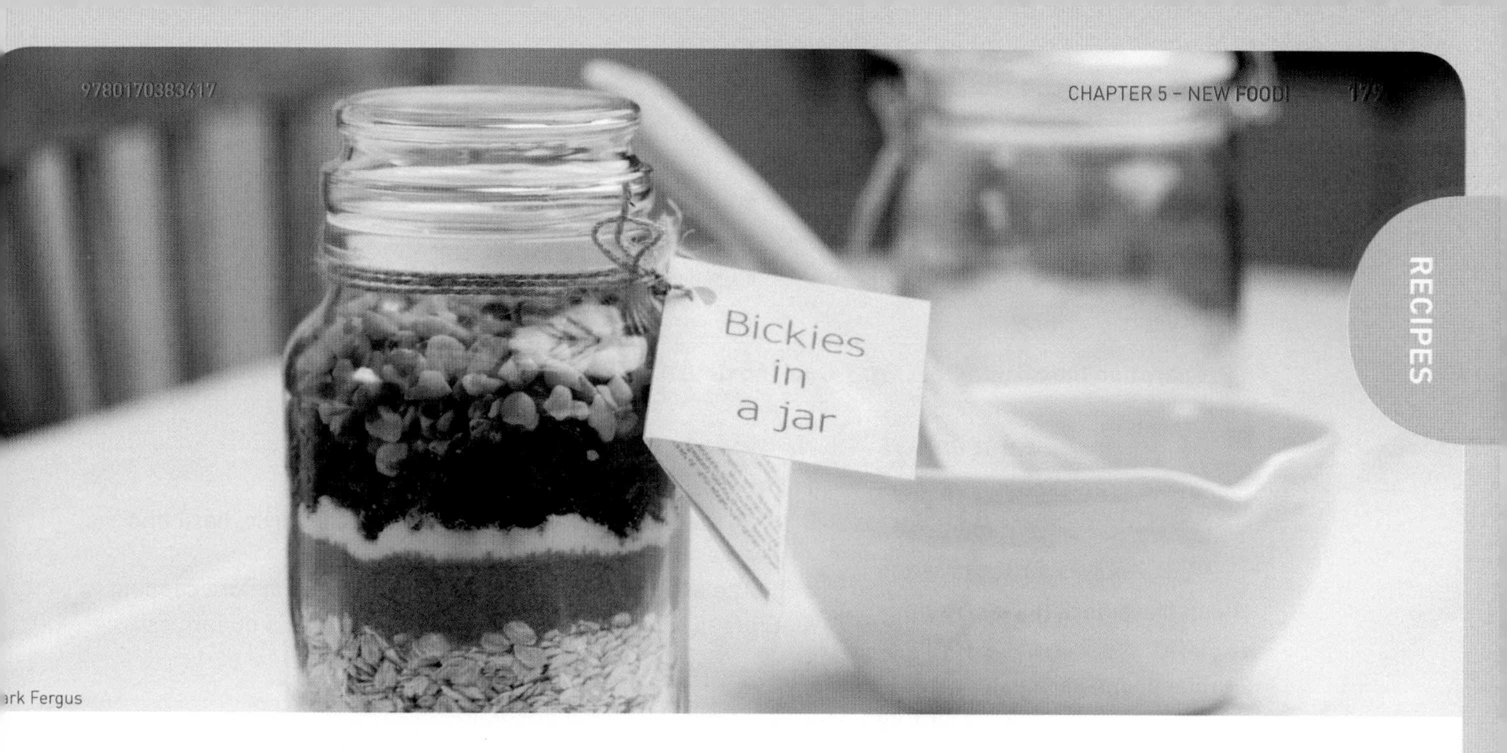

ark Fergus

RECIPES

Bickies in a jar

This is a popular gift and cake stall idea – with no cooking required!

Each student will need a clean, wide-mouthed large jar with lid, such as a large glass coffee jar. These can also be purchased relatively inexpensively from discount stores.

 **Preparation time** 20 minutes  **Cooking time** nil  **Serves** 1

INGREDIENTS

1 cup plain flour
½ teaspoon bicarbonate of soda
½ teaspoon baking powder
½ teaspoon salt
1 cup rolled oats
¾ cup packed brown sugar
1 teaspoon cinnamon
¼ cup caster sugar
1 cup raisins
1 cup chopped walnuts (optional)

Recipe adapted from http://www.homemadegiftguru.com/old-fashioned-oatmeal-cookie-recipe.html

METHOD

1. Stir together the flour, bicarbonate of soda, baking powder and salt.
2. Using a funnel, or a piece of A4 paper taped into a funnel shape, pour the flour mixture into the jar.
3. Press down the flour mixture until it won't pack any more.
4. If needed, clean the inside of the jar with a folded paper towel.
5. Add and press down oats tightly.
6. Mix the cinnamon into the brown sugar.
7. Spoon in the brown sugar and cinnamon, pressing layer firmly.
8. Pour in the caster sugar and press firmly.
9. Add the raisins and press firmly.
10. Add the walnuts and press firmly.
11. Seal with lid and decorate as desired.
12. Add label with these directions: Preheat oven to 180°C. Beat together ½ cup softened butter, 1 egg, 1 teaspoon vanilla and 1 tablespoon milk. Add Bickie Mix and blend well. Break up any clumps of mix. Drop tablespoon sized rounds of dough onto ungreased baking sheet. Bake 9–11 minutes or until edges turn golden brown.

QUESTIONS

1. Using the ingredient purchase receipts, calculate the cost of your 'Bickies in a jar'. Add the cost of packaging.
2. Set the sale price per jar. How much profit (after the ingredients and packaging costs are taken out) do you make?
3. Use word-processing software to label your jar attractively for a cake stall.

Pesto

Pesto is a product, like jams and pickles, that is made by specialty producers and sold in delicatessens and gift shops.

 **Preparation time** 5 minutes  **Cooking time** nil  **Makes** ½ cup

INGREDIENTS

- 1 tablespoon pine nuts
- ¼ cup olive oil
- 2 cloves garlic, crushed
- 1 cup fresh basil leaves, roughly chopped
- 2 tablespoons parmesan cheese, grated

METHOD

1. Process the pine nuts in a food processor.
2. With the motor running, add the olive oil, garlic, basil and parmesan cheese.
3. Bottle and cover with a thin layer of olive oil before capping.
4. Store in the refrigerator or freezer in useable quantities.

QUESTIONS

1. Why is this product popular?
2. Suggest a selling price and justify the price you have selected.
3. How could you promote this product?
4. Where could this product be distributed?
5. Design a label for this product to help make it more appealing than its competitors.

Mark Fergus

Frozen dessert

Some of the best-selling products are endorsed by famous personalities. Name this dessert after your favourite celebrity.

 **Preparation time** 15 minutes  **Cooking time** nil  **Makes** 6 muffin-sized frozen desserts

INGREDIENTS

300 mL cream
200 grams or ½ tin condensed milk
2 tablespoons coconut milk powder
125 grams frozen raspberries

METHOD

1. Line a muffin tin with 6 paper cases.
2. Combine the cream and condensed milk and beat with electric beaters until thick.
3. Divide the mixture in half.
4. Crush the frozen berries with a wooden spoon and stir into one half of the cream mixture.
5. Stir the coconut milk powder into the other half.
6. Spoon the white mixture into paper cases until half full. Finish with a layer of raspberry mixture to fill to top.
7. Freeze until set.
8. Garnish with flavoured ice-cream topping and fresh berries.

QUESTIONS

1. What other flavours could be used in this recipe? Suggest ingredients that could be substituted for the raspberries and coconut powder.
2. List any food products that you know of that are endorsed by celebrities.

CHAPTER 6

Mark Fergus

Is it good for me?

FOCUS AREA

FOCUS AREA: FOOD SELECTION AND HEALTH

This chapter looks at food and health, focusing on the role food plays in the body. You will investigate the functions and sources of food nutrients, our varied nutrient needs and factors that influence health. The chapter reveals the effects of poor eating and examines guides that are available to help Australians make decisions that aim to promote good health. You will study the factors that influence food habits and how governments, manufacturers and society can impact on nutrition levels. Foods with special health benefits will also be investigated.

CHAPTER OUTCOMES

In this chapter you will learn about:

- food and the body
- digestion
- food components
- nutritional needs
- factors influencing food habits
- nutrition and patterns of food consumption
- response to nutrition levels
- applications of food guides.

In this chapter you will learn to:

- outline functions of food
- describe digestion
- outline information on food components
- identify recommended dietary intakes throughout the life cycle
- select food for appropriate nutrients
- design and prepare foods to meet the needs of different groups
- recognise factors influencing food habits and explain their impact on food choices
- outline effects of inappropriate nutrient intake
- discuss responses to nutrition levels
- evaluate nutritional food guides
- analyse the nutritional value of foods
- modify foods to reflect food guides
- design, plan and prepare foods to reflect food guides.

RECIPES

- Plum pork and vegetable stir-fry, p. 210
- Quick vegetarian pizzas, p. 211
- Quinoa and rice salad, p. 212
- Hot, rich potato salad, p. 213
- Yakitori, p. 214
- Chunky polenta chips, p. 215

FOOD FACTS

A baby's body weight triples in its first year of life.

It takes approximately 1.5 hours for a meal to be digested in the stomach before it passes into the small intestine.

Until the 18th century, sailors were often ill during voyages because there was a lack of fresh fruit and vegetables available at sea. This resulted in a Vitamin C **deficiency** that caused scurvy, a condition that causes gums to weaken and shrink, teeth to fall out, and prevents wounds from healing. Many sailors died because of their poor health.

When villi are damaged, nutrients cannot be absorbed properly. People with coeliac disease are intolerant to gluten from wheat and have significant difficulty absorbing nutrients. Wheat irritates the villi, causing them to shrink back and flatten. Diseases of deficiency, such as **osteoporosis**, are commonly linked to this dietary problem.

Athletes fill up with huge quantities of carbohydrates, such as pasta and bread, before a marathon run. This is known as **carbo-loading** and it ensures that their energy supply will not run out during an event.

There are many different types of vegetarians. The strictest vegetarians are known as vegans. Vegans do not eat any animal products or animal byproducts, such as eggs or cow's milk. Many vegans do not consume honey.

Probiotics need cold temperatures to stay alive. If the temperature rises above 4°C, the good bacteria begin to die.

FOOD WORDS

carbo-loading ingesting large amounts of carbohydrate-rich foods in one sitting, usually before an athletic event

cholesterol constituent of some fats that sticks easily to the walls of the arteries

deficiency lack or absence of a particular nutrient

emaciated very lean or underweight through lack of food

enhance improve or build on

hormone chemical substance made by the body to perform a specific function; for example, to stimulate growth and development

hypertension very high blood pressure, often linked to a high salt intake

immunity ability to fight or resist disease or infection

legislation law passed by a government

manufacturer business involved in the making, processing or production of goods

micro-organism microscopic organism such as bacteria

osteoporosis condition of weak and porous bones due to a lack of calcium in the diet

policy management plan or strategy devised by a government or organisation

stunted restricted growth and development of the body

subsidise provide financial aid

sustenance nourishment in the form of food

6.1 Food and the body

Everybody needs to eat food to survive. Different foods give us the six nutrients that we need to maintain good health. They are the following:

1 carbohydrates
2 proteins
3 lipids
4 vitamins
5 minerals
6 water.

Without these food nutrients, you would die. All foods contain one or more of these nutrients. If you eat a balanced diet consisting of a variety of healthy foods, you are certain to receive these nutrients every day. Food has three main functions in the body:

- growth and development
- provision of energy
- repair and maintenance of the body's cells.

Growth and development

Babies, young children and adolescents grow at a rapid rate. In adults and the elderly, most growth has stopped and nutrients are mostly used for maintaining their bodies.

Your body cells must be able to grow and develop as you do, and food plays a major part in this. For example, protein is the building block for every body tissue cell, such as bone, teeth, skin and muscle. If a person is lacking protein in their diet, problems may occur, such as stunted growth.

Provision of energy

As a car needs fuel to work, your body needs food to help you carry out your daily tasks and regulate your body processes. Carbohydrates and fats, in particular, help to fuel the body, not only for physical activity but also for carrying out the functions needed for you to survive. These include digestion, breathing and the circulation of blood.

When the body runs low on energy, you feel fatigued and tired. In severe circumstances, a lack of energy may even lead to certain body processes shutting down, such as the reproductive system. On the other hand, excess energy causes fat build-up and can lead to obesity.

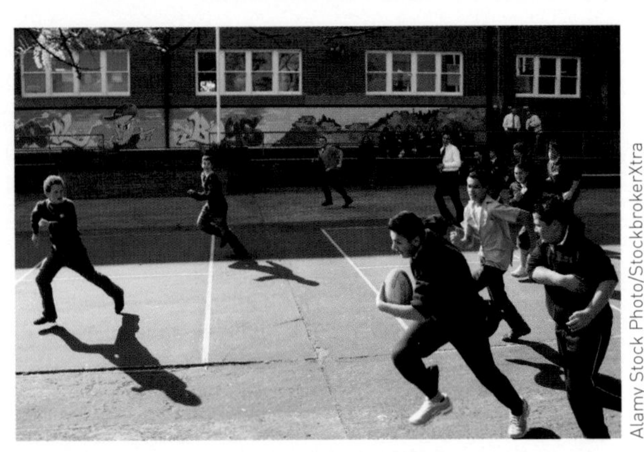

Alamy Stock Photo/StockbrokerXtra

Figure 6.1 Our bodies need fuel to help us carry out physical activity.

Repair and maintenance of cells

Achieving good health is one thing, but keeping it is another. You need to maintain your health, and you need healthy food in order to do this. For example, your skin is often cut or grazed, your hair falls out constantly and your red and white blood cells die on a regular basis. By consuming the right nutrients, your body will repair itself and stay healthy.

Similarly, you must be well enough to fight infection and disease. A healthy individual who eats well and exercises regularly is less likely to suffer from a cold or flu than a person who eats poorly and does not exercise.

UNIT REVIEW

LOOKING BACK

1 What are the three main functions of food?
2 List the six food nutrients.
3 What age groups grow at a rapid rate?
4 Which two nutrients provide most of our energy?
5 List three body tissues that are made from protein.

FOR YOU TO DO

6 Research three foods that contain carbohydrates.
7 Evaluate your diet. Identify the 'energy' foods that are always present.

TAKING IT FURTHER

8 In groups of two or three, plan and present a multimedia presentation of approximately 2–3 minutes that educates young people about the three functions of food. Use a storyboard to organise your research.

6.2 Digestion of food

The process of digestion

Digestion is the process whereby food is broken down into smaller chemical units that may be absorbed and used by the body.

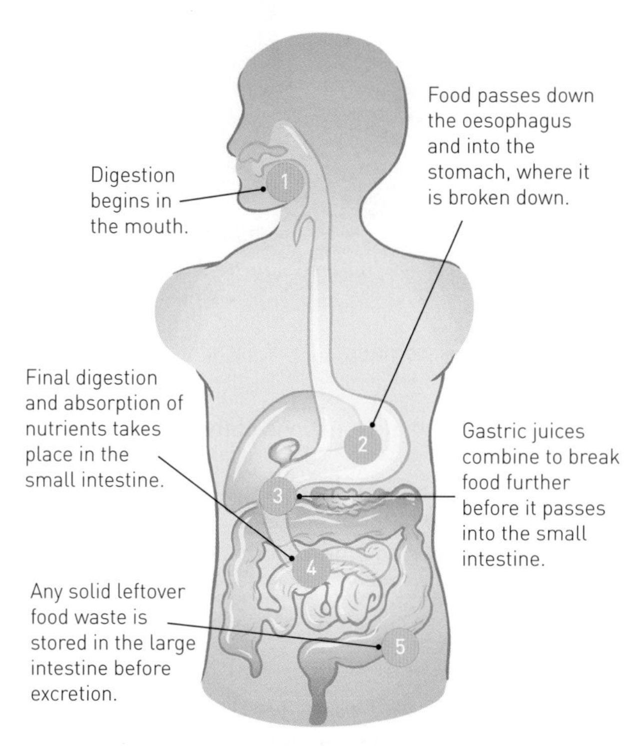

Figure 6.2 The process of digestion

The gastrointestinal tract

The gastrointestinal tract, also known as the GI tract, is where the digestion of food takes place. It begins in the mouth, proceeds through the oesophagus to the stomach, then through the small and large intestines and finally to the rectum. This entire tract measures around eight metres in length.

Absorption of nutrients

Most nutrients are absorbed in the small intestine. The villi that line the small intestine have a large surface area to allow the tiny units of each nutrient to pass through into either the bloodstream or the lymphatic system.

Carbohydrates are broken down into simple units of sugar called glucose. Protein breaks down into amino acids, and globules of lipids break down into fatty acids and glycerol. Everything is absorbed into the bloodstream, except lipids. These pass into the lymphatic system. When each tiny component of each nutrient passes into the bloodstream, they can be carried around the body to the cells that require them.

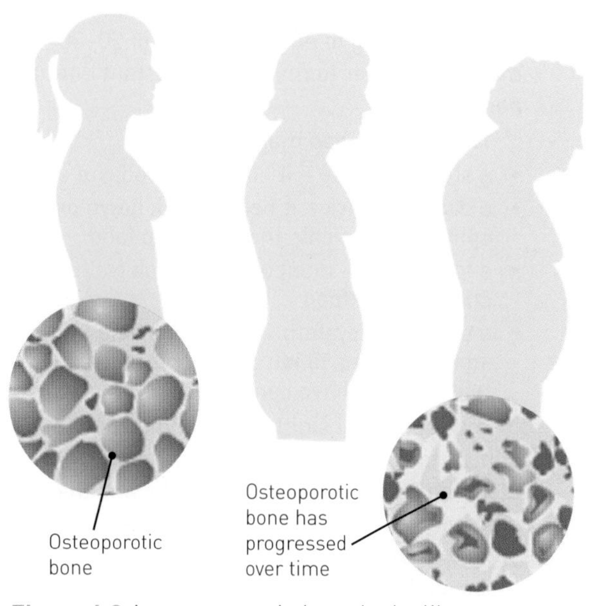

Figure 6.3 An osteoporotic bone looks like honeycomb and can lead to frequent breakages. Over time the spine may compress, fracture or crumble.

Metabolism

People digest and absorb food at different rates. The rate at which food is absorbed and digested is known as metabolism. Metabolism refers to the speed that food is broken down and used by the body. Every person has a different rate of metabolism. It can be affected by factors such as age, height, activity level and genetics. Each person must then eat food according to how quickly they digest and absorb it. Have you ever heard anybody complain of a slow metabolism?

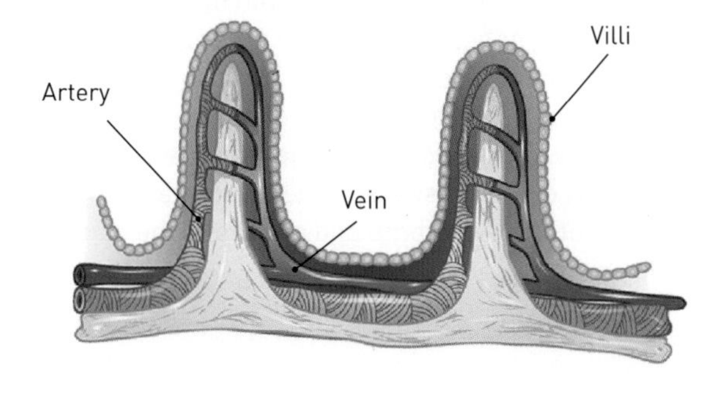

Figure 6.4 The intestine wall is lined with villi, which increase the surface area and assist with absorption.

FOOD IN FOCUS

HARD TO STOMACH

The stomach is a sack-like organ that not only breaks down the food you eat but also acts as a storage compartment. Your stomach can store two to three meals at a time. This prevents you from having to eat at intervals of 20–30 minutes in order to meet your constant energy needs.

The stomach has many functions:

- a storage device – it holds the food you eat
- a digestive sack – it helps break down or split the chemicals that make up food
- a food mixer – muscular stomach walls churn up the food
- a sterilising system – the strong acid in the stomach helps to kill any germs in the food that you may have consumed.

Your senses of sight, smell and taste can stimulate your digestive system, causing stomach acid to bubble and gurgle, and the stomach walls to contract or squeeze. The smell of food or the idea of eating causes a message to be sent to the brain, which in turn controls the release of more stomach acid.

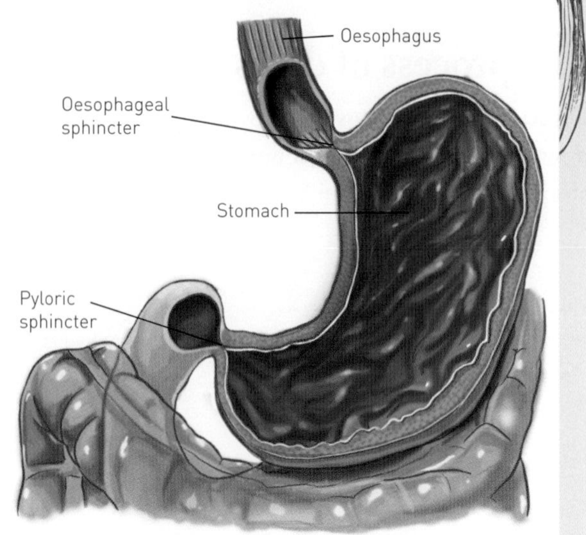

Figure 6.5 The stomach

ACTIVITIES

1. What are the main functions of the stomach?
2. Explain why it is helpful that the stomach can store two to three meals.
3. When you have an empty stomach, describe how you feel when you smell food cooking.

UNIT REVIEW

LOOKING BACK

1. How long is the GI tract?
2. Where does digestion begin and what happens at this stage?
3. At which stage does absorption take place?
4. What is metabolism?

FOR YOU TO DO

5. If someone claims to have a fast metabolism, what do they mean? Explain in your own words.
6. Pass a poster around the room with the heading 'Digestion', or get one student to act as a scribe. Provide your own definitions of digestion and display the poster for the class to see.
7. Interview someone with a digestive disorder such as coeliac disease (gluten intolerance). Find out the symptoms and causes of the disorder.

TAKING IT FURTHER

8. Select one stage of digestion.
   a. Using the internet, research and report in detail what happens to food at your selected stage of digestion.
   b. Provide your class with a summary. Use diagrams or illustrations to support your findings.
9. Evaluate the digestive process. Discover which digestive enzymes are responsible. For example, carbohydrate digestion begins in the mouth with an enzyme called amylase. Create a table with the headings 'Nutrient', 'Place(s) of digestion' and 'Enzyme(s) responsible'.
10. Research the role of the liver and pancreas in the digestive system. What do they do and where are they located?

6.3 Food components

Food nutrients perform special functions in the body. Different foods contain different amounts of each nutrient. Nutrition education teaches you what combinations of foods to eat in order to receive the correct balance of nutrients. Look at the table to see the functions and food sources of each nutrient.

Shutterstock.com/Jakub Zak

Figure 6.6 Athletes consume carbohydrate-rich foods to fill their cells with glucose energy.

Table 6.1 Nutrients

NUTRIENT CHARACTERISTICS	FUNCTIONS IN THE BODY	FOOD SOURCES
Protein There are two main types: 1 complete proteins 2 incomplete proteins. Complete proteins contain all of the nine essential amino acids required from food because the body cannot make them. Incomplete proteins lack at least one of the nine essential amino acids.	• Constituent of all body cells. • Building and repair of all hard and soft body tissue such as bone, teeth, muscle tissue and organs. • Required to make enzymes, antibodies, hormones and haemoglobin.	• Complete protein is found mainly in foods of animal origin: meat, fish, eggs, milk, cheese and yoghurt. • Vegetable sources mainly provide incomplete protein: nuts, pulses, seeds, wholegrain cereals and vegetables.
Carbohydrates There are three types of carbohydrates: simple sugars (monosaccharides and disaccharides), complex carbohydrates and dietary fibre. Simple sugars and complex carbohydrates are broken down into glucose during digestion. This is used by the cells of the body for energy.	• Source of energy in all cells of the body. • Glucose, which is not used by the body for energy, is converted into glycogen and is stored in the liver and muscles as an energy reserve. • Dietary fibre is not absorbed by the body. It adds bulk and helps to sweep the digestive tract. • Carbohydrates provide 16 kJ of energy per gram.	• Complex carbohydrates are usually high in dietary fibre, and include wholegrain breads and cereals, rice, pasta, breakfast cereals, vegetables, legumes and fruits. • Foods supplying simple sugars include cakes, pastries, desserts, confectionery and aerated drinks. These should be consumed in limited amounts.

NUTRIENT CHARACTERISTICS	FUNCTIONS IN THE BODY	FOOD SOURCES
Lipids Lipids (also known as fats) come from both animal and vegetable sources. Animal fats are usually solid, high in saturated fats and should be eaten in limited amounts because of the high level of cholesterol. Fats from vegetable sources are usually liquid (oils) and provide mostly monounsaturated and polyunsaturated fatty acids, which do not increase cholesterol levels.	• Lipids are the most concentrated source of energy, providing the body with 37 kJ per gram, more than twice the amount of energy provided by protein or carbohydrates. • Fats are an important source of fat-soluble Vitamins A, D, E and K. • Omega-3 fatty acids are needed for development of the brain and retina of the eye. • Omega-6 fatty acids are needed for the formation of all cells of the body.	• Saturated fats: butter, lard, meat, dairy products, palm oil, coconut oil and chocolate. • Monounsaturated fatty acids: olive oil, canola oil, avocado and nuts. • Polyunsaturated fatty acids: most vegetable oils, oily fish and fish oils. • Omega-3: soy bean and canola oil. • Omega-6: olive and sunflower oils.
Vitamins There are two main groups of vitamins: the fat-soluble Vitamins A, D, E and K and water-soluble Vitamins B and C. Most vitamins cannot be made by the body and must be found in food. Vitamins are micronutrients because they are only required in small amounts.	• Vitamins are involved in a range of essential functions, including normal cell division and growth, the absorption of other key nutrients, the production of hard and soft tissue, the production of energy, healing wounds and burns and the development of good eyesight.	• Fat-soluble vitamins are found in fish oils, milk, butter, cheese, eggs, vegetable oils, nuts and green leafy vegetables. • The B Group vitamins are found in dairy products, lean meat, fortified breakfast cereals and vegetables. • Vitamin C is found in citrus fruits, strawberries, kiwi fruit and capsicum.
Minerals Minerals are only required in minute amounts in the body. Some Australians' diets are deficient in iron, calcium and zinc.	• The main minerals required by the body are calcium, iron, chloride, phosphorus, magnesium, potassium, sodium, sulphur and zinc. • Minerals perform many functions in the body. For example, calcium is essential for the formation of bones and teeth, iron is essential for the formation of haemoglobin in the blood, and chloride acts with potassium and sodium to maintain fluid and electrolyte balance in the body.	• Calcium: milk, yoghurt, cheese, nuts, fish and soy beans. • Iron: red meat, oysters, fish, chicken, breakfast cereals, lentils, vegetables and dried apricots. • Magnesium: wholegrain cereals, nuts, pulses and green leafy vegetables. • Potassium: plant foods such as avocado, nuts, seeds and wholegrain cereals. • Zinc: shellfish, whole grains, lean meat, poultry, eggs and dairy products.

Source: Adapted from G. Heath, H. McKenzie & L. Tully, *Food by Design Third Edition*, Cengage Learning Australia, 2015, pp. 55–7

UNIT REVIEW

LOOKING BACK

1 List three protein foods.
2 What are the functions of vitamins?
3 List two important minerals and identify foods they are found in.
4 List three food sources of carbohydrates.

FOR YOU TO DO

5 Investigate two foods, other than those listed in this unit, that are good sources of Vitamin C.
6 Create a webpage on 'The six food nutrients'. Use software such as Adobe Dreamweaver or Microsoft FrontPage and work in pairs. Link at least one page to each nutrient and include a variety of information, pictures and interesting facts. Present this to the class using a data projector or save it on a local network for your classmates to see.

TAKING IT FURTHER

7 Research the effects of consumption of an inadequate quantity and/or quality of protein. Use a variety of nutrition-based websites to assist your research. TIP: Use the search term 'protein deficiency'.

6.4 Nutritional needs

Factors affecting nutritional needs

Each person requires different quantities of each nutrient, depending on factors such as age, gender, size, activity level, health and hunger.

Age

Children and adolescents are growing rapidly, so it is understood that they require large amounts of nutrients, such as protein for new body cells and carbohydrates for supplying energy. Elderly people must maintain their body cells, particularly their bones and muscles. In this case, minerals such as calcium are especially important.

Shutterstock.com/goodluz

Figure 6.7 Older people must consume calcium-rich foods to maintain bone density.

Gender

Men are generally more active than women and larger in size, so they require more energy foods that supply fat and carbohydrate. On the other hand, women menstruate every month and may undergo childbirth, so their iron requirements are often greater than those of men.

Size

Larger people require more nutrients in order to function normally. Smaller-framed people do not need to carry a heavy body around, so need less food. Think about a car and its petrol usage. A small hatchback requires only a small amount, while a station wagon requires a large quantity.

Activity level

In order to be active, people require food as fuel. A person who lives an inactive lifestyle requires fewer nutrients than a physically active person needs. If you were an Olympic swimmer, you would need to eat a lot more food to be able to meet the physical demands placed on you.

Health

People who have diet-related health concerns could require more or less of certain nutrients. For example, those who suffer from osteoporosis must consume a larger than normal intake of calcium. Those with anaemia must consume larger quantities of iron. Those suffering heart disease or high **cholesterol** must reduce their intake of saturated fat.

Hunger

The feeling of hunger is the signal that is sent from the brain to tell the body that it needs to receive **sustenance** in order to perform properly. When your body becomes hungry you begin to feel tired, sick, irritable and empty in the stomach. Every individual feels hunger at their own rate. Have you ever felt very hungry at morning tea and wanted to eat your lunch then? Hunger is not the same as appetite. Appetite is the desire to eat, even when you are not hungry!

Other special circumstances

In some circumstances an individual may need more or fewer food nutrients. Consider a pregnant woman who is required to provide nutrients for a growing foetus. The body provides this need by allowing for the delivery of nutrients to the baby via the mother's bloodstream. Calcium, protein, B Group vitamins and minerals are essential for the mother at this stage. She must ensure that she consumes the correct quantities of each nutrient to avoid deficiency in herself and her baby.

Recommended dietary intakes for various life stages

As you grow and develop you require different quantities of each nutrient. Your requirements depend not only on the factors already discussed, but also on the stage of the life cycle that you may be at.

People need to have a general idea of how much of each nutrient they should be consuming. The specific quantities of each nutrient needed for each group of people are called recommended dietary intakes (RDIs). These are the levels of daily intake of essential nutrients needed to meet the nutritional needs of the body. Look at RDI tables for a few selected nutrients. In which group do you belong? What are your particular nutrient needs?

The Energy RDIs shown in Table 6.2 are all based on an individual with moderate physical activity levels.

Note that the RDI is the average daily dietary intake level that is sufficient to meet the nutritional requirements of **nearly all** healthy individuals in a particular stage of life and gender group, including macronutrients, vitamins and minerals.

Table 6.2 Recommended dietary intakes

INFANT (7–12 MONTHS, ON A SOLID DIET)		
Protein	14 g/day (1.6 g/kg body weight)	
Energy	2500–3500 kJ/day (EER[1])	
Calcium	270 mg/day	
Iron	11 mg/day	
Magnesium	75 mg/day	
Vitamin B1	0.3 mg/day	
Vitamin C	30 mg/day	
YOUNG CHILD (4–8 YEARS)		
Protein	20 g/day (0.91 g/kg)	
Energy	6.1–8.2 MJ[2]/day	
Calcium	700 mg/day	
Iron	10 mg/day	
Magnesium	130 mg/day	
Vitamin B1	0.6 mg/day	
Vitamin C	35 mg/day	
ADOLESCENT (14–18 YEARS OF AGE)		
	Male	**Female**
Protein	65 g/day	45 g/day
Energy	11.9–14 MJ/day	10.3–10.9 MJ/day
Calcium	1300 mg/day	1300 mg/day
Iron	11 mg/day	15 mg/day
Magnesium	410 mg/day	360 mg/day
Vitamin B1	1.2 mg/day	1.1 mg/day
Vitamin C	40 mg/day	40 mg/day
ADULT (19–30 YEARS OF AGE)		
	Male	**Female**
Protein	64 g/day	46 g/day
Energy	11.6–15.2 MJ/day	9.2–12.5 MJ/day
Calcium	1000 mg/day	1000 mg/day
Iron	8 mg/day	18 mg/day
Magnesium	400 mg/day	310 mg/day
Vitamin B1	1.2 mg/day	1.1 mg/day
Vitamin C	45 mg/day	45 mg/day
PREGNANT (SECOND AND THIRD TRIMESTER) OR BREASTFEEDING (19–30 YEARS OF AGE)		
	Pregnant	**Lactating**
Protein	60 g/day	67 g/day
Energy	+1.4–1.9 MJ/day	+2.0–2.1 MJ/day
Calcium	1000 mg/day	1000 mg/day
Iron	27 mg/day	9 mg/day
Magnesium	350 mg/day	310 mg/day
Vitamin B1	1.4 mg/day	1.4 mg/day
Vitamin C	55 mg/day	85 mg/day

OLDER PEOPLE (70+ YEARS)		
	Male	Female
Protein	81 g/day	57 g/day
Energy	9.4–12.2 MJ/day	8.3–10.4 MJ/day
Calcium	1300 mg/day	1300 mg/day
Iron	8 mg/day	8 mg/day
Magnesium	420 mg/day	320 mg/day
Vitamin B1	1.2 mg/day	1.1 mg/day
Vitamin C	45 mg/day	45 mg/day

[1] EER = Estimated energy requirement, which can vary based on weight, height, activity level and so on.
[2] A unit of energy is the kilojoule (kJ) or megajoule (MJ). 1 MJ = 1000 kJ.

Source: Adapted from the National Health and Medical Research Council website: **www.nhmrc.gov.au**

People with special nutritional needs

Many people in society have special nutritional needs that do not necessarily correspond with the RDI tables in Table 6.2. Some special nutritional needs may be determined by factors such as health, beliefs or values.

Vegetarians do not eat flesh foods, such as red meat and poultry. The reason for their dietary requirements may be religious or based on their own beliefs, attitudes and health ideas. These people will have special dietary needs for protein, iron and B Group vitamins because flesh foods are completely absent from their diet.

Obesity sufferers have a high percentage of body fat and are required to cut down on food intake and increase exercise levels to try to overcome this nutritional disorder. These people may require a reduced intake of kilojoules in their diet.

HANDS-ON

PLANNING DIETS FOR DIFFERENT PEOPLE

PURPOSE

This activity is designed for you to use your research skills to investigate the nutrient needs of a particular group. Using this information you will then design, prepare and justify a nutritionally balanced meal.

MATERIALS

RDI tables

Recipe books

STEPS

1. Study the RDI tables for your selected group and determine their nutrient needs.
2. Plan a complete lunch or dinner menu. You should set two to three courses and include beverages.
3. Select one part of this menu to prepare in class. Fill out a food order and complete the practical in class time.

ACTIVITIES

1. In one to two paragraphs, summarise the nutrient needs of your groups.
2. Justify your choice of each food on the menu. Hand this to your teacher for marking.
3. Evaluate the success of your practical lesson. What went well? What looked good? What would you change or improve? Did you have any difficulties?

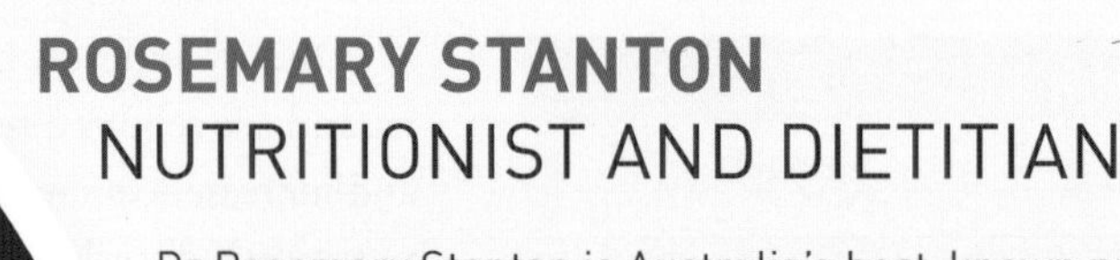

ROSEMARY STANTON
NUTRITIONIST AND DIETITIAN

Dr Rosemary Stanton is Australia's best-known nutritionist. She is the author of over 30 books, including nutrition textbooks as well as recipe books. She has also written over 3500 articles in magazines and newspapers throughout Australia and New Zealand, features frequently on radio and television and lectures to medical students, doctors, teachers, nurses, sporting groups and to the general public.

Rosemary has a science degree in biochemistry and pharmaceutical chemistry, postgraduate qualifications in nutrition and dietetics and a graduate diploma in administration.

As a practising nutritionist, Rosemary has a consulting business that services clients such as State and Commonwealth government departments, sports associations, primary industry groups, sections of the food industry and a major retailer. Her role within these groups is to further the aims of educating the public about food and nutrition. She has a close relationship with customers and their viewpoints, and is a spokesperson that is not beholden to any one interest group.

Rosemary Stanton was involved in one of the first school canteen surveys in NSW in the 1960s. She believes that school canteens are ideal to foster healthy eating habits in children.

With the growth in obesity levels in Australia, Rosemary is trying to educate people to stop eating so much processed junk, eat less and, of course, be more physically active. Rosemary doesn't agree with dieting, as she believes that a temporary diet or eating plan won't keep weight off in the long term, and that people need to change their eating and exercise habits permanently.

As a member of many professional associations such as the Dietetics Association of Australia, the Nutrition Society, the Australian Institute of Food Science Technology, the Gut Foundation and the Australian Health Communicators Association (just to name a few), it is Rosemary's main aim to change the poor eating habits of Australians so that people have healthier diets and eat more enjoyable foods that create minimal environmental damage.

Source: Adapted from information on the AIHW website and ICMI.com.au.
Photo: Fairfax Photos/Steven Baccon

ACTIVITIES

1. Give details of how Rosemary Stanton educates the public. Explain why her role is important.
2. Why do you think Rosemary does not believe in diets?
3. Interview friends and family to ask if and when they have heard of Dr Rosemary Stanton. Provide examples of food- and nutrition-related issues she has been involved in.
4. Create a report on one of the professional associations or consulting groups that Rosemary is involved with.

UNIT REVIEW

LOOKING BACK

1 List the seven factors that affect nutritional needs.
2 Explain how age affects our nutritional needs.
3 What does RDI stand for?
4 Give another term for lactating.
5 How much protein does a pregnant woman need to consume?

FOR YOU TO DO

6 Explain why women might need more iron than men.
7 Design and print an information sheet for a medical practice that will help vegetarians select a healthy, nutritious and balanced diet.
8 Create a table in your notebook showing each major life stage and a summary of the nutrient needs of people at that stage in life.
9 Analyse your current canteen menu. Write a proposal to send to the principal or canteen manager expressing the need to update the canteen menu. Make sure you comment on the poor nutritional value of any unhealthy foods that are sold and suggest healthy alternatives.

TAKING IT FURTHER

10 The RDI tables refer to 'EER'. Write an extended response to explain how and why EERs can vary from person to person, and why an exact requirement is difficult to determine.
11 Select 10–15 meals that would be nutritious enough to serve infants at a childcare centre. Plan a weekly lunch menu that may be used to provide these infants with a balanced, nutritious diet each day.
12 Plan a new canteen menu for your school. Ensure that foods are nutritious and exciting. Try not to include any junk foods.
13 Present an oral report on the special nutritional needs of someone with anaemia. Research this by using a variety of nutrition-based websites.

6.5 Factors that influence food habits

Many factors influence what and when we eat. Think about your own eating habits. What factors influence how you select food to eat? Discuss this as a class and brainstorm your ideas.

Social practices

The society in which you live has a great impact on your eating patterns and habits. Social influences can be seen in a variety of situations. For example, when your family eats dinner, you often eat what they eat and when they eat. Some families have take-away every Saturday night, while others have a baked chicken dinner early every Sunday evening. However, the trend of eating together as a family is changing because of our modern lifestyles and personal activities.

Peer pressure in society can affect when and what you eat. If your friends are eating something interesting from the canteen at morning tea, you may be influenced to purchase the same thing.

In many countries, people eat a large lunch that is followed by a nap or short sleep. In the evening, only a small meal is eaten.

Religion

Religious ties can have a direct impact on the food choices and habits of an individual. Religious restrictions or taboos may dictate food habits. For example:

- Jews do not consume meat and milk in the same meal
- Jews and Muslims do not eat any pork products
- Muslims fast for one month every year
- Hindus do not eat beef
- Christians do not eat meat on Good Friday.

Geographic location

Where you live has a great impact on your food habits. Consider the following examples:

- Certain foods cannot be grown in certain areas. Strawberries cannot be grown in areas that encounter cold frosts, as such conditions cause damage to the quality of the fruit. Wheat grows in hot, mostly dry conditions.
- Some isolated places do not get regular fresh food deliveries. If you lived in the desert of Central Australia, you would be lucky to receive a delivery of fresh fruit and vegetables once a fortnight!

- Climates in certain areas restrict food availability. Tropical fruits are available in warm conditions, but are hard to obtain at any other time of the year. It is unlikely that you would be eating fresh mangoes in the middle of winter, as the supply would be limited and the cost would be prohibitive. Drought or flood in an area can affect foods available. This could have a great impact on food supply and food cost.
- The landforms, soil types and frequency of rain in all continents of the world are different, and the conditions in each region will dictate which foods can be grown successfully.

Mark Fergus

Figure 6.8 Vegetarians consume various types of foods to meet their dietary needs.

Economic situation

The financial state of a country and employment levels for its population will determine the foods available to people. In some countries, some people can only afford to eat once per day, or even less, while in Australia we can generally afford to eat three good meals each day. The type of job you have and the amount of money that you earn will affect what you tend to eat. If you earn a lot of money you have the option to select fillet steak over chuck steak, brie over cheddar cheese and lobster over leatherjacket.

Some countries are unable to grow certain food crops, so must buy or import these foods from other countries. They are usually expensive to purchase.

If a country is at war, government money may be directed towards military uses rather than the provision of food. This then limits the food available to people.

iStock.com/FtLaudGirl

Figure 6.9 Mangoes grow best in warm, tropical climates. This is an example of a climate and geographic location that could influence food habits.

FOOD IN FOCUS

UNDERSTANDING WHAT REALLY MAKES US SICK

BIANCA NOGRADY

The lifestyle choices you make, such as diet (and) exercise, have a huge impact on your health. Yet most of us know nothing of the social factors that drive these.

When it comes to healing the sick, we look to doctors. When it comes to preventing us from getting sick in the first place, many say we should look to governments.

That's because the vast majority of our biggest killers such as heart disease, diabetes, cancer (etc.) are significantly affected by where we live, where we work, our income, our education, our socioeconomic status and our lifestyle.

These are called the 'social determinants of health' and leading health experts say they impact on our health more than anything else.

Wealth determines health - The World Health Organisation defines social

determinants of health as the conditions in which we are born, grow up, live, work and age, and these are shaped by the distribution of money, power and resources at global, national and local levels.

These are factors such as income, employment status, access to education, access to healthcare, access to affordable housing, transport, stress, age and disability.

Impact on risk factors - Social determinants affect your health by impacting on risk factors that can lead to chronic disease and poor health.

'People who are on lower incomes have higher rates of a whole lot of risk factors, be they tobacco, alcohol, substance abuse, obesity, these sorts of things.'

'Poor housing impacts overcrowding, so we see a greater incidence of skin infections and greater impacts of mental health and substance abuse...'

'Lack of education ... impacts directly on health but would seem to be related to poorer health outcomes generally, so access to employment etc.'

Understanding poverty's impact - There is no question poverty is bad for your health and in ways you might not always realise. For instance, it often means people can't buy healthy foods such as fresh fruit and vegetables. This in turn is associated with increased risk of diseases such as heart disease and some cancers.

Living in rural areas - Where we live also has a major impact on our health, and evidence shows that living in remote or isolated settings puts many of us at a significant health disadvantage.

Indigenous Australian living in rural and remote areas are at an even greater health disadvantage than non-Indigenous Australians.

Rural communities also have poorer access to health services.

Bad place to be - Being on the bottom rung when it comes to social determinants of health is a bad place to be, and it can be very difficult to climb out of the trap. The solutions are far harder than treating the end results.

'Simply telling people to do more [exercise or healthy eating] is not going to make a difference. The feasibility of doing more is driven by something they don't have as much control over, and here you need governments and business and policy all having to come together'.

Source: Understanding what really makes us Sick, Bianca Nogrady. From ABC Health & Wellbeing, 14 May 2015.

ACTIVITIES

1. What are some risk factors for people on lower incomes?
2. How might living in a rural area impact on your health?
3. Write an extended response addressing how a lack of education could have an impact on a person's health. Make sure you plan a structured response with an introduction, body and conclusion.

Technological developments

The level of technology that is available has a great impact on our food habits. Consider the following:

- Farming technology means that food can be grown, harvested and processed faster than ever before. Food is then easier to obtain and less expensive in the long run.
- With the introduction of microwave ovens and microwaveable meals, foods can be just heated and eaten. This technology has changed family eating habits. Can you see how?
- Appliances such as breadmakers, thermal cookers, coffee machines, blenders and sandwich grills enable you to produce fresh, restaurant-quality meals in your own home, quickly and easily. Why buy bread from the local supermarket if you can bake your own bread and have it fresh every morning without leaving home?
- Internet shopping technology has meant that people do not have to leave home to purchase food. You can select your favourite fresh and processed foods online and have them delivered to your door.

Individual preferences

Your individual preferences are shaped by many factors.

- **Your past experiences:** Past experiences can have a great impact on what you do and don't want to eat. Were you force-fed a certain food as a child? Do you like eating a particular food because it reminds you of an enjoyable event?
- **Your ideas and beliefs:** Some people do not eat meat because they have strong beliefs in animal rights. Others eat lots of fresh foods and limit fatty and sugary foods because they wish to keep their bodies in a state of good health.
- **Your health:** Many people suffer from allergies and intolerances, and this will in turn affect what they choose to eat. If you suffer from a peanut allergy, you would obviously not consume a spread such as peanut butter.
- **Your individual needs:** Every individual is different and has their own special needs. For example, some people need to consume more water and fibre than others in order to keep their bodies functioning.

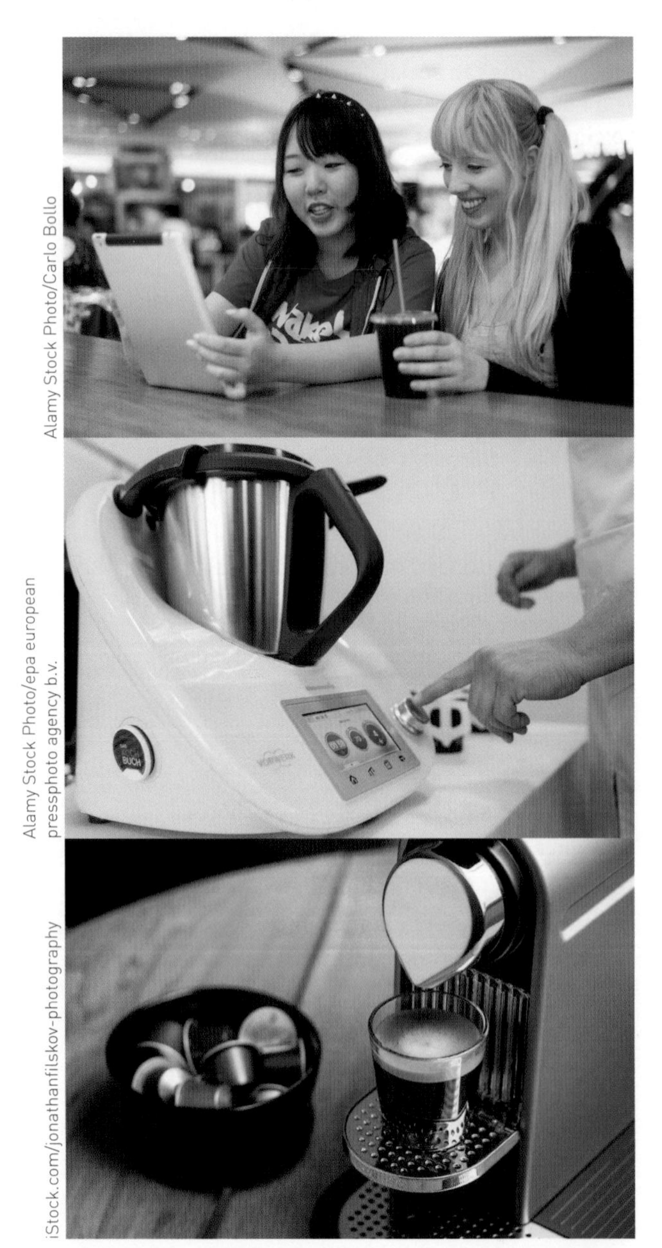

Alamy Stock Photo/Carlo Bollo

Alamy Stock Photo/epa european pressphoto agency b.v.

iStock.com/jonathanfilskov-photography

Figure 6.10 Technological developments have had a great impact on our food habits.

Mass media

The media and the companies who control them are responsible for advertising, marketing and reporting on issues in today's society. You cannot avoid being exposed to persuasive marketing through television, magazines, newspapers, radio and the internet. Slogans and jingles are always there to remind you of particular foods. Multinational food companies such as McDonald's advertise in public locations, billboards and bus stops, so that you are constantly aware that these foods are available.

Advertising makes use of unrealistic and manipulated body images, mainly in magazines and television targeted at teens and young adults. A slim model is often shown eating food such as pizza or fried chicken. Many models are slimmer or much slimmer than the average person, causing teenagers to compare themselves and become unnecessarily concerned about their body shape and size. Some may diet and cut out foods containing key nutrients, such as iron and carbohydrates, in an effort to lose weight.

iStock.com/Kevin Russ

Figure 6.11 Poor dietary habits are portrayed in the media, particularly in magazines and on television.

UNIT REVIEW

LOOKING BACK

1. List seven factors influencing food habits.
2. Give an example of how society influences food habits.
3. Explain how religion affects food habits.
4. Give an example of how individual preferences might influence eating habits.
5. Divide a poster into seven sections. Head each section with one of the seven factors influencing food habits. Under each heading, provide a description and one to two examples.

FOR YOU TO DO

6. Name one food that is not available in a certain geographic area.
7. Provide your own example of how you think the media might influence eating habits.
8. Create a collage depicting how the media presents distorted body images to young people.
9. Design a list of instructions to help an elderly person shop on the internet. You may need to first investigate an online supermarket such as Coles or Woolworths. Do not log on or register during this activity.

TAKING IT FURTHER

10. Investigate three other technological appliances that are now available to us that could alter our food habits.
11. 'Our food habits are determined by many factors. Every individual around the world responds to different influences that affect what, when and how they eat.' Write a two-page response to this statement. Make sure you plan your answer first. Find the key words, use examples to support your argument, and use your own words.
12. Investigate new technologies in food manufacture and food packaging. Select one relatively new technology for each and provide a one-page summary sheet to be handed out to your class. Use a variety of food magazines, journals and internet sites to assist you.

6.6 Nutrition and patterns of food consumption

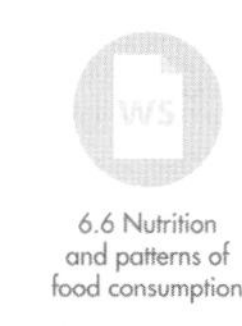

6.6 Nutrition and patterns of food consumption

Undernutrition and over-nutrition

We must be able to find the correct balance when it comes to food quantities. Having just the right amount of food is the key to good health. Too much or too little food can lead to some serious health issues.

Most people in our country have enough money to supply their basic dietary needs. Undernutrition in Australia is often the result of conditions such as anorexia nervosa, anaemia, osteoporosis and bowel disorders such as diverticulitis.

Undernutrition is commonly seen in developing countries, where people suffer from a lack of nutrients such as protein and energy from carbohydrates. They have difficulty meeting their dietary needs and are underweight and **emaciated**.

Alamy Stock Photo/René van den Berg

Figure 6.12 Supermarket aisles dedicated to snack foods are an indicator of our current food consumption patterns.

The main problem that Australians increasingly face today is excessive consumption of food, or over-nutrition. We are constantly bombarded with advertising for exciting, convenient foods to try. It is important to stay focused and take control of what you eat. Overweight, obesity, heart disease and diabetes are becoming serious health issues in Australia. Modern lifestyles, sedentary forms of entertainment such as gaming consoles and DVDs, cars, the internet, fast foods and a lack of education all contribute to the problems.

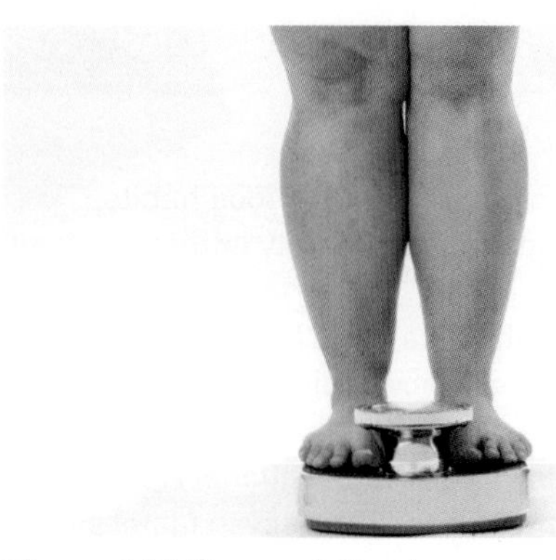

Figure 6.13 Over-nutrition is common in Australia today.

FOOD IN FOCUS

WHAT CAN PARENTS DO ABOUT CHILDHOOD OBESITY?

SARAH LE MARQUAND

Fat free. Sugar free. Sometimes foods. Forbidden foods. Encourage your kids to exercise and roam the neighbourhood – but do not let them out of your sight.

Is it any wonder the average parent is deeply confused?

Despite assumptions that tackling childhood obesity is simply a matter of 'common sense', research has found it is a little more complicated than that, and modern parents need guidance as they seek to instil healthy lifestyle habits in their offspring.

According to a recent review from the Murdoch Children's Research Institute, childhood obesity has doubled in prevalence since the 1980s, and now sits at more than 25 per cent.

'This is the first generation of children where the parents will outlive the children', says Murdoch Children's Research Institute Ambassador and board member, Sarah Murdoch. 'All of these diseases are preventable.'

The new research, published in the *Journal of Paediatrics and Child Health*, looks at obesity trends over the past 50 years and concludes that although obesity rates are stabilising in many developing countries, including Australia, an up-swing in the rates of severely obese adolescents is predicted.

'What I took from the paper was that we need to re-educate ourselves on what we feed our children and what we need to focus on,' says Murdoch. 'The good news is that it is plateauing so the awareness has come through. So how do we help children as they become adults and their bodies can be reset? The younger we get into it the better – the severity of this is scary.'

As to what this means for the well-meaning mums and dads, Murdoch is first to admit navigating a child through a balanced diet can be a minefield.

'There are so many myths at the moment and parents are so confused about what to feed their kids,' she says. 'We have to get people back on track with regard to food and portion size, but I think there is more to it than that. It is not just lazy children or people eating too much – it has become a disease and these kids are at risk of dying.'

Enter Step-a-thon, a national campaign targeted at primary school students that aims to educate children about the Importance of being active and – quite literally – making every step count as they strap on a pedometer to raise money for child health research.

'There are only 1 in 10 children getting the correct amount of exercise a day, so for us it is about getting kids to enjoy exercise,' says Murdoch.

Like many of the solutions put forward when addressing childhood obesity, the premise behind Step-a-thon appears deceptively simple. Deceptively so, that is, because the seemingly obvious need to remain active has become complicated in a time-poor, electronic-dependent era.

'I remember riding my bike around the street as a kid but now there are iPads, and trying to get them off them and outdoors is a struggle for parents,' says Murdoch.

Source: What can parents do about childhood obesity? Sarah Le Marquand, The Daily Telegraph 16 August 2015.

ACTIVITIES

1 What is the current rate of childhood obesity?
2 Explain the term 'portion size'.
3 What is a 'step-a-thon'?
4 How many children get the correct amount of daily exercise/activity?

HANDS-ON

OVER-NUTRITION

PURPOSE

To investigate dietary conditions relating to over-nutrition.

You will select one condition, such as diabetes, and prepare a PowerPoint presentation aimed at educating Australians on the condition, how it occurs and how it can be treated through good dietary practices.

MATERIALS

Textbook
Pen
Paper
Library nutrition resources such as books, magazines and journals, internet access and access to PowerPoint

STEPS

1 Select a condition of over-nutrition and research it using a variety of sources. Record the details of each source because you will need to prepare a bibliography at the end of your presentation.
2 Determine which information is interesting and informative enough to include in your presentation.
3 Start working in PowerPoint and use a combination of text and graphics. Experiment with sound and special effects.

UNIT REVIEW

LOOKING BACK

1 What is the main nutritional problem in Australia today?
2 Where is undernutrition mostly seen?
3 Why are many Australians overweight and obese?

FOR YOU TO DO

4 In pairs, design a brochure aimed at educating Australians on how to eat the right quantity of the right foods. Make it as interesting and colourful as possible. Print the brochure, in colour if possible, and display it in your classroom.

TAKING IT FURTHER

5 List some diet-related disorders that might be linked to over-nutrition.
6 What other health problems could occur if a person suffered from undernutrition?

6.7 Response to general nutrition levels

Social, political and manufacturing directions

Society's response

As a society, we must take responsibility for the foods we eat and our state of health. We ultimately decide what goes into our mouths. As the level of health is falling nationwide, we must try to be more conscious of what we should and shouldn't eat.

Society needs to become more educated about healthy foods. Speaking up about nutrition issues is a way to show that we are dissatisfied. If healthy food is not made available to us, we have no option but to select from unhealthy alternatives.

Consider the school canteen. If the foods available to students include fast foods such as meat pies, crisps, soft drinks and sweet pastries, good, wholesome foods are often crowded out of the canteen menu. Students have no other choice but to purchase these less nutritious items.

amanaimages/AGE/John Birdsall

Figure 6.14 Unhealthy snack foods are readily available to some children at school.

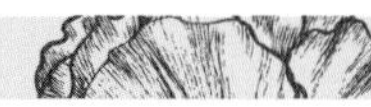
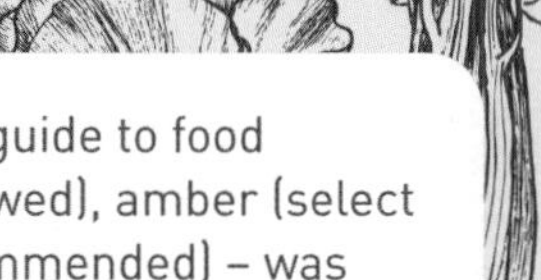

FOOD IN FOCUS

GOVERNMENT ORDERS REVIEW OF SCHOOL CANTEEN MENUS AMID OBESITY CRISIS

BRUCE MCDOUGALL

School canteen menus are set for a major overhaul amid revelations that many serve 'junk' food to children, including lollies, cakes and soft drinks, in defiance of state-imposed health guidelines.

A review has been ordered of the canteen food sold daily in schools across the state as a large number ignore bans on sugar-sweetened drinks and fat-laden offerings.

Meat pies, sausage rolls, pizza, chips and ice cream have also crept back into school canteens despite the war on obesity and healthy food rules imposed more than a decade ago.

The NSW Healthy School Canteen Strategy required all government schools to provide a healthy, nutritious canteen menu, but about 15 per cent of schools surveyed by a parent action group were found to sell soft drinks, while 82 per cent sold chips and more than a third still offered cakes and biscuits.

The collapse of the healthy food program has prompted a scathing attack by nutritionist Rosemary Stanton, who helped develop the menu guide for school canteens.

She described some food as 'junk' and said there was no enforcement of rules to ensure children were served only healthy items.

'One catering company has been doing meal deals with a sausage roll, chips and a soft drink, and that is really appalling', she said.

She said the traffic light guide to food groups – green (always allowed), amber (select carefully) and red (not recommended) – was helpful but needed enforcement.

There are currently no fines for canteens that continue to serve unhealthy food. Each principal is responsible for ensuring that their canteen is 'consistent with the Nutrition in Schools policy'.

Mandatory school canteen food guidelines were introduced at a cost of $750 000 more than a decade ago and soft drinks were banned in canteens and vending machines.

With canteen operations now turning up to $70 000 a year in public schools, the NSW Department of Education and NSW Health are 'considering evidence of best practice, available research and changes in the Australian Dietary Guidelines 2013'. The review could lead to tougher restrictions on foods in fat, sugar or salt.

The review is backed by the Healthy Kids Association, a not-for-profit group that aims to promote healthy food choices for children, and the Stephanie Alexander Kitchen Garden Foundation.

The Traffic Light Food Guide:

Green: Fill the menu. Contains a wide range of nutrients and generally low in saturated fat, sugar or salt.

Amber: Select carefully and in smaller servings.

Red: Not recommended. Should be sold not more than twice a term.

Source: Adapted from 'Government orders review of school canteen menus amid obesity crisis' by Bruce McDougall, *Daily Telegraph*, 1 May 2015.

ACTIVITIES

1. What percentage of schools are still selling soft drinks?
2. Can schools be fined for serving unhealthy foods?
3. Outline the Traffic Light Food Guide and provide examples of foods in each group.
4. Write a 3–4 paragraph response to explain the main issue relating to current canteens and what they are serving.
5. Research the Healthy Kids Association. What do they do?

Manufacturers' response

Food **manufacturers** must also take some responsibility. Society's concern over nutrition levels has led to improved food standards in the marketplace, such as:

- clear, informative and detailed food labels; for example, many now display the percentage of Recommended Daily Intake (RDI) of energy, fat, carbohydrates etc.
- more nutritious convenience foods sold through supermarkets; for example, pre-prepared salads and low-fat microwave meals
- modified versions of existing food products such as low-fat, salt and sugar; for example, low-salt margarines, low-fat yoghurt and reduced-sugar and reduced-fat chocolate bars
- controlled portion sizes; for example, snack-sized potato chips and smaller chocolate bars
- product and company information available over the internet; many companies will offer extra nutrition information via their website, particularly if the information does not fit onto a small food package
- the creation of company nutrition policies; these are often displayed on a food manufacturer's website
- improved menus at fast-food and take-away outlets; for example, McDonald's 'healthy choices' menu offers salad wraps, apple slices, bottled water and low-fat yoghurt.

6.7a Response to general nutrition levels

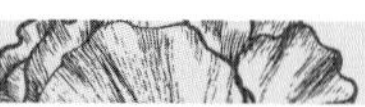

FOOD IN FOCUS

JUNK FOOD ADVERTISING TO KIDS: DO ADS MAKE KIDS FAT?

MIRANDA HERRON

How marketers influence children

Food and drink companies now have many ways to reach children as the line between entertainment and advertising is increasingly blurred.

It's not just the ads on TV. Children are targeted via:

- the internet
- social media
- viral marketing
- celebrity endorsements (particularly sports stars)
- product placement in TV shows and films
- competitions
- supermarket promotions and discounts
- smartphone 'adver-games' with embedded brand messages and licensed characters.

Experts call for regulation

'What the government's up against is a powerful, well-funded food and beverage industry,' says Kerin O'Dea, professor of population health and nutrition at the University of South Australia. She argues junk foods are highly profitable and it is not in the industry's financial interest to cooperate with public health initiatives.

'In the past, governments have legislated to force changes that are good for public health, like seatbelts in cars and smoking laws,' she says. 'But while Australia has been a world leader in the fight against big tobacco, we are quite timid when it comes to the processed food industry and the media companies that make money out of junk food advertising.'

What do the experts suggest?

As well as regulation of advertising and marketing, the health experts CHOICE spoke to want to see:

- state governments mandate, and enforce, healthy foods in school canteens and vending machines
- physical education teachers employed (rather than teachers taking sport) and the current requirement for schools to offer 120 minutes of moderate to vigorous exercise a week to be enforced
- mandatory kilojoule labelling in fast food stores extended nationally
- better labelling that is easier to understand.

Source: Adapted from Choice.com.au, 4 September 2014

ACTIVITIES

1. List three ways children are targeted by marketing.
2. Why might the junk food industry not want to cooperate with public health initiatives?
3. How would better labelling assist to enable better health for kids?
4. Brainstorm as a class – What can parents and families do to minimise the impact of junk food advertising and encourage healthy eating?

FOOD IN FOCUS

JUNK FOOD COMPANIES USE FACEBOOK TO GET AROUND CHILDREN'S TELEVISION ADVERTISING RESTRICTIONS

SUE DUNLEVY

Junk food companies are dodging television advertising restrictions by targeting kids on social media.

More than 13 million Australians a month are engaging in the Facebook pages of junk foods such as ice cream, chocolate, pizza, burger and fried chicken companies, according to a new study.

And unlike television ads, kids and teenagers are able to order some of the products online to be delivered to their door without even leaving their Facebook page.

The study by Sydney University's School of Public Health says such aggressive marketing could be behind rising childhood obesity.

There are currently voluntary restrictions on marketing junk food to children during programs aimed at children aged under 12.

Junk food brands used celebrity photos, videos, competitions, polls, quizzes and discount voucher to attract Facebook users to their Facebook page and promotional activities.

ACTIVITIES

1 What types of junk foods are commonly advertised via Facebook?
2 Can junk food manufacturers advertise during programs aimed at children under 12 years of age?
3 Have you experienced junk food advertisements while using social media, such as Facebook? Give details.

Ethical responsibilities of governments and manufacturers

The issue of responsibility is a controversial one. Nobody really wants to take responsibility for current nutritional concerns, but it seems that somebody must take charge! So who gets the blame?

Doing the right thing on behalf of society lies in the hands of the government and food companies. Governments determine which foods are imported, exported, bought, sold and **subsidised**, and are responsible for any **policy** or **legislation** related to food.

Nutritionists believe that governments must take a firm stance on nutrition levels and introduce special taxes on processed junk foods so that people find them more expensive to buy in relation to healthy foods. Other ideas include banning food advertisements aimed at children and including health warnings on food products.

Doctors in Australia have begun a new program to educate Australians by prescribing diet and exercise

NUTRITION INFORMATION (Average)

Serving Size: 30g (2 biscuits) Servings Per Pack: 24

	PER SERVE	PER 100g
Energy (kJ)	447	1490
(Cal)	107	355
Protein (g)	3.7	12.4
Fat, Total (g)	0.4	1.3
- Saturated Fat (g)	0.1	0.3
- Trans Fat (g)	0.0	0.0
- Polyunsaturated Fat (g)	0.2	0.8
- Monounsaturated Fat (g)	0.1	0.2
Carbohydrate, Total (g)	20.1	67.0
- Sugars (g)	0.9	2.9
Dietary Fibre (g)	3.3	11.0
Sodium (mg)	81	270
Potassium (mg)	102	340
Magnesium (mg)	32 (10% RDI)*	107

*Percentage of Recommended Dietary Intake (RDI)

Ingredients: Organic wholegrain **wheat** (97%), organic sugar, salt, **barley** malt extract.
Contains cereals containing gluten.

Bika BBQ Flavoured Corn Snack 70g

NUTRITION INFORMATION

Servings per package: 3
Serving Size: 25g

	Avg Quantity per Serving	Avg Quantity per 100g
Energy	515 kJ	2060 kJ
Protein	1.5 g	6.0 g
Fat - total	5.3 g	21.0 g
- saturated	2.5 g	10.0 g
Carbohydrate	17.5 g	70.0 g
- sugars	0.8 g	3.2 g
Sodium	206 mg	825 mg

Ingredients:
Corn (68%), palm oil, sugar, BBQ Seasoning 2.5% (contains wheat flour, hydrolysed soy protein, flavour enhancers 621, 631, 627, antioxidant 319, anti caking agent 551), flavour enhancer (621), salt, garlic, pepper, chili

CONTAINS: WHEAT, SOY

Imported by: Oriental Merchant Pty Ltd

10 Westgate Drive, Laverton Nth VIC 3026, AUSTRALIA, freecall 1800 806 842
Unit 10, 9 Manu St, Otahuhu, Auckland 2024, NEW ZEALAND, freecall: 0800 103 305

www.oriental.com.au

Best Before : See Pack
Product of Malaysia

Contains gluten containing cereals.

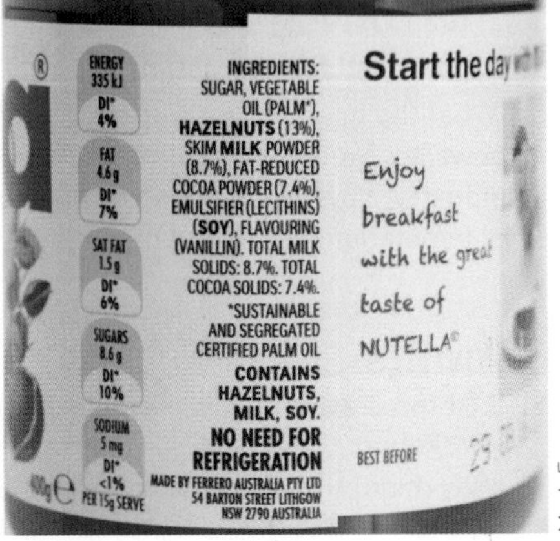

Figure 6.15 New food labels are clear and highly informative.

instead of drugs for those with diet-related conditions such as obesity or diabetes.

Food companies must do the right thing by listening to the needs and wants of consumers and providing healthy and nutritious foods that are low in fat, salt and sugar, but high in nutrients such as vitamins, minerals and fibre.

They must also consider the ethics of advertising their product. Manufacturers have to be careful that they do not mislead consumers or make incorrect claims about their products.

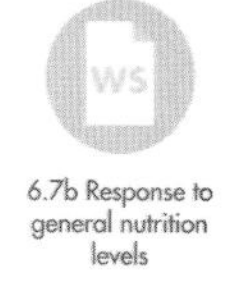
6.7b Response to general nutrition levels

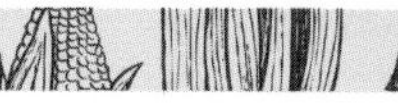

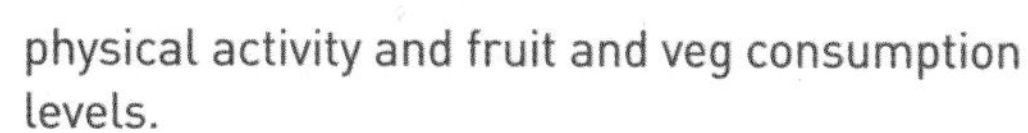

FOOD IN FOCUS

HEALTHY KIDS: HEALTHY LIFESTYLE PROGRAMS FOR HIGH SCHOOLS

In NSW there are a range of government programs available for schools to promote healthy lifestyles for students. The following programs encourage students to increase their physical activity and fruit and veg consumption levels.

1 NSW Premier's Sporting Challenge
2 Fresh Tastes @ school
3 Jump Rope for Heart.

Each program has support resources available for schools to make implementation easy.

Source: 2016 Healthy Kids, www.healthykids.nsw.gov.au

ACTIVITIES

1 Visit the Healthy Kids NSW weblink to investigate some of the available programs.

Healthy Kids

2 Does your school currently participate in any of these programs?
3 In a group of 3–4, devise your own program that might be used to increase health and wellbeing in high schools. Present it to the class.

HANDS-ON

WHO IS RESPONSIBLE?

PURPOSE

To make you aware of different people's opinions on who is responsible for nutrition levels in Australia.

STEPS

1 Interview four or five adults of different age and occupation and ask them the following questions:
   - What do you think society's role is in regard to Australia's poor nutrition levels?
   - What do you think your role is?
   - What do you think the government's role is?
   - What do you think the food manufacturer's role is?
2 Generate a summary sheet with each response shown clearly.
3 Discuss your findings with the class.
4 Write down your own responses to the questions.
5 What was the most interesting finding in your survey activity?
6 Did the class come to any conclusions about who is responsible?

UNIT REVIEW

LOOKING BACK

1 What strategies have governments taken to improve nutrition levels?
2 Who ultimately decides what goes into our mouths?
3 List one way a food manufacturer can address society's concerns for nutrition levels.

FOR YOU TO DO

4 How do governments determine which foods are available to us?
5 Imagine that the Department of Health has asked you to design a television advertisement to promote healthy eating and lifestyle. In groups of four or five, role-play a suitable advertisement. Make sure you video and upload the advertisement onto the school network for your class to watch and evaluate.
6 Think of another strategy that may improve poor nutrition levels. Give details to your class.

TAKING IT FURTHER

7 What is your opinion of taxes being levied on fast foods? Do you think it would help our nutrition levels?
8 Research 'Life. Be In It', '7-a-day' or similar campaigns. Write a short report on the campaign.

6.8 Using food guides for menu planning and food choices

Over the years, the Australian Government and other associations have provided guides designed to educate the Australian population. Some models have proven to be more successful than others. The most commonly used and well-known food guides in Australia are the Dietary Guidelines for Australians, the Australian Guide to Healthy Eating and the Healthy Eating Pyramid. Refer back through the text and, as a class, discuss each guide.

How can people apply these guides?

Each of these guides is aimed at different age levels or groups in society. Consider the Healthy Eating Pyramid and the Australian Guide to Healthy Eating. They both provide a basic diagrammatic representation of the foods that you should eat in the right proportions. These types of food guides would be helpful for young children or people who do not have a deep understanding of the English language. They can thus rely on pictures of foods and graphic representations to help them choose the right foods to eat.

On the other hand, the Dietary Guidelines for Australians are aimed at the average adult who speaks English fluently and can understand and act on each diet and lifestyle recommendation. The guidelines are based on the simple models described above, yet require a deeper level of understanding of nutrition and health.

Everybody should use these guides to assist in planning healthy diets every day. However, the type of guide each person uses will depend on who they are, how old they are and what level of understanding they have.

6.8 Using food guides for menu planning food choices

HANDS-ON

HEALTHY DIET PLANNING FOR TEENS

PURPOSE

To allow you to design, plan and prepare a nutritious menu for teenagers.

MATERIALS

Three nutrition guides
Recipes suitable for teenagers
Pen
Notebook

STEPS

1. Plan a suitable daily menu for a teenager.
2. Select one part of this menu to prepare in a practical lesson. Show it to your teacher for approval.
3. Fill out a food order and complete the practical in class time.
4. Have a friend taste and evaluate the meal while referring to the nutrition guides.
5. Complete a one-page justification of your food choices.

UNIT REVIEW

LOOKING BACK

1. Why are food models designed?
2. Who funds the development of these food guides?
3. List the Dietary Guidelines for Australians.
4. Name the food groups in the Australian Guide to Healthy Eating.

FOR YOU TO DO

5. Why might girls, women, vegetarians and athletes need more iron?
6. Why do you think you might need some fats and oils in your diet?
7. Plan a nutritious weekly dinner menu to be used at either a school cafeteria or a retirement village for elderly people. Remember to use the dietary guides.
8. Analyse the following menu for an 8-year-old child. Discuss good and bad points. Rewrite the menu to make it healthier.
   - Breakfast: 1 cup cocoa bubbles, ½ cup skim milk, 1 cup fruit drink
   - Morning tea: small packet plain potato chips
   - Lunch: cheese and tomato white bread sandwich, small chocolate milk
   - Afternoon tea: 2 sweet biscuits
   - Dinner: ¾ cup macaroni cheese, ¼ cup steamed vegetables
   - Dessert: 2 scoops vanilla ice-cream with canned peaches in syrup.

TAKING IT FURTHER

9. Construct a table displaying the three main dietary models used in Australia. List positive and negative points of each and include suggestions for possible improvements.
10. In groups of three or four, design a food guide aimed at teaching society about good eating. Prepare this strategy electronically (as a Microsoft Word document, PowerPoint presentation or web page) as though you were going to present it to the government.
11. Research and select a recipe for a meal that does not reflect the recommendations of the Healthy Eating Pyramid and Dietary Guidelines for Australians. Redesign the ingredients so that the recipe does follow the pyramid and guideline recommendations.

6.9 Additional content: active non-nutrients and health

iStock.com/ansonsaw

Figure 6.16 A variety of coloured vegetables are rich in phytoestrogens.

What are they?

Active non-nutrients fit into the family of functional foods. These are products or ingredients that provide an extra health benefit beyond the normal. Active non-nutrients are substances that are 'not nutrients' and are therefore not required in the diet, but may help with the functioning of body processes or assist to maintain or **enhance** good health.

Phytochemicals

Phytochemicals are special chemicals that come from plants and help reduce the risk of several diseases. Phytochemicals act as antioxidants in the body. Antioxidants are substances that protect the body against the harmful free radicals that are present in cigarette smoke, pollution and X-rays. Antioxidants also help to control blood cholesterol levels.

Phytoestrogens are a form of phytochemical. They produce a special **hormone** similar to one that the body produces called oestrogen.

Oestrogen is known to:

- help reduce the risk of cancer, **hypertension** and cardiovascular disease
- help to reduce hot flushes in menopausal women.

Mark Fergus

Figure 6.17 Hi-maize is a resistant starch found in many bread products.

Foods that contain phytoestrogens include soy bean products (such as tofu, soy milk and soy and linseed bread) and a variety of coloured vegetables (such as tomatoes, corn, oranges, celery, broccoli and carrots).

Probiotics

These are good **micro-organisms** that are found in dairy products. They:

- help to maintain the health of the intestinal tract
- fight intestinal disease
- help with gut functioning and digestion
- assist with **immunity** from diseases.

A probiotic product is a functional food that comes from a live organism known as a culture. Examples of friendly probiotics are lactobacillus and bifidobacteria. Too many unfriendly bacteria in the system can cause symptoms such as diarrhoea or abdominal pain. Most probiotics are found in foods like yoghurts, cheeses, fermented milk drinks such as Yakult and sour cream.

Prebiotics or modified resistant starches

Prebiotics are parts of food products that are not digested by the body. They resist digestion and help maintain good intestinal bacteria levels. Hi-maize is an example of a resistant starch. It acts like dietary fibre in the body and assists with regularity in passing stools. It is very fine and white, and mixes easily into breads and other baked products.

FOOD IN FOCUS

FREEDOM FOODS: ACTIVE BALANCE MULTIGRAIN AND CRANBERRY

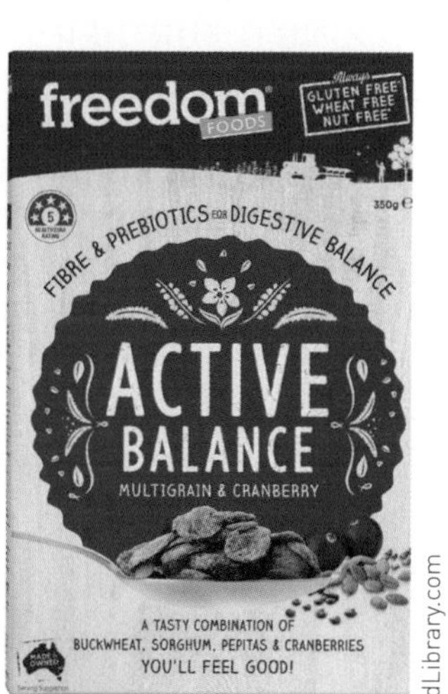

dLibrary.com

'Active Balance Multigrain and Cranberry offers 25% of your daily fibre intake, plus prebiotics to nourish the good bacteria in your digestive system and get everything balanced.'

Active Balance cereal improves digestion, is an excellent source of fibre, gluten, wheat and nut free, low in salt, no artificial colours or flavours, non-GMO, free from dairy, eggs and soy and has a five-star health rating.

Table 6.3 Nutritional information

AVERAGE QUANTITY PER 100 G	
Energy	1540 kJ
Protein	8.2 g
• gluten	Not detected
Fat, total	7.0 g
• saturated	1.2 g
Carbohydrate	59.9 g
• sugars	10.2 g
Dietary fibre, total	15.5 g
• insoluble fibre	7.0 g
• soluble fibre	2.5 g
• resistant dextrin	6.0 g
Sodium	90 mg
Potassium	260 mg

Source: Adapted from information on the Freedom Foods website. www.freedomfoods.com.au

ACTIVITIES

1. What percentage of an individual's fibre requirement is found in a bowl of Active Balance Multigrain and Cranberry?
2. Explain the term 'non-GMO'.
3. a How much protein is found in a 100 gram serve of this cereal?
   b How much protein is found in a 250 gram serve of this cereal?
4. Discuss as a class the differences between soluble and insoluble fibre.
5. Research the term 'resistant dextrin'. What is it and where is it found?

Potential health benefits of active non-nutrients

The main purpose of using functional foods is to try to improve the health of the population. Health care costs are increasing fast, so research into active non-nutrients is needed to help solve some of the problems.

Certainly these products give some form of relief and are generally widely accepted. However, they are not the only answer or resolution to the growing concerns over our nutritional problems. Some people may start to think that they can solve all of their dietary problems just by consuming these foods.

Functional foods can be used alongside a variety of other strategies. Educating consumers about good foods and the problems associated with fast, convenience and processed foods is very important.

UNIT REVIEW

LOOKING BACK

1 What are active non-nutrients?
2 Give one example of a phytochemical.
3 Name one probiotic food available on the market.

FOR YOU TO DO

4 Are functional foods the answer to our problems? Evaluate the health benefits of active non-nutrients and discuss your response as a class. Create a poster during the discussion. Divide it into two sections: 'Pros' and 'Cons'.

TAKING IT FURTHER

5 Summarise different types of active non-nutrients, functions and food sources.
6 Research one active non-nutrient by using a variety of sources such as books, the internet, journals and magazines. Provide an oral presentation (about three minutes). Use pictures and diagrams to assist you.

Chapter review

LOOKING BACK

1 What are the three main functions of food?
2 List the six food nutrients.
3 How is food digested?
4 What is 'metabolism'?
5 List two high-protein foods.
6 Name one vitamin and mineral and identify the main function of each.
7 List four factors that affect nutritional needs.
8 What does RDI stand for?
9 List four factors influencing our food habits.
10 Give an example of how our individual preferences might influence our eating habits.
11 Identify one food that is not available in one geographic area.
12 What is the main nutritional problem in Australia today?
13 Name two food models designed for Australians.
14 What is 'legislation'?
15 Define the term 'active non-nutrient'.

FOR YOU TO DO

16 Identify the protein, fat and carbohydrate foods present in your daily diet.
17 Describe lactose intolerance and provide examples of symptoms and foods to avoid.
18 Draw and label the GI tract.
19 Research the lymphatic system and provide information to describe its purpose.
20 List one day's worth of food choices for a strict vegetarian.
21 Provide a summary of each life stage and the nutrients required by each group.
22 Identify five tactics used by the media to influence food choices.
23 How might you educate society to eat the right quantity of the right foods?
24 Investigate the positives and negatives of functional foods.

TAKING IT FURTHER

25 In a detailed report, explain the role of the liver in human digestion.
26 Investigate protein/energy deficiencies such as kwashiorkor on the internet. Explain how and why they occur.
27 Design and produce six healthy snack options that could be supplied by school canteens.
28 Create a report on one recent technology and how it has affected both food availability and choices.
29 Investigate and report on one diet-related disorder that is linked to over-nutrition.
30 Evaluate the government's response to nutrition levels. What actions have they taken? What do they plan to do? Will this be adequate to improve the health of Australians? Submit an extended response of three to four pages.
31 How can each individual person take responsibility for their own nutrition levels? Prepare an oral presentation (3–5 minutes) to express your thoughts to the class.
32 Plan a dinner menu to be used in a hospital maternity unit. Remember to use the RDI tables and food guides.
33 Analyse the following diet of a 16-year-old boy:
- Breakfast: Milo, cereal, whole milk, 1 slice white toast with Nutella and water
- Morning tea: water, 4 milk arrowroot biscuits and an apple
- Lunch: chicken schnitzel and salad, white bread roll with mayonnaise and chocolate milk
- Afternoon tea: packet of M&Ms and water
- Dinner: spaghetti bolognese and orange cordial
- Dessert: 4 jelly snakes and tea with milk and 2 sugars.

34 Design, produce and evaluate a handout or e-brochure aimed at educating people about active non-nutrients.

RECIPES

Plum pork and vegetable stir-fry

This nutritious meal contains all the six food nutrients, is low in saturated fat and is tasty and easy to prepare.

 **Preparation time** 20 minutes  **Cooking time** 12 minutes  **Serves** 2

INGREDIENTS

- 250 grams pork fillet
- ½ tablespoon olive oil
- 80 grams vermicelli or other rice noodles
- 1 small onion, cut into wedges
- 1 teaspoon fresh ginger, grated
- ¼ cup plum jam
- 1 tablespoon soy sauce
- 100 grams snow peas, topped and tailed
- ½ large carrot, cut into slices or batons
- 1 small parsnip, cut into thin strips

METHOD

1. Prepare the vegetables as instructed, cut any fat off the pork and slice the meat thinly. Heat oil in a wok or frypan and cook the meat on a high heat for 3–4 minutes or until browned. Place on a paper towel and keep warm.
2. Place the vermicelli in a medium bowl and add boiling water to cover. Set aside to soften for 5–8 minutes.
3. Stir-fry the onion and ginger over a high heat for 1–2 minutes or until golden brown. Add the jam and soy sauce and stir for another 1–2 minutes. Add the snow peas, carrots and parsnips and stir-fry for 3 minutes.
4. Return the meat to the pan and cook on a high heat for 2–3 minutes or until it is heated through. Drain the vermicelli and arrange the pork and vegetables with noodles on a plate as desired.

QUESTIONS

1. How do you prepare the vegetables?
2. Which ingredient has the richest source of carbohydrate?
3. In this recipe, identify a source of protein, iron, lipids, vitamin and water.

Mark Fergus

Mark Fergus

Quick vegetarian pizzas

This is a tasty, healthy meal or snack that is free from any flesh foods. You can select your own ingredients and modify the pizza toppings to suit your own tastes.

 **Preparation time** 20 minutes  **Cooking time** 20 minutes  **Serves** 4

INGREDIENTS

- 800 grams spinach, washed, trimmed and roughly chopped
- 2 teaspoons garlic, crushed
- 300 grams button mushrooms, sliced
- 140 grams tomato-based pizza sauce
- 2 tablespoons tomato paste
- 4 large wholemeal pitta breads
- 200 grams low-fat ricotta cheese
- 80 grams low-fat cheddar cheese, grated

METHOD

1. Preheat oven to 220°C.
2. Wash and prepare the spinach, garlic and mushrooms.
3. Add the spinach to a large saucepan and stir over a low heat until wilted. Remove from heat and drain on absorbent paper.
4. Combine the pizza sauce, tomato paste and garlic in a bowl. Remove 2 tablespoons tomato mixture and mix it in a separate bowl with the mushrooms.
5. Place the pitta bread on a baking tray and spread with the remaining sauce mixture. Top with the spinach, then the mushroom mixture and finally the cheeses.
6. Bake for approximately 20 minutes or until topping is browned.

QUESTIONS

1. List four other vegetarian toppings that you could use as substitutes in this recipe.
2. What sorts of garnishes could be used with this recipe?
3. Design your own pizza. Give it a name and list the ingredients you would use.

9780170383417

Quinoa and rice salad

This is a modern, nutritious and colourful salad that everyone will love!

 **Preparation time** 40 minutes  **Cooking time** 20–25 minutes  **Serves** 2

INGREDIENTS

- 200 grams button mushrooms
- ½ cup white rice and quinoa (blended)
- 2 tablespoons French dressing
- ½ capsicum (yellow, red or a mixture)
- 1 small red onion
- 160 grams cherry tomatoes
- ¼ cup pitted Kalamata olives
- ½ cup flat leaf parsley leaves
- 1 tablespoon olive oil (extra virgin)
- salt and pepper

METHOD

1. Preheat oven to 200°C (fan forced).
2. Cook the rice and quinoa together based on packed directions. Stir and separate the grains and set aside to cool to room temperature.
3. Wash and halve the mushrooms. Place in a bowl with 1 tablespoon of salad dressing and toss to coat. Cover and set aside for 20–30 minutes.
4. Chop the capsicum and onion into wedges and place on a greased roasting pan. Spoon over the balance of the salad dressing and season with salt and pepper. Roast for 15 minutes or until tender. Add the tomatoes and olives and bake for another 5–8 minutes or until warmed through. Cool for 5 minutes.
5. Add the mushrooms, rice, quinoa and parsley to the roast vegetables in the pan and toss gently. Drizzle with olive oil and season to taste.

QUESTIONS

1. Re-read the recipe and list four things you can do to prepare for this recipe.
2. What other ingredients could be used? List foods you could substitute if needed.
3. Discuss different ways you could cook the rice and quinoa. How much water is required for each method?

Mark Fergus

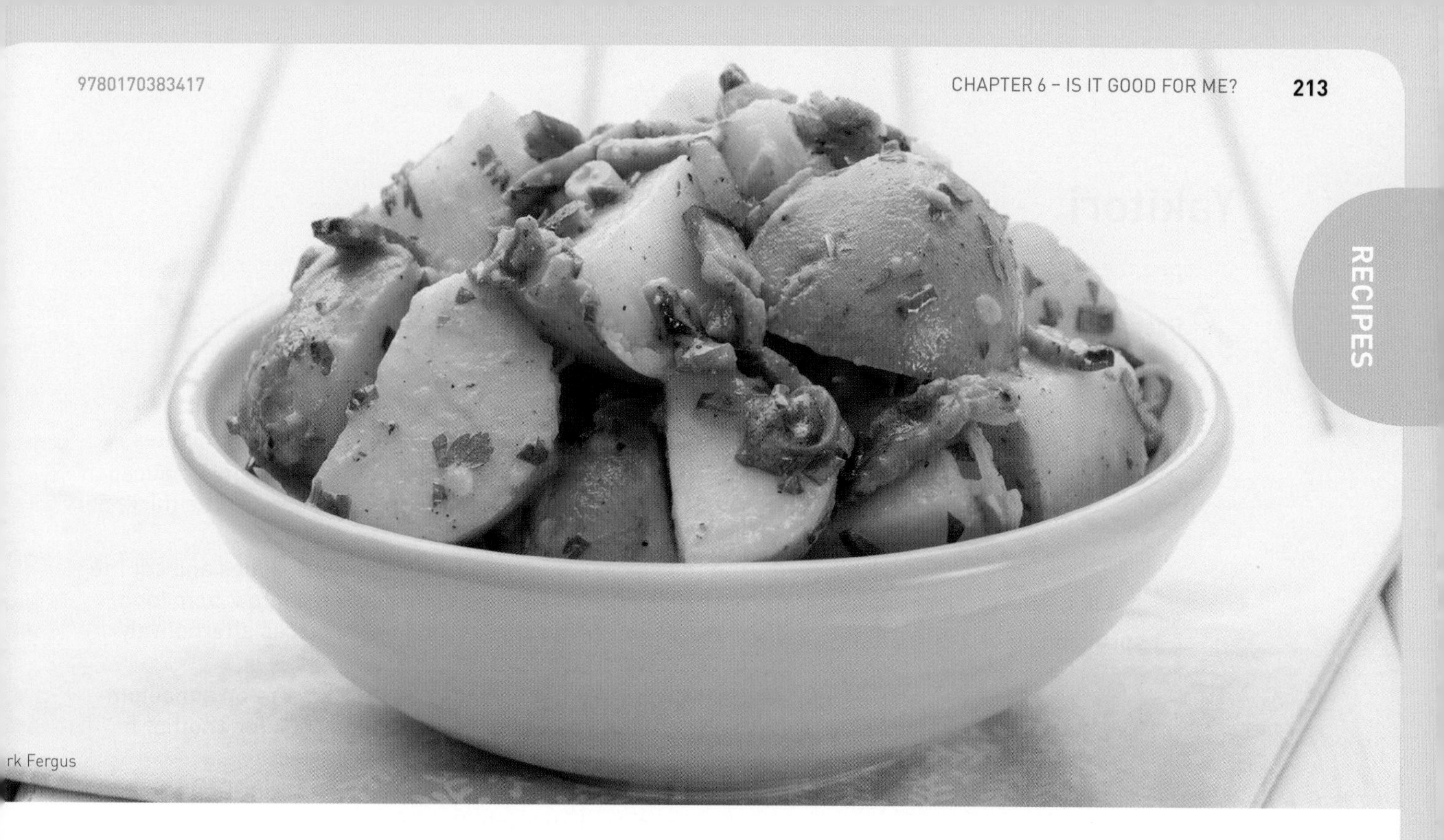
rk Fergus

Hot, rich potato salad

This warm winter recipe has a tasty, rich dressing. It is perfect either as an accompaniment to a meal or eaten on its own.

 **Preparation time** 20 minutes  **Cooking time** 25 minutes  **Serves 2–3**

INGREDIENTS

SALAD:

- 3 rashers bacon
- 750 grams small Desiree potatoes
- 2 shallots, sliced thinly
- 2 tablespoons parsley, chopped
- ¼ teaspoon salt
- pinch of black pepper

DRESSING:

- ⅓ cup melted butter
- 1 tablespoon Dijon mustard
- 2 tablespoons white wine vinegar

METHOD

1. Preheat oven to 180°C.
2. Trim the rind from the bacon and grill or pan-fry until crisp. Allow the bacon to cool, then chop it into small pieces.
3. Place whole potatoes into a saucepan of simmering water and cook until just tender. Do not let their skins break away too much. Drain and cool.
4. Make the dressing by whisking all ingredients together in a jug. Keep warm until required.
5. Cut the potatoes into quarters and place in an ovenproof dish with three-quarters of the bacon bits and the shallots, parsley, salt and black pepper. Pour in the dressing and toss gently. Sprinkle with the remaining bacon and place in oven for approximately 10 minutes or until warmed through. Serve hot.

QUESTIONS

1. How could you alter this meal to make it comply with the dietary guidelines?
2. This recipe uses Desiree potatoes. Go to the internet to find the names of three other types of potatoes.
3. List as many other ingredients as you can think of that can be included in a potato salad.
4. Design your own potato salad. Share your ingredients with your class.

9780170383417

RECIPES

Yakitori

In Japan there any many popular Yakitori bars for this tasty snack of skewered, seasoned chicken.

 **Preparation time** 30 minutes  **Cooking time** 10 minutes  **Makes** 5–6 skewers

INGREDIENTS

- ¼ cup soy sauce
- 2½ tablespoons sugar
- 2 tablespoons sake (optional)
- 3 tablespoons mirin
- 230 grams chicken thighs
- 1 spring onion or 2–3 shallots
- salt
- 2 teaspoons vegetable oil
- 5–6 bamboo skewers

METHOD

1. Soak the bamboo skewers in water for about 30 minutes
2. Mix the soy sauce, sugar, sake and mirin in a small saucepan and boil for 5–8 minutes until the sauce gets a little thick. Set aside to cool.
3. Cut the chicken thighs into 2.5 centimetre cubes and cut the spring onions or shallots (white sections) into 2.5 cm long pieces. Skewer the chicken and onion pieces alternatively onto skewers.
4. Heat the oil in a frypan and cook the skewers on a medium-high heat for 5 minutes, then turn and cook for another 5 minutes until cooked through and browned.
5. Place the sauce onto a shallow dish or saucer. Dip the chicken immediately into the sauce and serve.

QUESTIONS

1. Why do the skewers need to be soaked?
2. How should the chicken be cut for this recipe?
3. Select and list some accompaniments this Yakitori could be served with.
4. What garnishes can be used for this recipe?
5. Why is it important to ensure the chicken is cooked through to the centre of the skewer?

Mark Fergus

Mark Fergus

RECIPES

Chunky polenta chips

These unique chips are crispy on the outside and soft in the middle, and can be served either as a snack or an accompaniment to a meal.

 **Preparation time** 30 minutes  **Cooking time** 40-45 minutes  **Serves 2**

INGREDIENTS

- 1 cup polenta (yellow cornmeal)
- 1 cup chicken or vegetable stock liquid
- 1 cup water
- ¼ cup grated Parmesan cheese
- ½ tablespoon butter
- ¼ teaspoon salt
- cooking oil spray

METHOD

1. Preheat oven to 200° C.
2. Bring the water and stock to a boil in a saucepan.
3. Whisk in the polenta in a slow, steady stream.
4. Reduce heat and cook, stirring constantly for 8–10 minutes or until really thick. Be extra careful as polenta can erupt like a volcano if unattended and easily burn on the bottom of the saucepan.
5. Stir in the cheese, butter and salt.
6. Once the polenta is really thick then it is ready to pour into a square dish lined with cling wrap.
7. Allow to cool.
8. Once cool, turn out polenta and take off cling wrap. Cut into thick chips.
9. Place foil on a baking tray and spray with oil. Coat the chips with oil spray and bake for 20–30 minutes until crisp and brown. Turn once during cooking and spray the other side of the chips.
10. Serve hot.

QUESTIONS

1. When the polenta is boiled, what consistency should the final mixture be?
2. Why could hot polenta mixture be dangerous?
3. What modifications could you make to this recipe to add a different flavour to the chips?

CHAPTER 7

Mark Fergus

What's on the menu?

FOCUS AREA: FOOD SERVICE AND CATERING

FOCUS AREA

Throughout history, there have always been occasions when non-family members prepared meals for the family. Wealthy ancient Greeks and Romans employed chefs to prepare elaborate feasts. Travellers have always needed food and accommodation. Large cultural celebrations such as weddings have traditionally depended on the assistance of outside caterers.

CHAPTER OUTCOMES

In this chapter you will learn about:

- different food service and catering ventures and their value to society
- employment opportunities in the food/hospitality industry
- employer, employee and consumer rights and responsibilities
- menu planning
- recipe development
- purchasing systems
- food service and catering
- operating a small food business.

In this chapter you will learn to:

- examine the operations and contribution of different food service ventures
- conduct web searches
- outline employer, employee and consumer responsibilities
- identify and demonstrate safe work practices
- compare and develop menus and recipes for different catering events
- identify elements of a recipe
- design, plan and prepare food for functions
- set tables
- create a proposal for a small food business.

RECIPES

- Bruschetta, p. 244
- Sweet potato soup, p. 245
- Salt and pepper squid, p. 246
- Steak Diane, p. 247
- Mini baked cheesecakes, p. 248
- Thai chicken cakes, p. 249

FOOD FACTS

The word 'restaurant' derives from the French verb *restaurer*, meaning 'to restore'. It was first used in France in the 16th century to describe soups sold by street vendors that were advertised to restore your health.

The word 'bistro' comes from the Russian word *bystro*, meaning 'quickly'.

Casual workers make up just over half of workers in the food service industry.

There are many more male chefs than female chefs, yet in the home, more females cook than males.

Workers can choose whether to join a union or not; however, unions play a valuable role in protecting the rights of workers and improving their conditions.

The NSW Food Authority website lists names of businesses that have breached the *Food Act*.

Today, instead of using the term 'restaurant', food entrepreneurs employ terms like 'grill', 'bar', 'diner', 'shack', 'stand' and 'cafe'.

Famous French chef Fernand Point said, 'If the menu is not appealing, the people will lose their appetites and their desire to part with their money'.

Smorgasbord is the Swedish term for 'buffet'.

For safety reasons, dining venues have a set number of seats they can offer in both their indoor and outdoor dining areas. This is determined by the local council.

If cooked rice is left in the danger zone of 5–60°C for a few hours, it can give you food poisoning because the bacteria *Bacillus cereus* in rice multiplies rapidly in warm conditions.

FOOD WORDS

duty of care legal obligation imposed on an individual requiring adherence to a standard of reasonable care while performing any acts that could foreseeably harm others

employee a person working for another person or business for pay

employer a person or business that employs one or more people, especially for wages or salary

enterprise a company organised for commercial purposes

finger food small appetiser or sweet item that you can pick up and eat using your fingers

media forms of communication – radio, television, newspapers, magazines, social media, world wide web – that reach large numbers of people

portion size specific size, shape and weight of food to be served

stock control determination of the supply of goods kept by a business

7.1 Food service and catering ventures

While catering itself is all about the cooking of food, food service involves the serving of food to customers, patients, students and clients.

Profit and non-profit ventures

Most food service and catering ventures operate to make a profit (monetary gain). Some ventures are non-profit operations that support communities or provide an essential service.

Table 7.1 Profit-making and non-profit-making ventures

TYPICAL PROFIT-MAKING VENTURES	TYPICAL NON-PROFIT-MAKING VENTURES
Restaurants Cafes Bistros Hotels and motels – room service, bar service and restaurants Function centres Fast food and take-away stores Food stalls Mobile food vans and food trucks Caterers for: • public events • private functions • transport, such as airlines • film and television • canteens and cafeterias.	Canteens for: • schools, colleges and universities • sport • work. Fundraising food stalls Caterers in: • hospitals • nursing homes • prisons • defence service • boarding schools • childcare centres • charities • relief and emergency services.

Economic and social value

Economic contribution

Food ventures contribute to the economy as they pay government taxes that allow many community projects and services to be funded, such as education, welfare, highways and defence. Food ventures purchase food and supplies from other businesses, such as butchers, bakers and coffee merchants. They therefore assist other businesses to make money and employ staff.

Preparing and serving food requires staff. Workers use wages to purchase goods and services and this assists to promote economic growth. Workers also pay taxes to the government. Even customers pay taxes through the Good and Services Tax (GST) applied to food that is prepared.

Social contribution

Australians are relying more on food service providers. Working parents find that it often saves time and energy to eat out, buy take-away or organise a function outside the home. Australian households are also, on average, much smaller than in the past, and preparing meals for one or two people is considered more difficult than purchasing a meal already prepared.

Non-profit food service organisations provide essential social contributions. For example,

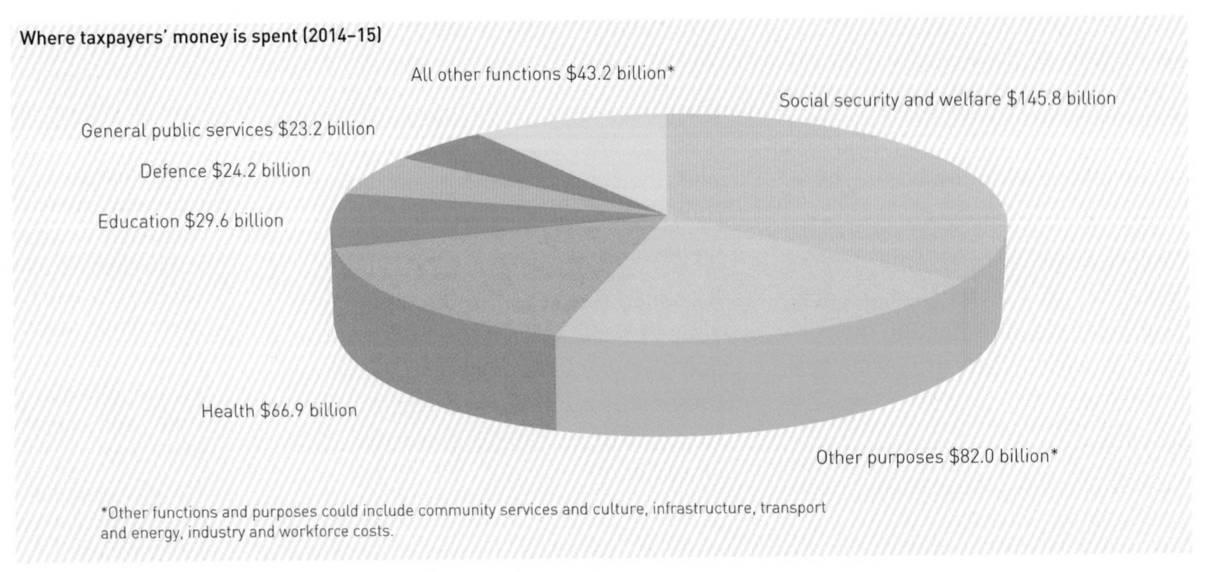

Figure 7.1 Where Australian taxpayers' money is spent

charities often feed people who are sick, frail and homeless, and others provide emergency services during disasters.

Many sporting and service clubs operate or lease ventures such as bistros and canteens to earn income to help provide facilities, equipment and services for their members, as well as for other groups in society. Some community organisations and charities hold food stalls to raise money to fund local community projects.

Alamy Stock Photo/MBI

Figure 7.2 Many families often enjoy sharing meals together at restaurants as occasional alternatives to eating at home.

FOOD IN FOCUS

KARIMBLA COMMUNITY RESTAURANT

The Karimbla Restaurant offers a variety of services to the South East Sydney Regions' frail aged, younger people with a disability and their carers. Some clients only need the service for a short time as they have just left hospital and/or are recovering from an illness. Others are more regular, particularly those who live on their own and find it is easier, more enjoyable and more affordable to eat at the restaurant than having to cook for themselves. Like with most restaurants, bookings are encouraged.

In a relaxed and friendly atmosphere, clients can enjoy a nutritious and affordable meal for $10.00, consisting of a warm bread roll and butter, a choice of two main meals followed by a choice of two desserts with an after-dinner mint, served with tea, coffee, hot chocolate or apple juice. Menus change twice a week to offer plenty of variety.

The restaurant also offers internet & computer classes to clients. Internet access is available for clients who would like to organise internet grocery shopping. Frozen meals are available to order to enable clients to have healthy, easily-prepared meals at home.

Source: Adapted from the Karimbla Community Restaurant website

ACTIVITIES

1 Who are the restaurant's clients?
2 What do the customers get for their money?
3 Besides a sit-down meal, what else does the restaurant offer?
4 Plan a menu for the restaurant. Give reasons for your food selection.

UNIT REVIEW

LOOKING BACK

1 Distinguish between the terms 'catering' and 'food service'.
2 Give three examples of profit-making food service and catering ventures.
3 What does the acronym GST stand for?
4 Identify three groups in society who may benefit from the services of non-profit food service and catering ventures.

FOR YOU TO DO

5 Write a paragraph about your favourite food service and catering establishment. Give reasons why you like the establishment.
6 If you were in charge of catering for a school or sporting canteen, list 10 healthy foods and drink items you would have on your menu.
7 Plan a dinner menu that could be prepared by an airline catering company for a flight from Sydney to Auckland. Make sure to include a dessert and drink.
8 Brainstorm at least seven ways the Australian government uses money gained from taxes.

TAKING IT FURTHER

9 Set up your own catering company and produce a social media page, leaflet or brochure providing prospective customers the following information:
- company name and contact details
- food services you provide, such as function and corporate catering
- examples of foods you offer, such as platters and finger foods
- appropriate images.

7.2 Employment opportunities in the industry

Food service and catering is a large, expanding and diverse industry that offers full-time, part-time and casual work opportunities.

Full-time **employees** work between 35 and 40 hours a week, while part-time employees work between 12 and 35 hours a week. Both are permanent positions that entitle staff to leave such as sick days, family leave and holidays. Casual employees work when required. Most casuals receive no or limited leave entitlements, but do earn a higher rate of pay.

Table 7.2 Personal requirements for working in food service and catering

REQUIREMENTS	BENEFITS AND OPPORTUNITIES
Communication skills	Select from a variety of careers
Good manners	Work in a variety of venues
Smart presentation	Work full-time or part-time
Team work	Learn about different cultures
Food skills	Travel
Dedication	Start at the bottom and work up to management level
Organisational skills	Learn skills in food preparation
Initiative	
Enthusiasm	
Fitness	

Back-of-house and front-of-house careers

In the hospitality industry, 'back of house' refers to where meals are prepared, meaning the kitchen. Those working 'front of house' are responsible for serving and greeting guests.

Table 7.3 Careers in hospitality: back of house and front of house

BACK OF HOUSE	FRONT OF HOUSE
Kitchen hand	Food and beverage attendant
Catering assistant	Bar staff
Counter hand	Barista
Sandwich hand	Customer service assistant
Cook	Supervisor
Chef	Host

Table 7.4 Career pathways in hospitality: front of house

CAREER PATHWAY OF FRONT-OF-HOUSE
Owner, operator, manager: Manages the whole operation – hiring staff, marketing and ensuring profitability
Head waiter (maître d'): Organises service staff, allocates tables for bookings, greets guests and compiles staff rosters
Food and beverage attendant: Takes orders, serves food and/or drinks to guests
Trainee attendant/runner: Assists food and beverage attendant by removing plates, setting tables and collecting glasses

Table 7.5 Career pathways in hospitality: back of house

CAREER PATHWAY OF BACK-OF-HOUSE
Owner, operator, manager: Manages the whole operation – hiring staff, marketing and ensuring profitability
Executive/head chef (chef de cuisine): Manages the whole operation of the kitchen such as ordering, stock control, planning menus and training kitchen staff
Sous chef: Second in command, manages day-to-day activities of the kitchen
Chef de partie: Specialist cook in area such as sauces, soups, fish, pastries, or grills and roasts
Commis chef: Has completed apprenticeship and works in any area of the kitchen preparing, cooking or finishing food
Apprentice: Chef who works under the direction of a chef. Apprenticeships last four years
Kitchen hand: Carries out cleaning duties and helps with simple food preparation

amanaimages/Milton Brown

Figure 7.3 Cooks have usually undergone some professional cooking training, but you cannot be classed as a chef unless you have completed an apprenticeship.

Management

Managers make decisions regarding the operations of an establishment. The food service and catering industry provides a good opportunity for employees to progress into management positions. It is not uncommon for chefs to open and manage their own food businesses after just a few years working in the industry, or for team leaders in some establishments to become managers. There are also tertiary courses where you can attain degrees in management skills specific to hospitality.

Delivery

Consumers like the convenience of prepared foods – from pizzas to diet foods – delivered to them, both at home and at their workplaces. For instance, many companies use a nationwide phone number or app for customers to place orders. The order is sent to a local store that prepares the food, places it in appropriate storage containers and then delivers it to the customer. Most food service establishments rely on suppliers to deliver ingredients and foods, which results in many employment opportunities in this field.

istock.com/mediaphotos

Figure 7.4 A barista is a beverage attendant who specialises in making coffee.

FOOD IN FOCUS

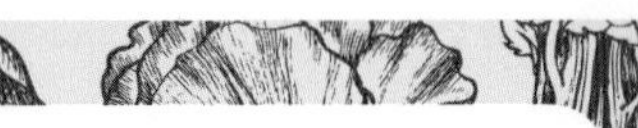

THE TRIPPAS WHITE GROUP

The Trippas White Group is a very large catering company operating restaurants, cafes and catering facilities across Australia. Their venues include Sydney Opera House Western Foyers, 360 Bar and Dining at Sydney Tower, Botanic Gardens Restaurant and Centennial Parklands Dining. Additionally, the company provides corporate catering services that include:

- first and business class lounge airport catering
- school boarding and staff catering
- boardroom catering
- office catering
- event catering
- cafe, kiosk and restaurant management.

The company employs more than 500 full-time, part-time and casual employees nationally. The company also claims to be an equal opportunity employer and is committed to providing a safe and supportive work environment for all staff. Jobs advertised on their website have included bartenders, wait staff, casual baristas, chefs, chefs de parte, catering assistants and casual kitchen hands, among others.

ACTIVITIES

1. Name two dining venues operated by the the Trippas White Group.
2. Give three examples of corporate catering the Trippas White Group provided.
3. What is the company committed to in relation to its workers?
4. Using the notes in the textbook, answer the following questions.
   a. What is the difference between part-time and casual employment?
   b. What is a chef de parte?
   c. What is the difference between a chef and a cook?

UNIT REVIEW

LOOKING BACK

1. List four back-of-house employment opportunities in the food service and catering industry.
2. List four front-of-house employment opportunities in the food service and catering industry.
3. Outline the role of management in an establishment.

FOR YOU TO DO

4. You are planning to operate your own cafe.
   a. Choose a location and give reasons why you have selected this location.
   b. List five dishes on your menu board.
   c. List five beverages on your menu board.
   d. Outline the roles of three staff members required to operate your cafe.
   e. Justify four qualities you would expect from staff.
   f. Write an advertisement for a staff position available at your cafe.

TAKING IT FURTHER

5. Plan and prepare a function for your teachers, parents or community members such as morning tea or a dinner. Your event could also help to raise money for a charity. Assign front- and back-of-house positions to different class members; for example, kitchen hand, food and beverage attendant, head chef and manager.
6. Visit a restaurant or cafe and interview the staff about their role in the operation of the venture.

KYLIE KWONG
CHEF

Kylie Kwong is a chef, author, television presenter and restaurateur. As a third-generation Chinese Australian she learned Cantonese cooking from her mother's side. She loved hanging around her uncles' noodle factory and enjoyed family gatherings involving banquets and steamboat parties.

Kylie undertook much of her apprenticeship at Neil Perry's Rockpool and later as head chef at Wokpool, a modern Asian noodle bar and restaurant. Later she became head chef at Bill's, owned by Bill Granger. In 1999, Kylie and Bill opened a restaurant together called Billy Kwong. It was Kylie's first restaurant and it was located in Sydney's Surry Hills.

In addition to her Chinese heritage, Kylie says her cooking style has been shaped by living in Australia, where we have access to some of the world's finest produce, including seafood from pristine waters and a bountiful array of Asian fruit, vegetables and herbs. According to Kylie, 'I have incorporated these ingredients into my food to create a distinctly Australian Chinese cuisine.'

She has also made a commitment to sustainability by only using organic and biodynamic food and Fairtrade tea, coffee and chocolate. Kwong explains, 'I wanted my work and social life to reflect my Buddhism. Offering my customers healthy, life-giving, precious food is the best way for me to help them.'

Kylie has also fused Indigenous food into her cooking. At Carriageworks Farmers Market in Sydney, Kylie sells dumplings and pork buns, pancakes with salt bush, and sticky rice parcels of macadamia and warrigal greens. Kylie says she has been inspired and influenced by some great cooks and food writers: 'From my mother to Alice Waters, Stephanie Alexander, Neil Perry, Stefano and Franca Manfredi, Maggie Beer, Elizabeth David, Fergus Henderson, Rene Redzepi and Alex Atala.' She is also inspired by her travels, which provide her the opportunity to learn more about food traditions worldwide.

Source: Adapted from information from the Kylie Kwong website. Photo: Newspix/John Fotiadis

ACTIVITIES

1 Where did Kylie undertake her apprenticeship?
2 At what restaurant was Kylie first in the position of head chef?
3 Kylie specialises in what cuisine?
4 How does Kylie commit to sustainability?
5 Give an example on how Kylie has incorporated Indigenous food into her cuisine.
6 Name five people who have influenced Kylie.
7 How has travelling inspired Kylie?

7.3 Employer and employee rights and responsibilities

Employers and employees have specific rights and responsibilities to follow in order to ensure safety and fairness at work.

Work health and safety

The food service and catering industry is fast-paced and involves heat, water, tools and machinery. Common injuries include:

- slips and falls
- sprains and strains
- scalds and burns
- cuts and abrasions.

To prevent injury, it is important that food service and catering staff:

- use safety signage
- wear protective clothing and shoes
- have easy access to first aid and emergency contact numbers
- are trained in safety
- keep work areas clean and clear
- store chemicals appropriately
- lift objects correctly.

Table 7.6 Employer **duty of care** and employee rights and responsibilities

EMPLOYER DUTY OF CARE	EMPLOYEE RIGHTS AND RESPONSIBILITIES
Ensure a safe work environment	Look after your own personal health and safety
Ensure safe systems at work	Help other employees' health and safety, where possible
Provide WHS information and training to staff	Complete all WHS training
Ensure and supervise the safe use of equipment and products	Follow safety policies and procedures, including personal protective clothing
Provide emergency procedure training	Follow emergency procedures (e.g. fire evacuation)
Keep records of incidents and safety checks	Report unsafe or unhealthy conditions and report injuries
Monitor employees' health and wellbeing	Follow laws about alcohol and drugs

Shutterstock.com/Di Studio

Figure 7.5 Chefs wear protective clothing and shoes.

The *Work Health and Safety Act 2011* (NSW) aims to protect the health, safety and welfare of people at work. It covers employees, employers and self-employed people. Failure to provide safety for workers can result in heavy fines, business closure or imprisonment of offenders. The government organisation WorkCover can inspect premises at any time to check on health and safety issues. It must close and inspect all workplaces where a death has occurred as a result of an accident.

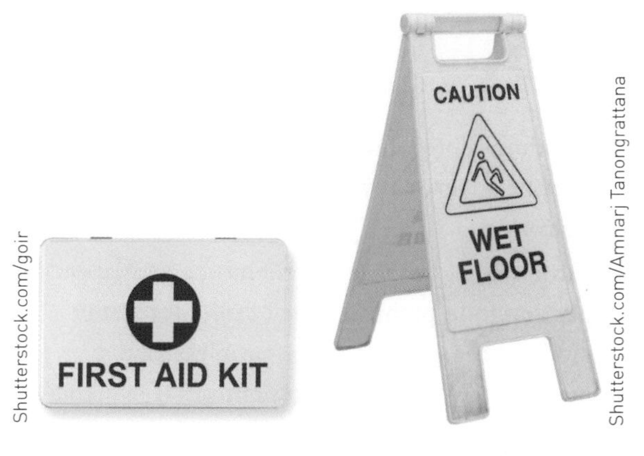

Figure 7.6 Signs help with safety.

Industrial legislation

Industry awards

Awards are the minimum working conditions of employees in a particular industry or occupation. They are negotiated between unions and employers and are permanent until new improved awards are agreed to. Employers who provide less than award conditions can be prosecuted.

Awards usually deal with:

- minimum wage rates
- leave entitlements such as holiday and sick leave
- hours of work
- penalty rates, overtime and casual rates
- allowances, such as for travel or uniform.

Enterprise agreements

Under **enterprise** agreements, employees or their unions can negotiate with employers to gain conditions better than the award. Once an agreement has been reached, the improved working conditions are then written down and cannot be breached.

Anti-discrimination legislation

Under anti-discrimination legislation, employees and customers cannot be discriminated against on the basis of:

- age
- gender
- religion
- race
- disabilities
- sexual preference
- marital or parental status.

Equal employment opportunity principles

Equal employment opportunity (EEO) means that all people, regardless of gender, race, age, marital or parental status, sexual preference, disability or religious belief, have the right to be given fair consideration for a job, or other job-related benefits such as staff training and development, if they have the qualifications, skills and experience for the employment opportunity.

UNIT REVIEW

LOOKING BACK

1 Name the legislation that ensures workers have a right to safety.
2 Name four injuries common to the food service and catering industry.
3 What are three areas covered by industry awards?
4 Employees and customers cannot be discriminated against on the basis of several personal aspects of their lives. List five of these.

FOR YOU TO DO

5 Produce a safety sign that can be displayed in your school kitchen.
6 What personal protection items are worn in commercial kitchens and why are the items necessary for safety?

TAKING IT FURTHER

7 Investigate current award wages and conditions of different careers in hospitality, using the internet to access government and union websites. Use keywords in your search such as Australia, awards or chefs.
8 Examine the first aid kit for your food technology classroom and state how the items can be used in relation to first aid.

7.4 Consumer rights and responsibilities

When you purchase goods or use a service you become a consumer. Consumers have a right to:

- safety
- choice
- be heard
- satisfaction of basic needs such as food and shelter
- redress (refund, repair or exchange on faulty goods)
- consumer education
- a healthy environment.

Consumers also have a responsibility to:

- use products wisely
- choose carefully
- make use of available information
- seek redress with faulty goods or express dissatisfaction with poor service.

Safety and hygiene

In NSW, the *Food Act 2003* and *Food Regulation 2010* are the primary laws protecting the right of consumers to clean and safe food. According to these instruments:

- food has to be fit for human consumption
- foods must meet certain standards; for example, meat pies must contain a certain percentage of meat
- consumers must not be deceived; for example, beef meat pies must contain beef
- fair trading is promoted; for example, a food cannot be advertised with false and misleading information.

Environmental health officers enforce legislation by inspecting food premises. They can enter at any time and have the power to take samples, issue warnings, and fine or close food premises if hygiene standards are not met.

Under this legislation, food for sale must be prepared in a commercial kitchen where materials and equipment meet set standards. Most food ventures need council approval and inspection before they can operate. Food businesses also need to have a supervisor in charge of food safety in order to implement practices that prevent food poisoning.

Under the law, establishments serving alcohol require a licensee and staff trained in Responsible Service of Alcohol (RSA). Public liability insurance is also compulsory in case customers are injured or fall sick as a result of an establishment's operations.

Under the *Smoke-Free Environment Act 2000* (NSW), smoking is banned both indoors and outdoors at commercial dining areas such as restaurants and cafes. Smoking is also not permitted near a food stall. The law protects the rights of non-smokers as passive smoking is a health risk.

iStock.com/TkKurikawa

Figure 7.7 Smoking is prohibited both inside and outside dining areas.

iStock.com/shalamov

Figure 7.8 Staff need to be trained in responsible service of alcohol (RSA) before they can serve alcohol.

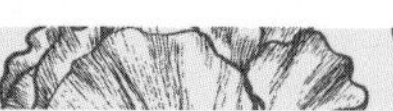

FOOD IN FOCUS

FOOD POISONING ON THE RISE

The number of Australians struck down by food poisoning has leapt almost 80 per cent in a decade and the number of outbreaks linked to restaurants has more than doubled, according to the latest government statistics. The figures capture only a fraction of infections since most victims don't go to a doctor, experts say.

People cook less and eat out more, say public health experts. 'Traditionally, food is prepared and eaten immediately, but now food might be prepared and left longer before eating. Also, the exposure is greater; if someone accidently contaminated food in their home and fed it to their family, the organism is only being exposed to five people rather than 500', says CSIRO food microbiologist Cathy Moir.

The dramatic increase of salmonella poisoning is largely attributed to a shift in preference towards chicken, with contamination by raw or undercooked chicken a major cause. However, raw or minimally cooked eggs are the largest cause of food-borne illness in Australia, particularly with the popularity of raw egg-based dishes like aioli.

There have also been outbreaks of salmonella in fresh produce such as rockmelons and cucumbers as health-conscious consumers favour salads, raw vegetables and minimally processed foods with lower salt and fat contents. 'The majority of dishes are safe and don't have salmonella on the surface, but once you start handling them – and handling them in a way that doesn't cook them – you're allowing more opportunities for salmonella to spread.' Food Safety Information Council CEO Juliana Madden says following one simple habit would go a long way. 'Honestly, a lot of the problem would be fixed if we just washed our hands'.

An ageing population is also 'a huge challenge', says food industry consultant Patricia Desmarchelier. Older people are particularly susceptible to illness and reduced mobility could mean that elderly people shopped less often, stored food for longer and ate more pre-prepared food, which could put them at risk.

Source: Adapted from 'Food poisoning on the rise' Inga Ting, Goodfood.com, 4 March 2014.

ACTIVITIES

1 Read the text and produce a mind map of the causes for the rise in food poisoning.

2 What advice would you give an elderly person who lives on their own to avoid food poisoning?

Value for money

Consumers know that it costs more to purchase a meal than to make it at home, but they still have a right to receive value for money. Consumers expect fresh, high-quality, nutritious ingredients and adequate, consistent **portion sizes**. They also expect friendly and efficient service. At restaurants, consumers often tip staff if the food is appealing and the service is exceptional.

By law, food service providers must provide menus where items are correctly priced and adequately described. Consumers also have the right to complain and be compensated if the food or service is poor.

Food labelling and marketing

Food service providers are not required to label their products like packaged foods, but they must still inform the consumer of the main ingredients, cooking method and portion size of the dishes available. Some providers use photographs and displays to inform customers, while others describe their menu items verbally; for example, 'Chicken Montrose (grilled chicken breast served with a fresh tomato and basil sauce)'.

A consumer has the right to ask if a dish has ingredients such as peanuts to which they may be allergic. Many food service providers now label vegetarian and gluten-free options on their menus to cater to a wider clientele.

The food service and catering industry is very competitive. Common marketing practices employed to entice customers include **media** advertising, sponsorships, competitions, food deals, memberships and giveaways.

UNIT REVIEW

LOOKING BACK

1 List four rights of consumers.
2 Name two pieces of legislation that the food service and catering industry must abide by when preparing food.
3 Describe the role of an environmental health officer.
4 What marketing practices do food service providers use to entice customers to purchase from them?

FOR YOU TO DO

5 Search for images of poor kitchen hygiene and make a list of any rules that have not been followed in one of the images you have found.
6 A restaurant reviewer has given a restaurant one star out of 10 for value for money. List factors that may contribute to poor value for money in food service.
7 Write informative but short descriptions for each of the following menu items:
- a seafood family feast
- a chocolate lover's dream
- healthy tropical chicken.

TAKING IT FURTHER

8 Produce a short video or other form of visual presentation on how to maintain food safety for the food industry.

7.5 Menu planning considerations

A menu is a list of dishes available to customers, which are ordered and usually presented in courses. Many factors need to be considered when planning menus.

Scale of function

Functions can involve a small or large number of guests. With smaller functions it is easier to provide more courses and options than it is when dealing with larger numbers.

Typical courses in a menu are as follows.

- Appetiser: Tasty food that is not too filling and stimulates the desire for more food
- Entrée (or first course): Usually a fish, soup, or salad in small portions
- Main course: Usually meat, fish or poultry with vegetable accompaniment
- Dessert: Sweet dishes, fruit or cheese.

WS 7.5 Menu planning considerations

Table 7.7 Types of menus

TYPES OF MENUS	CHARACTERISTICS
à la carte	• Lists all the dishes available, arranged in courses and each priced separately • Provides an extensive choice of menu items • Allows the customer to choose the number and type of dishes • There food is cooked to order so there is a waiting time • Can be expensive because skill is required to prepare each dish individually
Table d'hôte	• A restricted menu • Offers a small number of courses – usually two or three • Offers limited choice within each course • Set price • Faster service • Controls catering costs
Function menu	• For special occasions such as weddings and High Teas • Usually a fixed menu with little or no choice • Customer usually selects a menu package • Priced per head • Choice of sit-down meal, buffet or finger food service
Du jour menu	• Changes daily • Presented on a blackboard or verbally to the customer • A useful way to accommodate seasonal produce, test new recipes or use excess ingredients
Cyclic	• A series of fixed meals (usually breakfast, lunch and dinner) that rotates over a period of time, such as a week, fortnight or month • Provides only a few choices to pick from for each meal • Usually well-balanced nutritionally • Commonly used by schools, hospitals, nursing homes, camps and airline catering
Degustation	• Sampling many courses of small portions of food that show off the chef's specialities

Facilities, staff, time and money

An establishment plans menus according to the facilities, staff, time and money they have available. For example, an à la carte menu requires chefs with skill. It also requires a lot of equipment to prepare the varied dishes and is costly because many ingredients are required. Offering a table d'hôte menu limits choice and may help an establishment to save time and money.

Some caterers, such as those in nursing homes, need to feed a large number of people at one time. They usually have limited funding and facilities, and many dishes on their menus contain inexpensive ingredients and methods of cooking such as baking, which allows a lot of food to be cooked at once.

Arnold's Riverside Restaurant à la carte menu

Garlic and herb damper	$3.70
Soup of the day served with Turkish bread	$5.00
Wedges served with sweet chilli sauce and sour cream	$5.00

Garlic prawns served on a jasmine rice timbale	$11.90
Caesar salad – cos lettuce, egg and croutons served with chef's own Parmesan and anchovy dressing	$8.90
South Australian oysters (½ dozen)	$11.50
Chicken skewers – two skewers served on cos lettuce with a creamy sweet chilli sauce	$8.90
Penne pasta Dijon – chicken tossed in cream, sundried tomatoes and Dijon mustard and served with penne pasta	$9.20

Chargrilled prime eye fillet – 350 grams, with your choice of sauce and cooked to your liking	$21.00
Chicken filo – chicken breast fillet pocketed with sundried tomatoes and ham, wrapped in filo pastry and baked	$21.00
Fettucini puttanesca – olives, chilli and anchovies, tossed in tomato concasse and served with Parmesan cheese	$16.00
Honey and soy duck – pan-fried with crispy skin with honey and soy jus, served on bok choy	$18.90
Fisherman's catch – grilled barramundi, prawns, oysters, scallops and calamari, with lemon and mango mayonnaise	$21.00
Herb-crusted lamb racks – local lamb, oven-baked and served with couscous and a basil pesto sauce	$19.90

All main meals served with your choice of seasonal vegetables or a fresh garden salad

Chocolate bavarois served with raspberry coulis and fresh chantilly cream	$6.90
Banana crêpes served with banana and mango parfait	$6.90
Bailey's cheesecake – homemade cheesecake served with Tia Maria sauce	$7.20

Figure 7.9 An à la carte menu

$45 THREE COURSES
$39 TWO COURSES

APPETISERS

Chicken Caesar – chicken, bacon, croutons, cos lettuce and anchovy dressing
Deep-fried salted calamari served with a marinated green paw-paw and mint salad
Vegetarian nori roll plate – vinegared rice, pickled and fresh vegetables, wasabi and soy

MAIN COURSES

Chargrilled Atlantic salmon served with baby spinach, roasted red onion and tzatziki
Free range chicken breast served with grilled sweet potato, chive and coriander salad
Penne pasta – tomato, pesto, eggplant and kalamata olives

DESSERT

Fresh seasonal fruit plate
Chocolate pudding served with double cream
Ice-cream terrine

All mains served with bread rolls, chips, salad
Coffee or tea also provided

Figure 7.10 A table d'hôte menu

LILLYPILLY'S COUNTRY COTTAGE FUNCTION MENU $60.50 PER PERSON

PRE-DINNER STARTER

Homemade dips served with crudités and a selection of crackers

Entrée Savoury pancake topped with roasted Mediterranean-style vegetables and rustic salsa

Rib fillet of beef served with a mushroom and green peppercorn glaze

Chocolate, date and almond torte served with chocolate sauce and double cream Tea and coffee

Figure 7.11 A set function menu

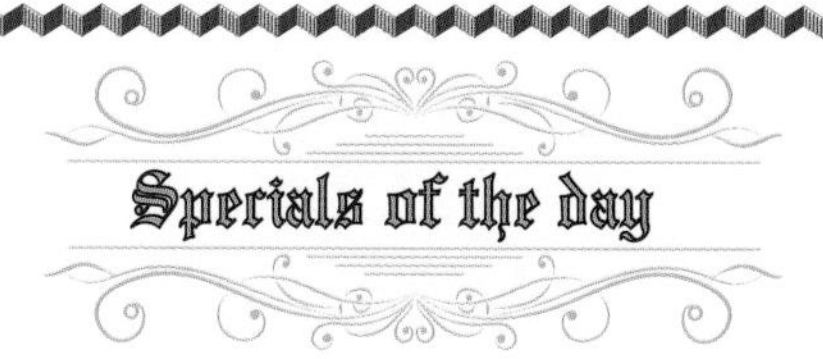

Herbed eggplant, cheese and tomato stacks	$7.00
Creamy gnocchi with prosciutto	$8.00

Seared fish cutlets with creamy pesto	$13.00
Chargrilled Mexican chicken	$16.00

Coconut and almond pudding	$7.50
Zabaglione	$6.00

Figure 7.12 A du jour menu

Meal	Monday	Tuesday	Wednesday	Thursday	Friday
Breakfast	• Fried eggs • Grilled tomatoes • Banana pancakes • Fresh fruit platter • Cereals/milk /yoghurt • Fresh juices	• Scrambled eggs • Bacon and sausages • Cinnamon waffles • Fresh fruit platter • Cereals/milk /yoghurt • Fresh juices	• Omelette • Breakfast burrito • Miso and rice • Fresh fruit platter • Cereals/milk /yoghurt • Fresh juices	• Frittata • Bacon and sausages • Hash browns • Fresh fruit platter • Cereals/milk /yoghurt • Fresh juices	• Fried eggs • Bacon and sausages • Sautéed mushrooms • Fresh fruit platter • Cereals/milk /yoghurt • Fresh juices
Lunch: Main	• Lemon couscous • Moroccan lamb • Harissa chicken • Steamed vegetables • Roast beef with Yorkshire pudding • Fresh garden salad Tabouli	• Vegetarian jambalaya • Cajun chicken and dumplings • Blackened snapper and tofu • Beef and sausage gumbo • Steamed seasonal vegetables • Oven-roasted chicken • Windsor court salad • Tossed green salad	• Greek spiced rice pilaf • Moussaka • Chicken kebabs • Spanokopita • BBQ marinated calamari • Roast lamb with mint sauce • Greek salad • Green salad	• Herb-tossed noodles • Beef stroganoff • Portuguese chicken • Eggplant marsala • Tomato and rosemary quiche • Steamed vegetables • Fresh garden salad • Potato salad	• Fragrant rice • Lamb rogan josh • Tandoori chicken • Pumpkin curry • Roast pork with apple • Pappadums • Banana and coconut salad • Fresh green salad
Lunch: Dessert	• Chocolate pudding • Ice-cream • Fruit salad	• Pecan pie with cream • Ice-cream • Fruit salad	• Chocolate mousse • Ice-cream • Fruit salad and toppings	• Sticky date pudding • Ice-cream and caramel sauce • Fruit salad	• Crème brûlée • Praline icecream • Fruit salad
Afternoon tea	• Vietnamese spring rolls • Meat and cheese platters • Fresh fruit	• Hot dogs • Vege dogs • Fresh fruit	• Pasties • Potato wedges • Fresh fruit	• Assorted wraps • Hot chicken kebabs • Fresh fruit	• Sushi • Sausage rolls • Fresh fruit

Figure 7.13 A cyclic menu used for film catering to feed cast and crew. Buffet service is provided for crew to select meals from. When filming at night, dinner options are available from a menu.

Figure 7.14 A degustation menu involves many small portions of foods to taste.

Year, time and occasion

Menus are often planned with consideration to the time of year. In winter, more soups and warm desserts are on a menu, while more salads and cold desserts are available in summer. In spring, lots of fresh vegetables are offered.

Depending on the time of day, many establishments will offer a breakfast, lunch or dinner menu. Breakfast menus and lunch menus are often lighter, quicker to prepare and include fewer options than dinner menus.

Different occasions also influence menus. For example, a wedding commonly uses function menus. Barbecues are often used to feed a large number of people for occasions such as school fundraisers.

Types of customers

The health, occupation, gender, age, preferences and number of people a food service provider has to cater for affect menu planning. For example, hospital caterers must account for the health, gender and age of patients in order to provide a nutritious meal. They consult with dietitians to plan suitable meals. Men, for instance, often have a higher energy requirement than women, and those recovering from heart disease may require low-fat meals. To cater for personal preferences, they often include several options as part of their menus, such as vegetarian or gluten-free items. Today, strict nutritional guidelines exist for childcare and school catering services.

UNIT REVIEW

LOOKING BACK

1 List three courses on a menu.
2 Explain the difference between an à la carte and a table d'hôte menu.
3 List the factors you need to consider when planning menus for different customers.

FOR YOU TO DO

4 Plan a blackboard menu of four breakfast options for cafe.
5 Plan a three-day menu (breakfast, lunch and dinner) for Year 9 and 10 students at school camp. The food will be prepared and served in a mess hall. Make sure you do not include nut products.

TAKING IT FURTHER

6 You are catering for a school formal. Prepare a professional-looking three-course function menu for the occasion. Set a price for the menu.
7 Find an example on the internet of the following menus:
- à la carte
- table d'hôte
- buffet function menu.

7.6 Recipe development

Many food service and catering establishments develop new recipes or modify existing recipes to add interest to their menus or to suit the needs of diners. For example, a restaurant chef may modify a basic recipe for apple crumble and turn it into a deconstructed apple crumble, while a hospital cook may produce a low-fat, low-sugar version of this dish to offer patients with heart disease.

Once they are proved reliable, recipes in the food service and catering industry become known as standard recipes. These recipes make it easier to cater for both large and small functions.

Ingredients

Ingredients in a recipe must always be of high quality, and preferably fresh, but in large-scale production some prepared ingredients may be more viable; for example, frozen vegetables. When modifying ingredients to provide healthier options, many low-fat, low-sugar and high-fibre ingredient alternatives are available.

Method of preparation

The method in a recipe provides instructions on preparing the ingredients.

When serving a large number of people at one time, baking and roasting are more suitable methods than frying or grilling, as they require less constant attention. These methods are also more appropriate than frying when developing low-fat recipes. Dishes that can be made in advance, such as chilled desserts, save time when preparing food for functions.

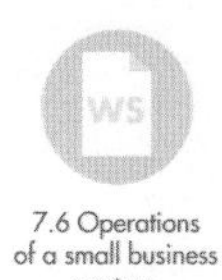

7.6 Operations of a small business venture

Portion size

A standard recipe indicates a number of servings, or the number of people it should feed. It also assists with ordering. For example, if a recipe serves 10, and 100 portions are required, then the quantity of each ingredient is multiplied by 10.

Cost per portion

The cost per portion is useful in assisting to set menu prices. It is also useful when selecting recipes to suit a budget.

Measuring

No matter how small or large the quantity of ingredients that are required in a recipe, they must be measured accurately. Too much or too little could spoil the appearance or flavour of a dish.

Recipe writing

When a recipe has been developed, it is important that it is written as procedural text so that others can follow it and produce a quality product.

HANDS-ON

DESIGN A CANAPÉ

PURPOSE

To engage in developing and creating canapés. Canapés are popular finger foods that are easy to prepare.

MATERIALS

Cream cheese
Mini toast
Choice of ingredients cut to size, such as stuffed olives, gherkins, smoked salmon, cold meats or caviar
Finely chopped herbs such as dill, parsley or chives
Piping bag

STEPS

1. Whip cream cheese.
2. Pipe onto mini toast.
3. Decorate the toast with a topping of your choice so that the canapés look colourful and appealing. You can garnish with herbs if necessary.
4. Take a photo of your design.

Tuna quiche

Name

Preparation time: 10 minutes
Cooking time: 40 minutes
Cost per serve: $1.90 Cost per portion

List of ingredients to be used

Quantities required

Serves 10	Serves 25	Serves 50	Ingredient
10	25	50	eggs
940 mL	2.3 litres	4.7 litres	reduced-fat milk
2½ cups	625 grams	1.25 kilograms	reduced-fat cheese, grated
2	6	12	onion, chopped
460 grams	1.15 kilograms	2.3 kilograms	tuna, drained and flaked
1¼ cups	400 grams	800 grams	wholemeal plain flour
2½ teaspoons	1½ tablespoons	3 tablespoons	baking powder

Method

1 Beat eggs and milk together.
2 In a separate bowl, mix together the cheese, onion, tuna, flour and baking powder.
3 Add the egg and milk mixture to the flour mixture.
4 Pour into a lightly greased dish.
5 Bake in 200°C oven for 35–40 minutes.

Method of preparation

Figure 7.15 The elements of a recipe

HANDS-ON

PRODUCE YOUR VERSION OF SCONES

PURPOSE

In this activity you will find and practise a scone recipe then change it to make it your own recipe, such as pizza scrolls.

MATERIALS

Cookbooks or recipe websites
Ingredients for scones
Other ingredients you want to use

STEPS

1 Find a basic scone recipe that uses Australian standard measurements.
2 Practise this recipe.
3 Come up with a variation for your scone, such as other ingredients or changes in shape.
4 Write up your idea as a recipe.
5 Trial the recipe.
6 Evaluate the recipe.
7 Rewrite the recipe with any changes you made.

Lemon and passionfruit butter

Goal or aim

Ingredients

2 egg yolks
1 teaspoon lemon rind, grated
2 dessertspoons fresh lime juice
125 grams softened butter
1 tablespoon passionfruit pulp

Equipment

double saucepan
whisk
paring knife
chopping board
wooden spoon

Materials/resources

Method

1 *Place* egg yolks, lemon rind and juice **in the top of a double saucepan**.
2 *Stir* with a wooden spoon over simmering water for 1 minute.
3 *Cut* softened butter **into small pieces**.
4 *Whisk* into mixture **gradually**.
5 *Continue* whisking **until sauce thickens**
6 *Remove* from heat **immediately**.
7 *Stir* in passionfruit pulp and *cool* to **room temperature before serving**.

Steps in order
Italicised text indicates verbs or adverbs, which usually begin sentences containing instructions.

Bold text indicates words that qualify given information – about how, where, when and for how long.

Delicious served on toast. Optional serving suggestion

Figure 7.16 Procedure text is used for writing recipes.

UNIT REVIEW

LOOKING BACK

1 Identify the five elements of a standard recipe.
2 Explain the importance of cost per portion in a standard recipe.
3 Explain the importance of measuring in recipe.

FOR YOU TO DO

4 Here are the ingredients for 20 serves of fruit muffins:
- 2 cups flour
- 1 teaspoon baking powder
- ⅓ cup caster sugar
- 1 teaspoon nutmeg
- 2 tablespoons oil
- 1 cup creamy yoghurt
- 1 egg
- 1 cup fruit, such as fresh blueberries or strawberries.

Using your knowledge of food and nutrition, suggest how a caterer at a sports and recreation camp could modify the below ingredients list so that the recipe:
- feeds 100 athletes
- is low in cost
- is highly nutritious.

TAKING IT FURTHER

5 Watch a cooking show. Report to the class the following information:
- the name of the chef or presenter
- the name of the dish they prepared
- a description of how the dish was prepared and what the chef did to make the dish interesting and appealing.

6 In groups, produce your own five-minute cooking segment of an interesting recipe you have developed or modified.

7.7 Purchasing systems

Ordering

In the food service and catering industry, food is usually ordered in bulk quantities, often from food wholesalers. Some establishments have contracts with suppliers to ensure that they receive consistent and exclusive products. Some caterers and chefs do their own shopping and will attend fresh food markets, selecting the best quality produce for their establishments. A business will telephone, email or use the internet to place orders.

Receiving

When receiving supplies of food, staff members are responsible for checking that:
- orders are correct
- food and packaging are undamaged
- ingredients are fresh.

If the stock meets the standards, it is quickly stored in its correct location to prevent risks to hygiene and safety. Incorrect or poor quality stock is not accepted.

Controlling

Stock control is very important as it ensures that there is sufficient good-quality stock when required, but not so much stock that there is waste or a problem with storage.

Some stock is checked daily while other stock is checked weekly or monthly. For example, cafes check milk supplies daily and coffee supplies weekly. Supplies of items such as straws may also be checked monthly.

There are different systems used in stock control. Some establishments use computer programs and others use a card system, while some simply list foods that are low in stock on a board or piece of paper.

Issuing

In most food establishments, food is collected from the dry, cool or frozen storage area as required. In larger establishments, food required may be requested using a requisition sheet and then issued by a stock controller.

The first-in, first-out (FIFO) system is employed when issuing food. When new stock arrives, the fresher ingredients go to the back while the older stock moves to the front so that it is used first. This helps to prevent food wastage.

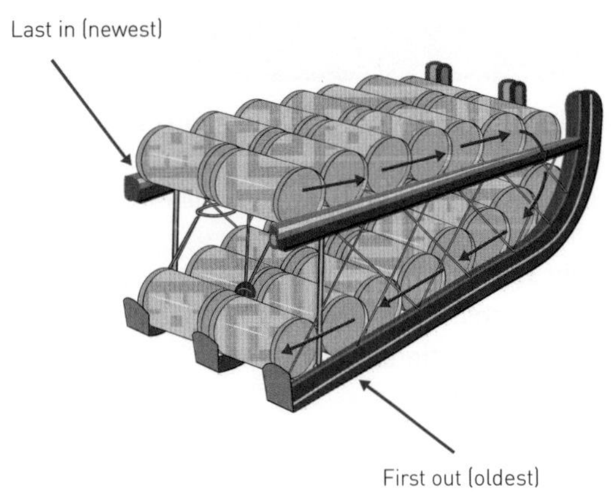

Figure 7.17 A FIFO food rotation system

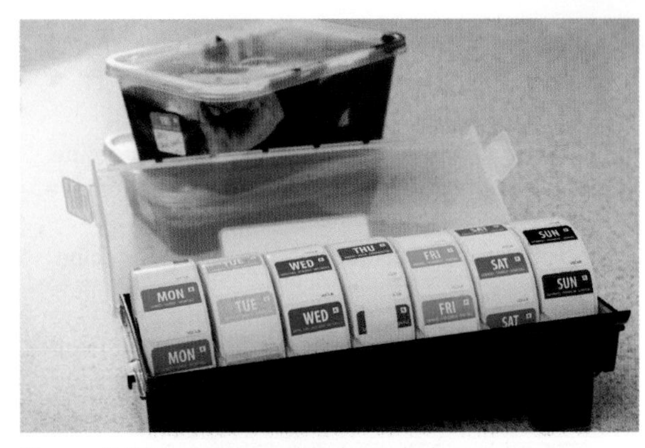

Figure 7.18 Labelling with dates helps to use up older stock first.

UNIT REVIEW

LOOKING BACK

1 What is checked when a food is received?
2 Why is it important to quickly store food after it has been delivered?
3 What procedures are involved in FIFO?

FOR YOU TO DO

4 Consider a take-away shop located near a busy workplace.
   a Estimate how often (daily, weekly or monthly) you think they would need to order:
   - bread
   - frozen chips
   - fresh tomatoes.

   b For each food in part **a**, describe how you would judge the freshness of food delivered.
   c What drinks may a take-away order from a supplier?

TAKING IT FURTHER

5 Visit your school canteen and interview the manager about their food purchasing system.
6 Find wholesale food suppliers in your area and produce a database to store the information about three of these suppliers.
7 Practise ordering food supplies over the internet from one of the supermarket websites.

7.8 Food service and catering considerations

When serving food, a number of factors need to be taken into consideration.

Plating food

Depending on the occasion and the establishment there are different ways of plating food, but no matter what form of plate service is used, good presentation of food is essential as first impressions count. So when plating foods, remember to:

- control the portion so that everyone gets the same amount but neither overload nor have too little on the plate
- arrange foods so that they look colourful
- use garnishes and decorations to add interest to dishes but ensure that they complement the food
- use clean and unchipped plates and wipe away any spills with a clean cloth
- place the best side of the food facing up.

Style of meal

Styles of meals can range from casual to formal. Take-away food is served packaged and taken away, or eaten at the venue on uncovered tables with plastic cutlery. For many casual meals, such as those served in bistros, plates and cutlery are provided, and again tables are commonly uncovered.

A popular style of meal is the buffet, where diners select foods from a variety of prepared dishes on display either on a table or from behind a counter. Providing **finger foods** for a guest to select at a function is also considered self-service. With more formal meals, the food and beverage attendant will be more involved in serving the guests and the table setting will be more structured.

Number of courses

Buffet and self-service work well when there are only a few courses to be served. As courses increase in number, it may be easier to provide plate service using traditional formal table setting.

When setting a table for formal service:

- allow 60 cm from the centre of one place setting to the centre of the next
- use clean, crisp tablecloths with dry, shiny, clean and undamaged glassware, cutlery and crockery
- working from the outside in, arrange cutlery 2 cm from the table edge in the order of courses with the knife blades facing the plate (in à la carte service, only the main course knife and fork are set; the rest are laid out after orders have been placed)
- dessert cutlery may be presented with the dessert or laid on the table before the meal, either next to the other cutlery or horizontally above the plate; if horizontal, the dessertspoon handle faces right and the dessert fork is placed below the dessertspoon with its handle facing left
- clear all plates, condiments and unused cutlery prior to serving desserts
- place glasses at the top of the dinner knife; in formal settings there are different glasses for different drinks
- place serviettes in the centre of the plate, to the left of the forks, above the plate, on the bread-and-butter plate or in the wine glass
- use attractive and appropriate table decorations that do not interfere with service or communication between guests.

Customer requirements

Families, couples, groups, business people, young children and seniors all require different styles of food service. For example, families with young children often prefer low-cost dining with prompt service and a children's menu.

Cost

Service where dining facilities are provided, as well as waiting staff to serve food, is more costly than self-service, buffet or take-away service. Usually, the more a customer pays, the more an elaborate and attentive service can be expected.

Time available

When a meal needs to be eaten quickly, take-away service or self-service works well. When more time is available, a more formal table service may be appropriate.

Table 7.8 Styles of plating food

SERVICE	DESCRIPTION
Plate service	All the food is put on a plate before being served.
Part plate service	The main item is put on the plate while the vegetables are served in a separate container from which customers help themselves.
Semi-silver service	The main item is put on a plate while vegetables are placed onto the plate at the table by a food attendant.
Silver service	An empty plate is presented to the customer and the food attendant serves all the food using a fork and spoon.
Guéridon service	The food attendant prepares all or part of a dish at a small table or trolley beside the customer's table.

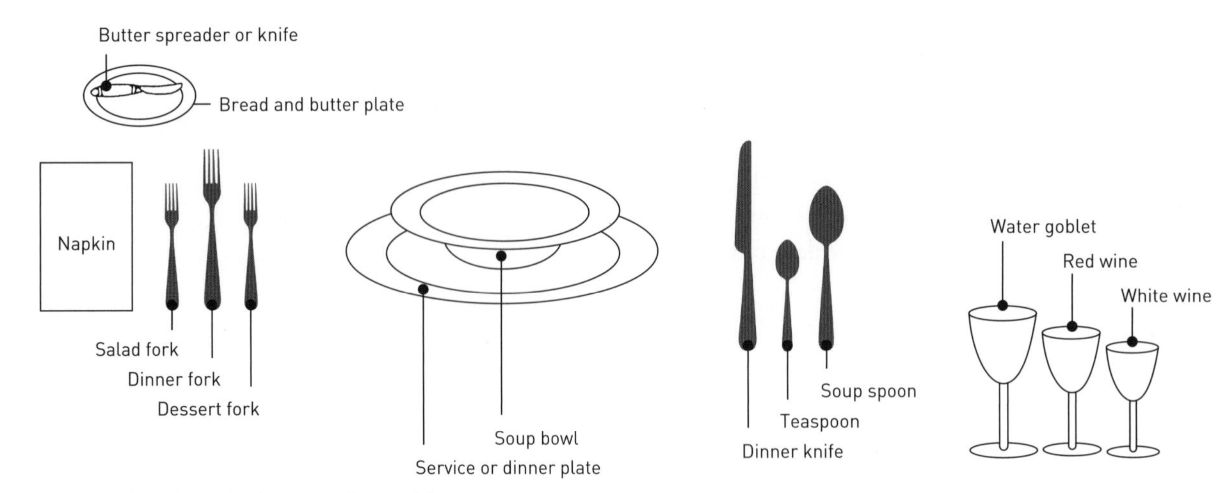

Figure 7.19 A formal place setting guide

FOOD IN FOCUS

BOURKE STREET BAKERY

Paul Allam and David McGuinness were bakers and chefs who shared a love of good food and an appreciation of all things hand-made. In 2004, they opened the Bourke Street Bakery in edgy Surry Hills as a little corner store bakery cafe where all their produce was freshly made on the premises.

While undertaking the fit-out of the bakery, the pair would hand out samples of test-bakes. By the time the store opened, the neighbours were hooked. Soon, customers would queue for their rustic sourdoughs, hazelnut and raisin loaves, pork and fennel sausage rolls, chicken pies, *pain au chocolat*, flourless chocolate cake and their renowned ginger *brûlée* tarts.

Over the years, Bourke Street Bakery has grown to over 100 staff and expanded its locations throughout Sydney. They have also expanded their business with some venues offering restaurant service, while some others offer sourdough bread-making classes and catering menus for both adults and children. The business has increased so much that now their pastries are made off-site at their Marrickville premise.

As a separate entity to Bourke Street Bakery, the owners established the Bread and Butter Project, Australia's first social enterprise bakery, where they teach refugees the art of baking in order to help them find jobs.

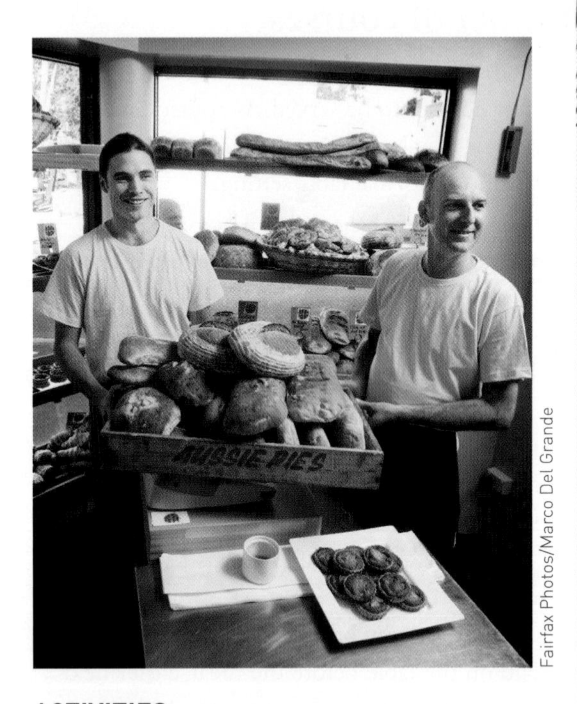

Fairfax Photos/Marco Del Grande

ACTIVITIES

1. In two words, describe what type of food establishment Bourke Street Bakery was when its first store was opened.
2. How did they promote their first store before it was opened for business?
3. List four items you would find at Bourke Street Bakery.
4. How has the Bourke Street Bakery expanded their business?
5. Complete a mind map on how Bourke Street Bakery contributes to the community.

UNIT REVIEW

LOOKING BACK

1 Compare two styles of plating food for service.
2 List five tips for plating food attractively.
3 Identify four different types of customers common to the food service and catering industry.

FOR YOU TO DO

4 You are a wedding planner. Make some suggestions to a couple with 100 guests on:
- styles of meal service
- number of courses
- costs per head
- reception time
- decorations.

5 Practise setting tables for the following function menu (or sketch the table setting):
- spring vegetable soup
- rack of lamb served with puréed kumara
- lemon ricotta tart.

TAKING IT FURTHER

6 Prepare a dish suitable for a Valentine's Day dinner in class and then set the table appropriately for the occasion.

7.9 Additional content: operations of a small business venture

A small business usually has fewer than one hundred employees. Many restaurants, cafes, take-away outlets and function caterers are run as small businesses.

Most owners of small businesses claim that working in their own business is enjoyable and rewarding but it is also time-consuming and hard work. To ensure success, small food ventures are encouraged to write proposals prior to establishing new businesses.

A proposal usually includes:

- a description of the business – where, what and why
- a market analysis – the opposition and the opportunities, strengths and weakness of the business
- a marketing plan – how the business will be promoted
- a financial plan – where finances will be obtained and how income is expected.

Table 7.9 Factors to consider when establishing and operating a small business

ECONOMIC	LEGAL	ENVIRONMENTAL	COMMERCIAL
• Equipment and decor • Menus • Advertising • Rent • Utilities such as water and gas • Labour costs • Food • Insurances • Cleaning • Taxes such as GST • Loans	• Businesses registration • Council approval • Legislation – health and hygiene, occupational health and safety, fair trading, industrial	• Environmentally friendly packaging • Recycling glass, paper, water and grease • Energy-saving techniques	• Producing goods and services that consumers want • Setting prices to cover costs and that consumers are willing to pay • Appropriate location and operating times

FOOD IN FOCUS

FOOD TRUCKS

Rafael Rashid operates two food trucks that operate every day at various Melbourne locations as well as at music festivals. Beatbox Kitchen serves up burgers and fries while Taco Truck sells the Mexican favourite.

While it might seem like a low-stress, low-maintenance business, that's not necessarily the case, says Rashid. 'The hardest part of the job is debunking the myth that street food is low quality, and perfecting the right balance of freshness and speed. You have to have the right menu, with simple and great food.'

Rashid says, 'There's much more to operating a food truck than just driving from venue to venue, dishing out simple but tasty meals. There are serious costs involved too; for instance, you have to pay high rent at some music festivals. We rent a warehouse to store supplies; it has a kitchen where we prep food, which allows us to get out a lot more often than if we just had the truck.'

Fairfax Photos

Sydney's Eat Art Truck has a barbecue theme. Favourite menu items include pulled pork in a bun with barbecue sauce; shichimi wings (Japanese pepper seasoned barbecue chicken wings); and beef ssam, a Korean-style lettuce wrap with twice-cooked beef. The flashy van is covered with canvas, painted by a different street artist each month.

Operator Stuart McGill agrees that the costs of operating a food truck aren't to be dismissed. 'We use a prep kitchen, and then we've got wage costs for myself and two other chefs. The labour cost is definitely our biggest expense. We have a 15 kVA generator we have to fill with premium petrol and that takes a tank every three hours. Running costs aren't minimal.'

The Eat Art truck has a basic kitchen that comprises two combi ovens, a chargrill, a deep fryer and a touchscreen point-of-sale system. 'It's essentially like a commercial kitchen, just on the back of a truck,' McGill says. 'We had to abide by the same standards and build it to the criteria that any other restaurant would have to abide by. So we had to think about exhaust fans and things like that. We obviously have to have hot running water and we have to deal with waste water, all those sorts of things that you wouldn't automatically think of, but a crucial part to you operating in a normal manner.'

Source: 'The Business of Street Food', by Danielle Bowling, *Hospitality Magazine*, 23 August 2012. The Intermedia Group

ACTIVITIES

1 List four foods that are sold from food trucks, as described in the article.
2 Using the article, make a list of costs involved in operating a food truck.
3 Predict other costs that may be involved in operating a food truck.

UNIT REVIEW

LOOKING BACK

1 Define the term 'small business'.
2 Give three examples of small businesses that operate in the food service and catering industry.
3 List five economic costs in operating a business.
4 List two legal issues that need to be considered when starting a business.

FOR YOU TO DO

5 Make a proposal for a small food business that you would like to operate. Include the following information in your proposal:
- the type of business you would like to start
- where your business will be located and why
- the foods that will be sold
- what equipment your business will need and why
- what staff will be required and why
- how you will promote your business.

TAKING IT FURTHER

6 Visit a small food business in your community and report on the business; for example, products sold, opening hours, location and staff duties.

Chapter review

LOOKING BACK

1 List three profit-making food service ventures.
2 Describe the role of two back-of-house career opportunities in the food service industry.
3 Describe the role of two front-of-house career opportunities in the food service industry.
4 Describe the role of a manager in the food service industry.
5 List three requirements for working in the food service and catering industry.
6 Outline two responsibilities of employees.
7 What is an award?
8 Outline four consumer rights.
9 What does the acronym FIFO stand for?
10 Identify four parts of a recipe.

FOR YOU TO DO

11 **a** Find examples of an à la carte menu and a table d'hôte menu, and compare them.
b Discuss the advantages and disadvantages of both menus.
12 Explain how a menu may be different for a sit-down function than for a function where the guests stand.
13 Make a list of different ways to modify a basic muffin recipe.

TAKING IT FURTHER

14 Plan and prepare a Mother's Day menu suitable to serve on a tray. Justify your food selections.
15 **a** Look up employment opportunities in the food industry on the internet.
b Find a front-of-house and a back-of-house opportunity.
c Identify the qualities and qualifications required for the job.

Bruschetta

This appetiser is popular at many dining establishments.

 **Preparation time** 10 minutes  **Cooking time** 5 minutes  **Serves** 2

INGREDIENTS

- 2 slices of Italian or Vienna bread
- 1 tomato
- A few basil leaves
- ¼ Spanish onion
- 1 tablespoon olive oil

METHOD

1. Toast the bread.
2. Wash, dry and cut the tomato in half.
3. Deseed the tomato. Square off and then dice.
4. Dice the onion.
5. Chop the basil.
6. Combine the tomato, onion, oil and basil.
7. Divide onto the toast just before serving.

QUESTIONS

1. From where does bruschetta originate?
2. Why should the tomato be washed?
3. What other breads could be used for bruschetta?
4. What are some other popular appetisers and cocktail (finger) foods?

Mark Fergus

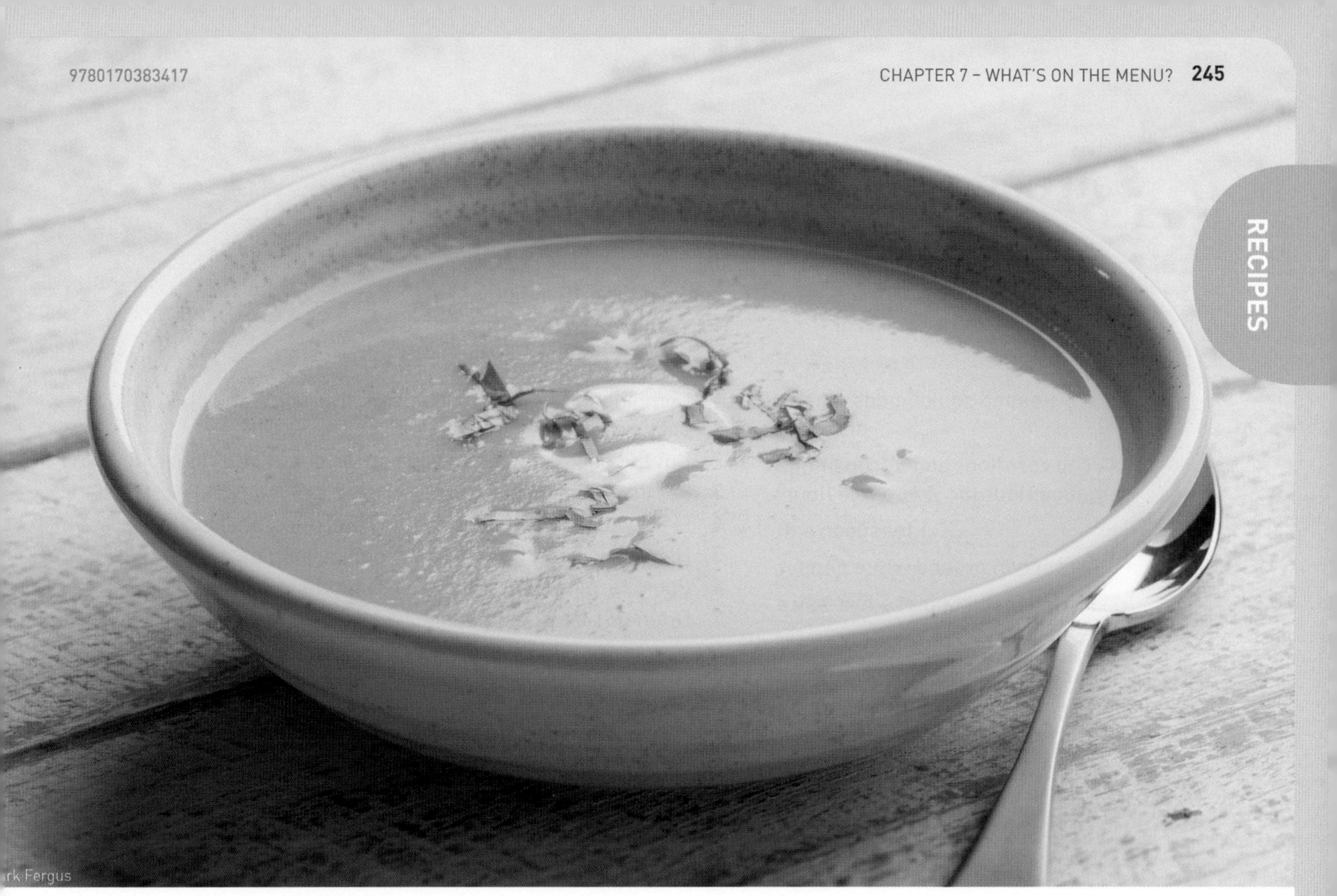

rk Fergus

Sweet potato soup

When garnished, soup makes an appealing winter entrée.

 **Preparation time** 10 minutes  **Cooking time** 15 minutes  **Serves** 2

INGREDIENTS

- 1 teaspoon olive oil
- ⅓ onion, chopped
- ¼ teaspoon crushed garlic
- 1 teaspoon curry powder
- 300 grams orange sweet potatoes (kumara), peeled, cut into 2 cm cubes
- 400 mL chicken stock
- ¼ cup coconut cream

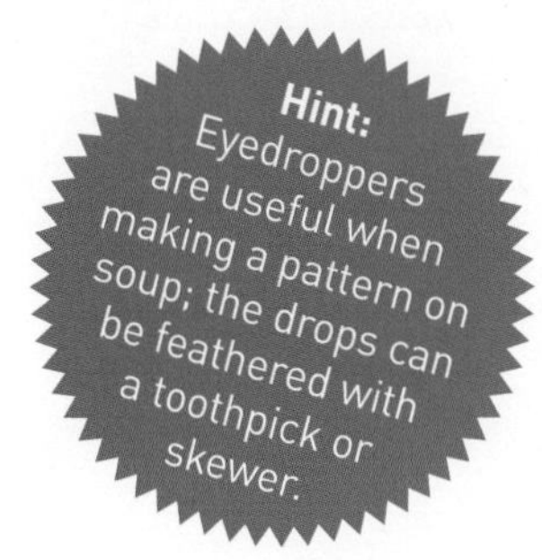

METHOD

1. Heat the oil in a large saucepan over medium-high heat.
2. Sauté the onion and garlic.
3. Add the curry powder, stirring for 1 minute or until aromatic.
4. Add the sweet potato and stock. Stir to combine. Bring to the boil.
5. Reduce heat to low. Simmer, partially covered, for 15 minutes or until the sweet potato is tender. Remove from heat.
6. Using a food processor or hand blender, puree the soup in batches until smooth.
7. Return the soup to saucepan over low heat. Stir in the coconut cream (do not allow to boil). Add more stock if soup is too thick. Season.
8. Serve.

QUESTIONS

1. What is the difference between simmering and boiling?
2. What does it mean to season?
3. How could you garnish the dish?
4. What are some other types of soups?
5. How many millilitres would ¼ cup coconut cream be?

9780170383417

Salt and pepper squid

Calamari has become a staple on many dining menus.

 **Preparation time** 10 minutes  **Cooking time** 15 minutes  **Serves** 2

INGREDIENTS

- 2 cleaned squid hoods – fresh or frozen
- ¼ cup cornflour and ¼ cup of fine semolina flour (or ½ cup rice flour)
- 1 teaspoon salt
- 1 teaspoon ground white pepper
- ½ teaspoon Chinese five-spice
- ¼ teaspoon chilli powder
- Vegetable oil, for shallow frying

METHOD

1. Combine the flours, pepper, salt, five-spice and chilli in a shallow dish.
2. Cut the squid into 5 mm wide strips or cut into rings.
3. Pat dry with a paper towel.
4. Lightly toss the squid in the flour mixture, shaking off excess.
5. Heat the oil in a frypan over medium-high heat.
6. Cook, turning, for 2 minutes or until light golden.
7. Drain on absorbent paper.
8. Serve with lemon, tartare sauce or aioli.

QUESTIONS

1. How much oil is used when shallow frying?
2. How can you tell when the oil is ready to add food?
3. Why should you not overcrowd the pan with food?
4. What types of oil are suitable for frying?

Mark Fergus

9780170383417

ark Fergus

Steak Diane

The steak in this popular dish from the 1970s can be easily replaced with chicken slices.

 **Preparation time** 10 minutes  **Cooking time** 15 minutes  **Serves** 2

INGREDIENTS

- 15 grams butter
- 1 tablespoon oil
- ½ cup corn flour or plain flour
- 2 tablespoons olive oil
- 2 slices of minute steak
- ½ teaspoon minced garlic
- 1 teaspoon Dijon mustard
- 1 teaspoon Worcestershire sauce
- ¼ cup cream
- 1 tablespoon finely chopped flat-leaf parsley

METHOD

1. Coat the steaks in flour.
2. Heat the butter and oil in a frypan over medium heat.
3. Cook the meat for 2–3 minutes on each side until golden and remove from pan.
4. Add the garlic to the frypan and cook for 30 seconds.
5. Add the meat and then stir in the mustard, sauce and cream.
6. Simmer on medium heat for 2–3 minutes until a thick sauce covers the meat.
7. Stir in the parsley.
8. Serve with cooked vegetables.

QUESTIONS

1. What is the role of the butter in this recipe?
2. What is the role of the flour in this recipe?
3. Where does Dijon mustard originate from?
4. What are two ingredients in Worcestershire sauce?

9780170383417

RECIPES

Mini baked cheesecakes

Desserts are often prepared well in advance of the main meal. Have fun decorating this dessert.

 **Preparation time** 15 minutes  **Cooking time** 25 minutes **Chilling time** 3 hours  **Makes** 2

INGREDIENTS

125 grams cream cheese
¼ cup caster sugar
¼ teaspoon vanilla extract
1 teaspoon lemon juice
1 egg
2 Arnott's Granita biscuits
2 paper muffin cases

METHOD

1. Preheat oven to 180°C. Line a muffin tin with paper cases.
2. Beat together the cream cheese, sugar, lemon juice and vanilla until fluffy.
3. Mix in the egg.
4. Pour the cream cheese mixture into the muffin cups, filling each ¾ full.
5. Top the filling with the biscuits.
6. Bake for 15–20 minutes or until firm.
7. Refrigerate the cheesecakes until chilled. Remove paper cases.
8. Ensuring biscuit is the base, decorate the cheesecakes with canned peaches, passion fruit, berry coulis or chocolate shavings.

QUESTIONS

1. Predict the equipment required for this recipe.
2. What berries could be used to decorate the cheesecakes?
3. How many times would you need to multiply this recipe if you had to cater for 30 people?

Mark Fergus

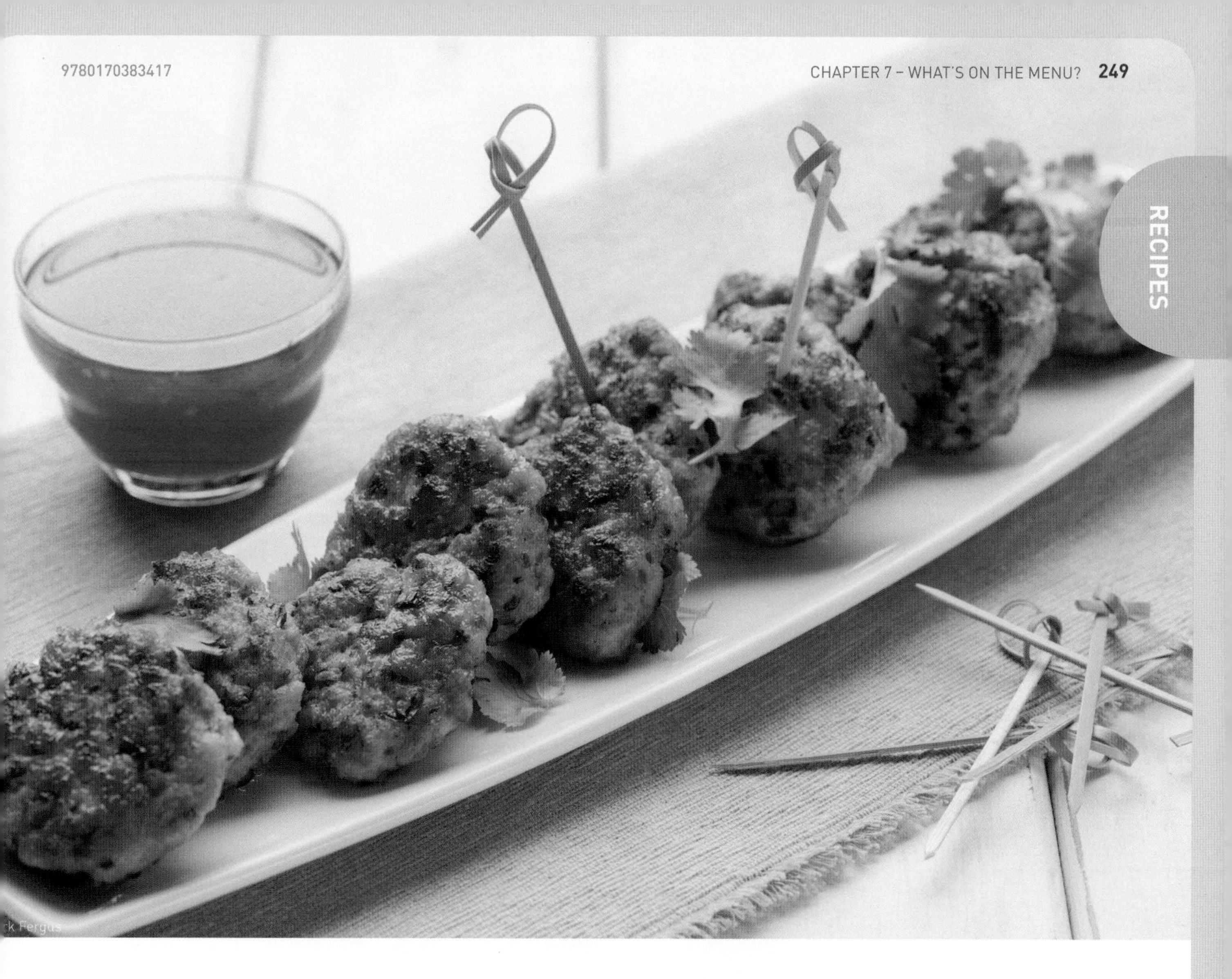

Thai chicken cakes

These are great to serve as finger food for a function.

 **Preparation time** 25 minutes  **Cooking time** 10 minutes  **Serves** 2

INGREDIENTS

250 grams chicken mince
1 tablespoon of beaten egg ⅓ cup fresh breadcrumbs
1 teaspoon fish sauce
1 teaspoon red curry paste
1 tablespoon finely chopped coriander
⅛ teaspoon of chilli powder
1 shallot, finely chopped
¼ cup vegetable oil for frying
sweet chilli sauce for serving

METHOD

1 Place the mince, egg, breadcrumbs, fish sauce, curry paste, coriander, chili and shallots in a bowl and use your hands to combine.
2 Using a tablespoonful of mixture, shape into 12 small patties, pressing to flatten slightly. Transfer to a large baking tray and chill for 10 minutes.
3 Heat oil in a frypan over medium-high heat. Cook the patties for 2 minutes each side or until cooked through.
4 Drain on absorbent paper.
5 Serve with some toothpicks and sweet chilli dipping sauce.

QUESTIONS

1 How do you make fresh breadcrumbs?
2 What is the main ingredient in fish sauce?
3 What are other suitable finger foods that could be served at a large function?

CHAPTER 8

Mark Fergus

Food for life

FOCUS AREA: FOOD FOR SPECIAL NEEDS

FOCUS AREA

The food needs of individuals are determined by a number of factors, including cultural influences and location, as well as changes in age, health and lifestyle. This chapter investigates the special food needs of individuals. The recipes included in this chapter are quick to make and nutritious, and can be easily adapted to meet the specific needs of all individuals.

CHAPTER OUTCOMES

In this chapter you will learn about:

- reasons for special food needs
- support available for people with special food needs
- special foods for special needs
- planning special food needs.

In this chapter you will learn to:

- identify why people may have special food needs
- examine support for people with special food needs
- prepare and assess the suitability of foods for dietary disorders
- address special food needs
- research the circumstances of a particular group and organise a dietary plan.

RECIPES

FOOD FACTS

The recommended weight gain during pregnancy is 10–13 kg.

The sense of smell is intensified during pregnancy. This is why some pregnant women suffer from nausea when they smell certain foods.

Approximately 25–30 per cent of Australian children are overweight or obese. Australia has one of the highest rates of childhood obesity in the world and it is rapidly increasing. Overweight and obese children have an 80 per cent risk of becoming overweight or obese adults.

10–20 people in Australia die each year from anaphylaxis, an extreme allergic reaction where the mouth, tongue and throat swell and constrict breathing. Many more have severe allergic reactions.

Food intolerances affect 5–10 per cent of the adult population, while food allergies affect 1 per cent.

Results from the 2011–12 National Health Survey run by the Australian Bureau of Statistics were released in 2013. The results showed that:

- 62.8 per cent of Australian adults were overweight or obese
- more men (69.7%) than women (55.7%) were overweight or obese
- approximately 25 per cent of all children were overweight or obese
- slightly more girls (25.7%) than boys (24.6%) were overweight or obese.

Athletes in heavy training are at risk of iron deficiency because they lose iron through the sweat and tissue damage that occurs when the body is injured during exercise. Female athletes are particularly at risk.

FOOD WORDS

anaemia condition resulting from an inadequate intake of iron, causing the sufferer to feel weak, tired and fatigued

antibody protein produced by the body to fight foreign particles

balanced diet nutrition intake that provides energy, nutrients, dietary fibre and fluids in amounts appropriate for good health

colostrum first breast liquid produced by a mother after childbirth; it is high in proteins, minerals and antibodies

convalescence period of recovery of health and strength after injury or illness

food allergy abnormal, unpleasant or adverse reaction to food by the body's immune system

food intolerance abnormal, unpleasant or adverse reaction to food caused when the nerve endings in different parts of the body become irritated

lactation production of milk for breastfeeding

life cycle stages through which people pass that begin when a female egg is fertilised and end at the time of death

logistics planning and carrying out of any complex or large-scale activity

perishable having a short shelf life; spoiling quickly and needing careful storage

wean introduce semi-solid foods into the diet of the infant and reduce milk intake

8.1 Special food needs

The stages of the human life cycle are infancy, childhood, adolescence, adulthood and older age, as well as pregnancy and lactation. Nutritional requirements are different at each stage of the life cycle (see Chapter 2).

Pregnancy

In pregnancy, a woman's diet must provide all the nutrients required for the development of the embryo and foetus. This is why it is important for the mother to have a well-balanced, nutritious diet established prior to becoming pregnant.

A developing foetus does not use a lot of energy, so a large amount of extra food is not necessary.

A pregnant mother requires:

- a good supply of protein for the growth of new cells
- additional calcium, phosphorus and Vitamin D to assist bone and teeth formation
- B Group vitamins, which aid the metabolism of carbohydrates
- iron and folate to prevent anaemia
- increased fibre intake to prevent constipation
- six to eight glasses of water per day.

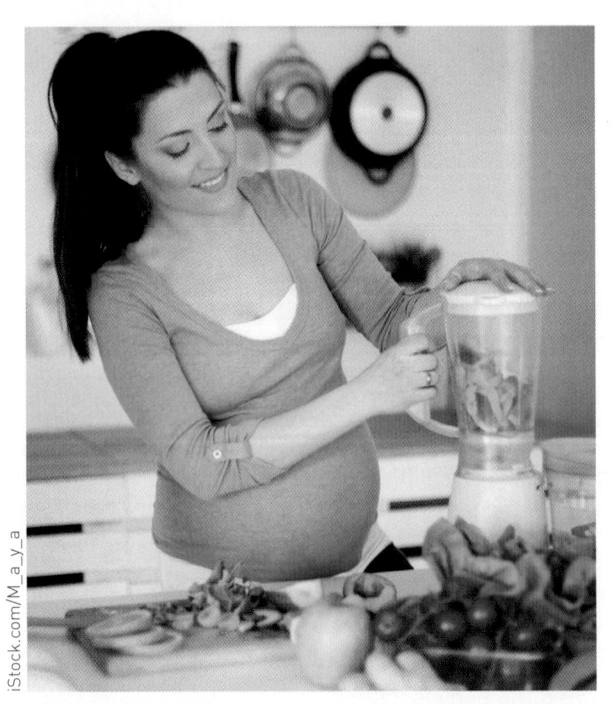

iStock.com/M_a_y_a

Figure 8.1 Pregnancy requires a nutrient-rich diet.

Alcohol consumption should be restricted during pregnancy, as it may cause foetal alcohol syndrome. Smoking and the consumption of illegal substances are not recommended at any stage of the life cycle and during pregnancy they can reduce the supply of oxygen, which significantly affects the growth of the baby.

Lactation

When a woman is breastfeeding she is said to be lactating. An increase in nutrients, specifically protein, vitamins and minerals, is essential at this time as the mother is producing the food requirements for her newborn child. This production of breastmilk is known as lactation. Breastmilk is high in energy and a complete source of nutrition for most babies up until the age of six months.

Breastmilk provides perfect nutrition and important substances for protection against infection, and is easily digested by an infant. The first fluid to come from the breast is a substance called colostrum, which is rich in antibodies, proteins and minerals but has less sugar and fat than the milk that follows.

Infancy

Infancy is the stage of the life cycle from birth to 12 months of age. Nutrition is important during this time, as rapid growth and development take place. A well-balanced diet is required for the total development of the child. After about six months of thriving on breast or formula milk, a baby may need some solid food for extra nutrients. From about six months of age, babies also need texture in their food to encourage them to chew and develop the muscles that will be required for speech. From the very first time a baby is offered food other than milk, the process of weaning has started. It is important not to introduce solid food too early as the digestive system and kidneys are not developed enough to cope with food.

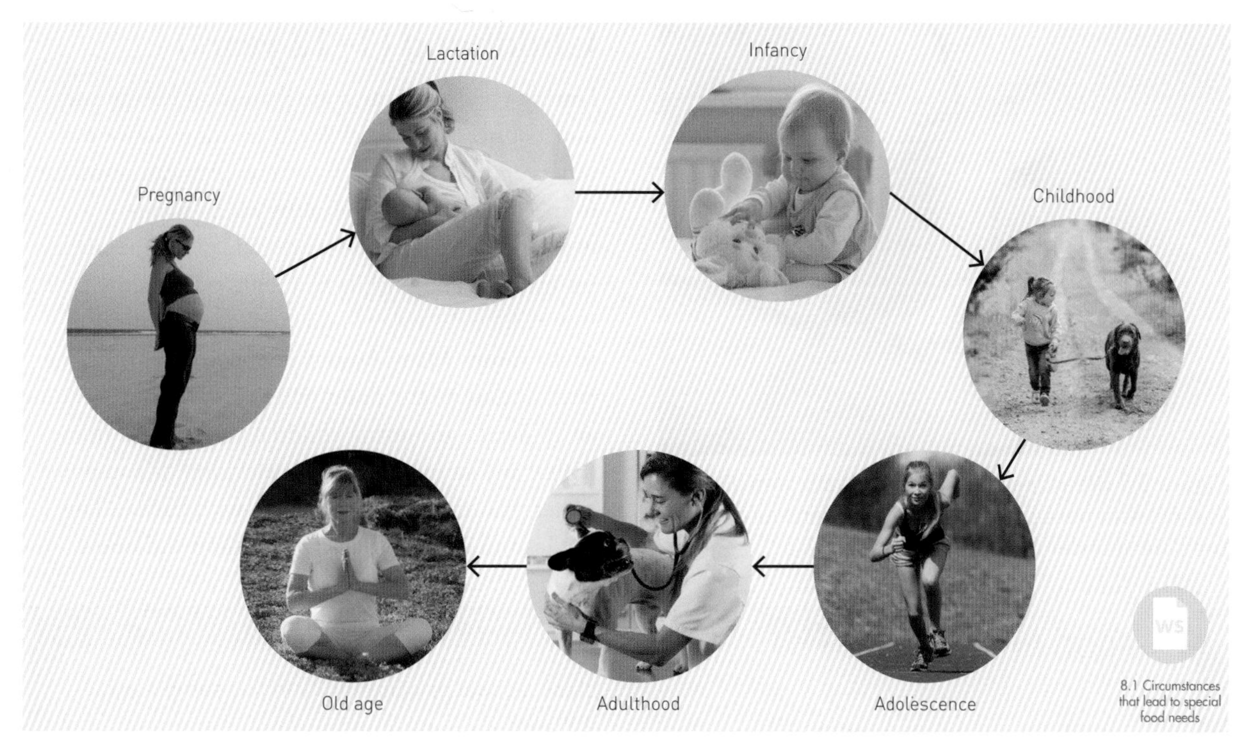

Figure 8.2 Stages of the life cycle

Clockwise from top left: Getty Images/Stephen Mallon; Getty Images/KidStock; Shutterstock.com/denniro; Shutterstock.com/pliona; Getty Images/Westend61; Shutterstock.com/135pixels; Getty Images/Elfi Kluck

FOOD IN FOCUS

BREAST IS BEST FOR SIX MONTHS

Mothers will be encouraged to feed their babies only breast milk for at least six months as part of an ambitious new national breastfeeding policy.

The goal of this policy is to increase breastfeeding levels, as currently only 14 percent of mothers are fully breastfeeding their babies to six months. Nearly half of mothers have abandoned breastfeeding after three months.

Federal and state health ministers have endorsed the new strategy, which calls for more community acceptance of breastfeeding in public, more support and training for mothers before and after delivery, and increased access to parental leave.

Breastfeeding expert Jennifer James said government leadership was vital to making breastfeeding more socially acceptable in Australia. 'Considering over a quarter of Australians think that breastfeeding in public is unacceptable, we know there is a long way to go.' Dr James said it was particularly worrying that recent research had shown many women aged 18–25 disapproved of breastfeeding in public.

Dr James welcomed the new strategy's recommendation for more practical support for mothers to begin breastfeeding and support for workplaces to adapt to the needs of breastfeeding mothers. Health Minister Nicola Roxon said Australia did not have high levels of breastfeeding compared with many countries, even though there was strong evidence of health benefits for babies and mothers.

Source: Adapted from 'Breast is best for six months' by Mark Metherell, *Sydney Morning Herald*, 14 November 2009.

ACTIVITIES

1 Identify why the national breastfeeding policy was developed in 2009.
2 Why do you think many 18 to 25-year-old women disapprove of breastfeeding in public? Do you think older women also find it socially unacceptable? Discuss your thoughts with the rest of your class.
3 Explain how community acceptance and increased parental leave would support breastfeeding mothers.
4 Use the internet to research a country with high levels of breastfeeding. Compare this country with Australia and identify the similarities and differences that are evident.

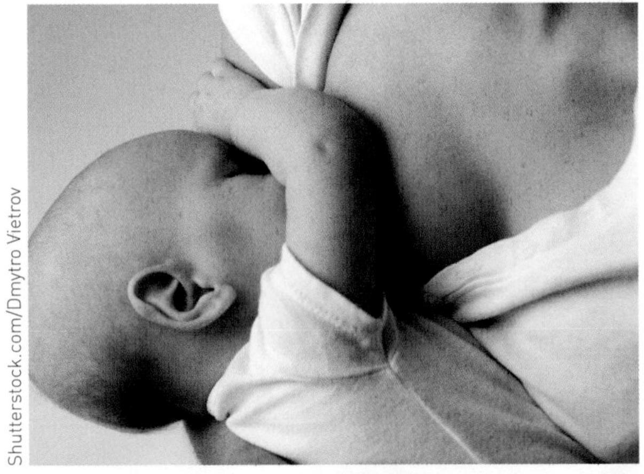

Figure 8.3 Breastfeeding supplies a baby with all food and nutrient needs. Mothers are being encouraged to breastfeed under new government policies.

Childhood (1–12 years)

During childhood there are increasing energy and nutrient needs. Some children are more active than others and may require more complex carbohydrates to meet energy needs. Healthy eating habits developed in early childhood will lay the foundation for healthy eating habits in adulthood. Healthy eating may reduce the risk of nutrition-related disorders in the future. Many Australian children are suffering from obesity to the point where it has almost become an epidemic. Too many children come home from school and sit in front of the television or computer rather than run and play outside with friends. This can lead to children being overweight and obese.

FOOD IN FOCUS

CHILDHOOD OBESITY RATES COULD BE SLASHED WITH SIMPLE CHANGES, REPORT SAYS

Childhood obesity rates could be slashed if families made small changes such as 15 minutes of exercise and cutting one small chocolate bar from their diet each day.

Research from Canberra University's Health Research Institute found one in four Australian children were overweight, although this number could be halved with simple changes.

The research, conducted with Robert de Castella's charity SmartStart for Kids, was based on data collected from a sample of 31 000 Australia students since 2000. It found the prevalence of obesity could be reduced markedly within 15 months.

'We asked what it would take to reverse the increase seen in the prevalence of overweight in Australian school children over recent decades,' said professor of public health Tom Cochrane.

'We were surprised by the answer: just a small daily dietary restriction equivalent to just one treat-size bar of chocolate and about 15 minutes extra of moderate physical activity per day.'

'Overweight and obesity is widespread in Australian school children,' the report said. 'No class, school or community in our sample was immune.'

The report found advertising and exposure to junk food were obstacles to reform along with limited access to healthy, affordable foods in schools.

'Clearly, this is a societal problem related to the way that children's lives are lived today compared with the way they would have been lived 40 years ago, when the so-called problem of childhood obesity did not exist at the epidemic levels seen today.'

Mr de Castella said childhood obesity was causing significant health challenges and required renewed attention.

'Each day kids are exposed to a series of factors which affect their weight, such as the availability and exposure to food and drink; loss of opportunities for play; and the uptake of television and other media

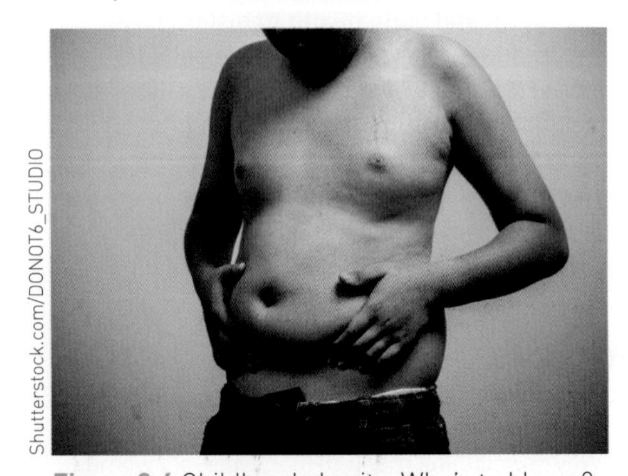

Figure 8.4 Childhood obesity: Who's to blame?

>

which encourage sedentary lifestyles,' Mr de Castella said.

'However, this research demonstrates that child obesity can be addressed and throws down a challenge for Australia to do better for its children and their future health and wellbeing.'

Source: 'Childhood obesity rates could be slashed with simple changes, report says', by Henry Belot, *The Age*, 11 November 2015.

ACTIVITIES

1 Outline why childhood obesity is so widespread in Australian school children.
2 Use the internet to research strategies that could be used to reduce current childhood obesity rates. Discuss your findings with the rest of the class.
3 Plan a one-day menu that a school-age child would enjoy eating, which could be used to help them lose weight.
4 Research childhood obesity rates in other countries. Compare and contrast these findings.
5 Discuss strategies that could be implemented to prevent children becoming overweight or obese.

Adolescence (12–20 years)

The teenage years are a time of rapid growth and development, when adolescents need more energy. Rapidly growing bones also need plenty of calcium. The increase in body tissue and hence blood volume in boys, and the onset of menstruation in girls, mean that extra iron is needed by both sexes. There is an increased need for:

- energy for functioning of the body
- protein for growth and repair
- vitamins for maintenance, health and development
- minerals, especially calcium, phosphorus and iron.

Adulthood

At this stage in the life cycle, growth has usually stopped and activity levels have stabilised or declined. If a balanced diet is consumed, all nutrients necessary for good health will be obtained. Adults need to ensure that their energy intake balances their energy output so that weight gain is avoided. By middle age approximately 50 per cent of Australians are overweight, due to a combination of high kilojoule intake and decreased physical activity. The dietary guidelines for Australian adults have been developed to encourage healthy lifestyles that minimise the risk of developing diet-related disorders. Refer to these dietary guidelines in Chapter 2.

The aged

Australians are living longer than ever before. In 2008, around 13 per cent of Australia's population was aged 65 and over; by 2056 this figure is expected to be between 23 and 25 per cent. Good nutrition and an active lifestyle are vital in maintaining the wellbeing and health of older people.

Older Australians, particularly those living alone, can lose the motivation to prepare meals for themselves and this can lead to malnutrition. While appetite and energy requirements generally decrease, a balanced diet is important. Attention should also be paid to specific nutrients such as protein, calcium, iron and Vitamin C. Being overweight or obese may become a problem unless sufficient exercise is undertaken.

iStock.com/FredFroese

Figure 8.5 Good nutrition and an active lifestyle are vital in maintaining the wellbeing and health of older people.

The nutritional needs for older people have been recognised through the development of the following specific guidelines for healthy independent Australians aged 65 years and over:

1 Enjoy a wide variety of nutritious foods.
2 Keep active to maintain muscle strength and a healthy body weight.
3 Eat at least three meals every day.
4 Care for your food – prepare and store it correctly.
5 Eat plenty of vegetables (including legumes) and fruit.
6 Eat plenty of cereals, breads and pastas.
7 Eat a diet low in saturated fat.
8 Drink adequate amounts of water and/or other fluids.
9 If you drink alcohol, limit your intake.
10 Choose foods low in salt and use salt sparingly.
11 Include foods high in calcium.
12 Use added sugars in moderation.

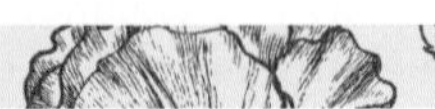

FOOD IN FOCUS

TEENAGERS FAVOUR FAT OVER FRUIT: SURVEY

The teenage appetite for greasy food has been confirmed by new Australian research showing one in four adolescents ignore fruit but eat fast food daily.

The results from a survey of 3800 teenagers has revealed what many adults already suspected – the healthy, balanced diet has eluded many young people.

'Our study found that most teenagers are far from having diets that will provide their growing bodies with the nutrients they need to ensure their long term health and wellbeing,' said Professor David Crawford, who headed the Deakin University study.

The survey of 12 to 15-year-olds found that 90 per cent were consuming unhealthy foods like fast foods, energy-dense snacks and sugar-sweetened drinks on a daily basis.

A quarter admitted to eating fast food every day, while more than a third said they hardly ever ate fruit.

Only one third of teenagers ate a balanced diet – at least one food from each of the five food groups every day.

Prof Crawford said it was clear a significant number of adolescents fell short of the recommendations outlined in the Australian Guide to Healthy Eating.

'The daily inclusion of fast foods coupled with the omission of a variety of healthy foods is setting many teenagers up for serious health problems,' he said.

The most prominent was obesity and 'spin-off' conditions like type two diabetes, poor heart health and depression.

The results, to be published in the Asia-Pacific Journal of Clinical Nutrition later this year, show that bread and cereals were the most commonly consumed of the food groups, followed by vegetables, dairy foods and then protein-rich foods like meat, eggs, nuts and legumes. Fruit came last.

Teenagers in regional areas tended to eat more vegetables and less fast foods than their metropolitan counterparts.

Girls' diets included more fruit and less fast food and sweetened drinks than boys, and boys consumed more meat.

Source: 'Teenagers favour fat over fruit: survey', *The Age*, 11 April 2007. The Age.

ACTIVITIES

1 Define a healthy, balanced diet.
2 Revisit Chapter 2 and find the recommended dietary intake (RDI) for children and adolescents. List five of your favourite foods and investigate whether the sodium, sugar and saturated fat levels are higher than Food Standards Australia recommends.
3 Visit the Food for Kids website and select a food product that you consume regularly. Does it meet the RDI levels set down for your gender and age? Discuss your findings with the rest of your class.

Food for Kids

4 Identify an unhealthy food that you consume regularly. In a group, discuss healthier foods that could be used as an alternative.
5 Unhealthy eating does not only result in obesity. Explain two other conditions that can result from poor eating habits.
6 Evaluate why teenagers in rural areas tend to consume greater amounts of fresh fruit and vegetables compared to those living in metropolitan areas.

UNIT REVIEW

LOOKING BACK

1 Why is it important for pregnant women to consume sufficient dietary fibre?
2 What does the term 'lactation' mean? Explain why the mother needs a good supply of protein during lactation.
3 Why is it important to establish good food habits during very early childhood? Explain how the diets of parents can influence what their children eat.
4 Compare the Dietary Guidelines for Children and Adolescents in Australia to the Dietary Guidelines for Australian Adults. Outline their differences and similarities. (The guidelines appear in Chapter 2.)
5 How can adolescents be encouraged to take responsibility for their own nutrition?
6 Explain how the nutritional requirements of adults are different from adolescents.
7 Why is it important for the elderly to select nutrient-dense foods rather than energy-dense foods?

FOR YOU TO DO

8 Breastfeeding offers nutritional advantages. Research other reasons for breastfeeding.
9 Research the nutritional composition of infant formula. Compare the nutritional composition of breast milk with infant formula.
10 Refer to the Australian Guide to Healthy Eating:
  a Compare the recommended food serves for pregnant women with those of adult females. Outline any similarities and differences.
  b Design a daily menu that would be appropriate for a pregnant woman.
  c Identify foods in the menu that provide iron, calcium, Vitamin C, folate and protein.
11 Identify your 10 favourite foods. Why do you like these foods? How many of these foods are nutrient-dense?
12 Outline the major faults in the following meal pattern of some teenagers: no breakfast; a packet of chips at recess; a sausage roll, soft drink and chocolate bar for lunch. How can such eating habits be changed?
13 Suggest five nutritious meals that are suitable for adults. Explain why the meals are nutritious.
14 Prepare an easily made dinner that would be appropriate for a frail older person. Use your imagination to make it nutritious and delicious, yet relatively inexpensive. Visit a supermarket and cost the ingredients used to prepare this meal.

TAKING IT FURTHER

15 Investigate the problems that teenage mothers may face in meeting their own nutritional needs as well as those of the growing baby.
16 In class, debate the following topic: 'Breastfeeding or bottle-feeding: Which is best?'
17 Design a brochure promoting nutritious diets for children between the ages of 2 and 5 years. This brochure could be handed out to adults at preschools or displayed in doctors' surgeries.
18 You have the chance to make a New Year's resolution to improve your eating habits. Using no more than 60 characters, send a text message to a friend, highlighting the changes you intend to make.
19 Plan a national television advertisement or presentation to improve the health of Australian adolescents. Use a desktop publishing program to develop ideas and plan a storyboard to show you would present your ideas. Consider:
  - the information you want to get across to your audience
  - the methods you will use to make adolescents aware of your campaign.
20 Jack is an 80-year-old pensioner who lives alone in a one-bedroom apartment on the third floor of an inner-city block of flats. Jack has no family and cannot be bothered to cook for himself. Jack has a small refrigerator, a stove and an electric jug. His diet is not well-balanced and his health is poor. Discuss how Jack's diet could be improved and write down a plan that he could follow.

8.2 Health status

Food allergies and food intolerances affect nearly everyone at some point in their life. You may have had an unpleasant reaction to something you have eaten and wondered whether you have an allergy or intolerance to that particular food. Such reactions, if severe enough, can cause devastating illness and, in some cases, can be fatal.

Allergies

A **food allergy** occurs when your immune system responds to a food it mistakenly believes is harmful. When you eat it, the immune system immediately releases huge amounts of chemicals into your body to try to 'protect' it against the food consumed. These chemicals then trigger allergic symptoms that can affect your breathing, skin, gastrointestinal tract and cardiovascular system.

Common triggers for food allergies include:

- seafood
- gluten (wheat, rye, barley and oats)
- cow's milk
- nuts – especially peanuts
- soy beans and soy products
- sesame seeds.

8.2 Health status

Intolerance

Food intolerance occurs when the nerve endings in different parts of the body become irritated, causing symptoms such as stomach and bowel troubles, headaches, swelling or hives.

Telling the difference

Food allergy and food intolerance can be difficult to distinguish, because the symptoms are very similar. The main difference is that a reaction to a food intolerance takes a lot longer to appear, often several hours or a day or two later. The body reacts only when a food chemical (natural or added) builds up in sufficient quantities to cause a reaction.

Common triggers for food intolerances include:

- milk
- eggs
- nuts
- fish/shellfish
- wheat/flour
- chocolate
- artificial colours
- pork/bacon
- chicken
- tomato
- soft cheese
- yeast.

istock.com/Diane Labombarbe

Figure 8.6 Foods produced from soy (such as soy sauce, tofu, soy beans and soy milk) are lactose-free.

FOOD IN FOCUS

RILEY'S STORY

Riley is almost 17 and has suffered from food allergies and intolerances all of his life. Riley was first diagnosed with an allergy to dairy foods when he was six months of age, and six months later he was also diagnosed with an allergy to egg. Severe wheat intolerance occurred when he was two years old, and at age 6 he suffered a serious reaction to red meat, requiring immediate hospitalisation and the administration of adrenalin; this was his first anaphylactic reaction.

With a reaction as serious as this, appointments with specialists and dietitians became more frequent. Under instruction from an immunologist, the services of an accredited dietitian were sought, as Riley had to undertake an elimination diet. All known allergens and food chemical triggers were removed from his diet. Then, once Riley was symptom-free, foods were then reintroduced in stages to determine tolerance levels, symptoms and their severity. The outcome of this testing indicated that Riley must make sure that the following foods are completely eliminated from his diet:

1 wheat and wheat-containing foods
2 all artificial colours and preservatives
3 all amine-containing foods
4 some foods that contain salicylates.

As a result of these food allergies and intolerances, Riley has to maintain a very strict diet. If he deviates from this diet then he and those around him have to suffer the symptoms that occur – severe hyperactivity, extreme fatigue and lethargy, vomiting, aggressive and violent behaviour, inability to concentrate, rashes, reflux and insomnia.

Riley is still able to maintain a very happy, normal life, despite the fact that he is required to carry antihistamine medication and an EpiPen at all times and is unable to indulge in a slice of pizza, packet of chips or many foods that others can tolerate. Riley has competed at state level in swimming, athletics and cross country, and he continues to be dedicated to training, excelling in middle-distance running and all aspects of swimming – he has his sights set firmly on becoming a triathlete in the future!

Riley and his family are lucky to have an active support network in their town, with Riley's school also being very supportive and proactive in educating other children in his peer groups about allergies and ensuring that they realise that Riley is no different.

Riley believes that the main thing to remember is that a food allergy or intolerance is not a disability; it just requires some special attention and care to ensure that sufferers are accommodated in all social situations. It is not necessary to exclude a person from a situation because of fear; people need to simply educate themselves to ensure safety in all situations.

ACTIVITIES

1 Explain the difference between food allergy and food intolerance.
2 Define the term 'anaphylactic reaction' and outline the cause and symptoms.
3 Find out more about one of Riley's intolerances. Identify foods that should be avoided by people affected by this intolerance. List alternative food sources.
4 Make a list of commercial food products that cater for people affected by this intolerance.
5 If you were asked to plan a diet for Riley, what difficulties would you face? How can these difficulties be overcome?
6 List the different ways that you could make a person with an allergy and/or intolerance feel welcome and included at a social function you were hosting.

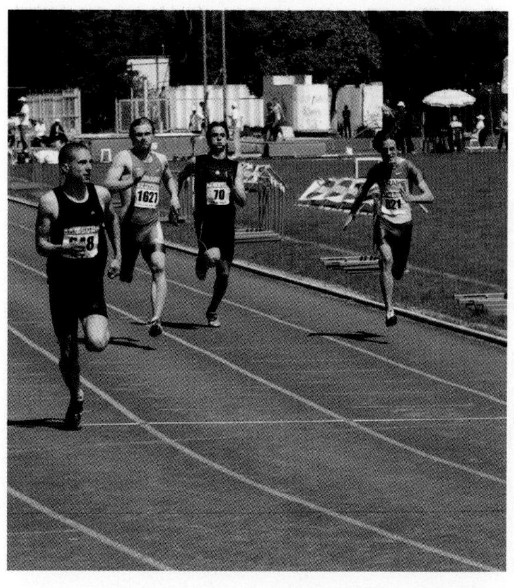

Dreamstime.com/Denys Kuvaiev

Recovery from illness or injury

When people are recovering from an illness or an injury they are **convalescing** or getting better. Injury and illness create stress on the body. Adequate nutrition can really make the difference in the recovery of people with severe injury or illness.

Injury increases the body's metabolism, and larger quantities of protein, vitamins and minerals are especially important. However, hunger may decrease, because illness or injury tends to reduce the appetite. A patient may not feel well enough to eat a high-protein meal such as fish and vegetables, yet will be able to benefit from an intake of protein by consuming a protein-rich drink made from milk and eggs, or by eating yoghurt.

Small, regular meals play an important role in the recovery from illness or injury. It is important to remember that all meals do not have to be soft for them to be easily digested, unless the patient has no teeth or is too ill to chew food. Foods that are more difficult to digest should be avoided; for example, fried foods, pastries, cakes, nuts or any fatty foods.

UNIT REVIEW

LOOKING BACK

1 Outline the causes, symptoms and treatment of a person suffering from a food allergy or intolerance.
2 Explain the cause and symptoms of coeliac disease.
3 Define the term 'convalescence'.
4 Outline the special nutritional needs of convalescent people.
5 Suggest ways of making a convalescent person's meals attractive.

FOR YOU TO DO

6 Interview your family and friends to find out if any of them suffer from a food allergy or intolerance. Make a list of the foods they must avoid. Suggest suitable alternatives for each of the foods.
7 Plan a day's menu suitable for a teenager with an allergy to milk and gluten. Explain whether you found this task difficult.
8 Interview a relative or friend who has been hospitalised for surgery. List the types of foods served during their hospital stay. Suggest three recipes that would assist recovery after surgery.
9 Think of an occasion when you were ill and recovered slowly. Which foods did you consume? Did the doctor recommend any special foods? Why?

TAKING IT FURTHER

10 Design an information pamphlet for parents whose child has a particular allergy or intolerance. The pamphlet should include an explanation of the allergy/intolerance symptoms. You should also include menu suggestions for breakfast, lunch, dinner and snacks. It may also have a feature recipe of a food product or meal you have tested that would be suitable.
11 Plan a day's menu for:
   a a person who has facial injuries and is finding it difficult to chew
   b a person who has a knee injury but no difficulty in chewing.
12 Explain why you chose these menus. Are the same nutrients provided in both menus?

8.3 Diet factors

The major causes of death in Australia are linked to diet and lifestyle. Some people consume a well-balanced diet while others do not. Poor nutrition and an unhealthy lifestyle are related to many disorders, including alcoholism, anorexia nervosa, bulimia, dental cavities, diabetes, heart disease and obesity. Making the right food choices can significantly reduce the risk of developing diet-related disorders.

Diet-related disorders

Alcoholism

Alcoholism is the regular consumption of large quantities of alcohol over a long period of time, which can lead to serious health problems, such as liver disease, disease of the pancreas and some types of cancer. Alcoholic drinks also provide many extra kilojoules and often can lead to obesity.

9780170383417

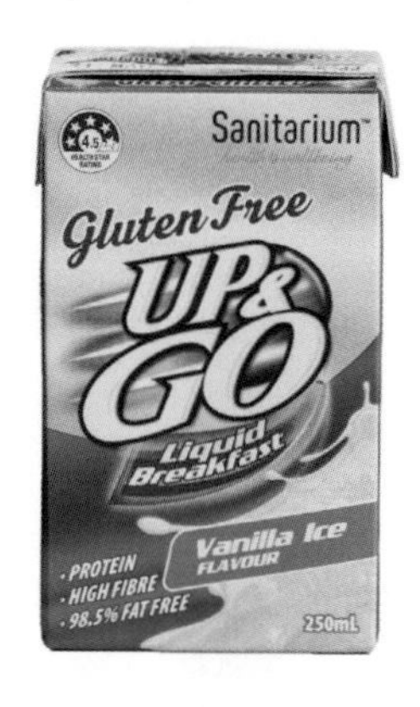

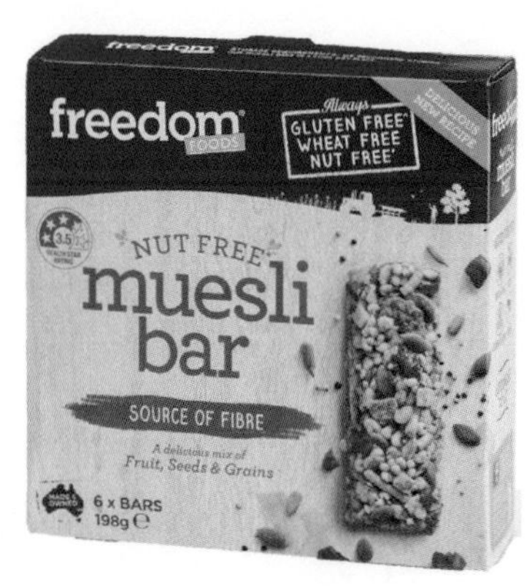

Mark Fergus

Figure 8.7 Supermarkets now stock a variety of gluten-free products.

Anaemia

Anaemia is a blood condition and is the most common deficiency disease in Australia, particularly in females. Estimates suggest that around one in five menstruating women and half of all pregnant women are anaemic. Anaemia has a wide range of causes, including certain diseases, conditions and medications. Iron deficiency is the most common cause.

Anorexia nervosa

One of two common eating disorders, anorexia is a condition in which a person refuses to eat sufficient food to maintain a minimum normal weight for age and height. The consequent wasting away has serious effects on many body systems, and may result in death.

Bulimia

Bulimia is an eating disorder marked by cycles of binge eating of excessive quantities of food, followed by purging through vomiting or use of laxatives or diuretics. The purging can seriously damage health. Unlike anorexia, a person with bulimia is rarely grossly underweight.

Dental cavities

The incidence of dental cavities in the Australian population has reduced over the past decades, mainly due to improved nutrition, fluoridated water supplies and improved dental hygiene. Dental cavities occur when bacteria in the mouth convert sugars and refined starches from foods and liquids into acids. The acid that is produced in the mouth can dissolve the outer layer of tooth enamel, resulting in a cavity.

Diabetes

Diabetes can be life-threatening if not carefully controlled. It is expected to affect one in four Australians within the next few years. Diabetes occurs when there is too much glucose in the blood. If blood glucose levels are constantly higher than the recommended range, health problems can develop.

Coronary heart disease

Coronary heart disease kills one in three adult Australians and is the leading cause of death in Australia. It occurs when the arteries in the heart become clogged with fatty deposits that restrict the amount of blood that can be pumped through them – sometimes they even become totally blocked. This makes the heart work harder to pump blood around the body and places extra pressure on the heart.

Obesity is a very serious problem in Australia – approximately two-thirds of the adult population are overweight and 27 per cent are classified as obese. It is estimated that, if the current trend continues, by 2020, 70 per cent of Australians will be above their healthy weight range. Too much food and too little physical exercise result in an increase in weight. Obese people are those who are 20 per cent or more above their recommended weight for height. People who are obese are also at a higher risk of developing other diet-related disorders.

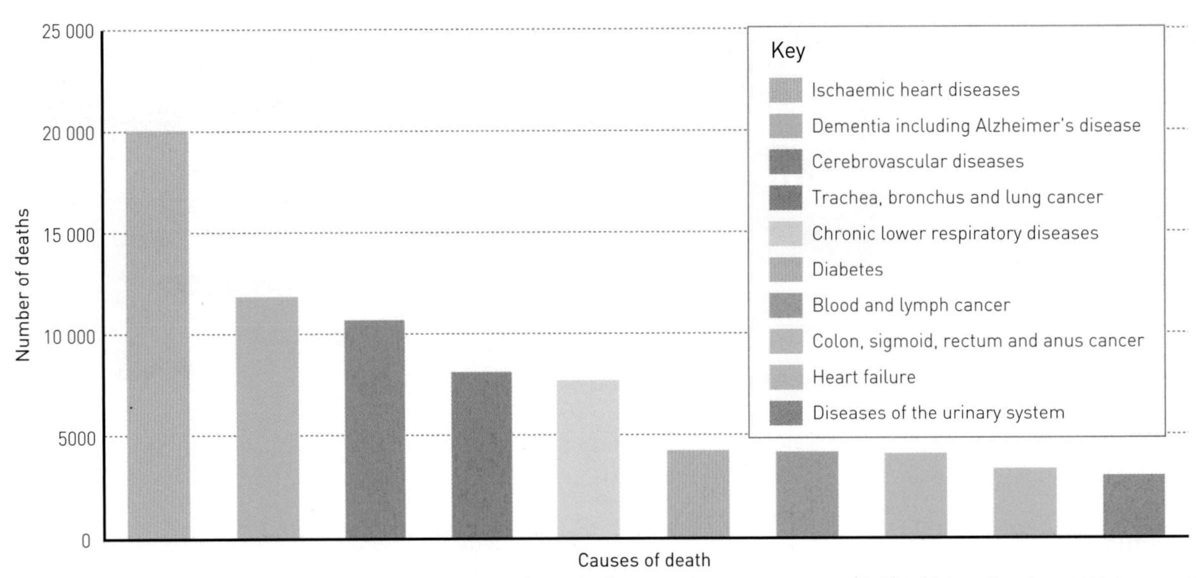

Source: Australian Bureau of Statistics, 2014 (3303.0). Creative Commons (CC) BY Attribution 2.5 Australia

Figure 8.8 Top 10 major causes of death in Australia, 2014

FOOD IN FOCUS

CORONARY HEART DISEASE

Garry Cornius, a 64-year-old man, is one of the increasing number of Australians who have experienced coronary heart disease. He is now learning to modify his lifestyle, with the assistance of many support agencies, so that he can continue to lead a happy and healthy life.

In 2009, Garry was enjoying a movie when he started to experience indigestion, a shortness of breath and pains in his arm, neck and chest. He was close to his local hospital and decided, luckily, to visit the emergency department to see what was causing these symptoms. Garry was quickly admitted. After many blood tests, X-rays and other procedures, Garry was asked to undergo a stress test on a treadmill. This stress test caused Garry's heart rate and blood pressure to elevate rapidly and induced a heart attack. He was medicated and rushed to Royal North Shore Hospital in a critical care ambulance. On arrival, Garry was immediately worked on by a cardiac surgeon, who discovered that Garry's main artery to his heart was 99 per cent blocked.

After undergoing surgery, spending one week in intensive care and another six weeks in recovery, he began the slow process of making changes to his lifestyle. Through this process Garry has discovered that exercise, rehabilitation and lifestyle modification are beneficial in preventing and controlling coronary heart disease.

Risk factors that can be easily modified are:

- tobacco smoking
- physical inactivity
- alcohol misuse
- poor nutrition.

Garry continues to access a wide network of agencies who provide support and information for people with special needs. These networks include his family, friends, private organisations and government agencies, all of which were invaluable to his recovery and continued good health.

ACTIVITIES

1 Define the term 'coronary heart disease'.
2 Visit the Virtual Medical Centre website. View the video titled 'Cardiovascular Disease Prevention' and design a flyer that could be used to educate people on the ways in which the disease can be prevented.
3 Visit the Heart Foundation website and outline the dietary advice you could provide to a person who was at risk of developing coronary heart disease.
4 Use the recipe finder on the Heart Foundation website to plan a day's menu that would meet Garry's special dietary needs.

The Virtual Medical Centre – Cardiovascular Disease Prevention

The Heart Foundation

Source: Australian Bureau of Statistics, 2013

FOOD IN FOCUS

A SURGE IN CASES OF EATING DISORDERS

A report to be released today, conducted by Deloitte Access Economics and commissioned by eating disorder group The Butterfly Foundation, estimates there are 913 986 people with eating disorders, about 4 per cent of the population.

It also calculated there were 1829 disorder-related deaths this year, far higher than the 14 recorded by the Australian Bureau of Statistics in 2010.

More than a third (36 per cent) of all sufferers are male, while binge eating disorder is the most common subtype across both men and women.

Anorexia, bulimia and eating disorder not otherwise specified (EDNOS) are also more common than previously thought.

Among those with EDNOS, about half, or 214 000, are morbidly obese.

The paper found the estimated cost to the health system is $100 million, while the productivity impact on the economy was close to that of anxiety and depression - about $15.1 billion.

Butterfly Foundation chief executive Christine Morgan said the figures, derived from recent population-based surveys in South Australia, New Zealand and the US due to a lack of Australia-wide data, were still 'conservative'.

'The numbers didn't surprise me, but it affirmed for me this is really serious and it affirmed for me (that) we can't be quiet about this,' she said. 'We've got to start pushing to have this properly understood as an illness in this country.'

Mortality rates are almost twice as high for people with eating disorders as those without, jumping to six times for those with anorexia nervosa, from which an estimated 25 753 people suffer.

The report also concluded there was a lack of treatment options, with just two public hospital beds dedicated to patients in NSW, five in Queensland and 19 in Victoria.

Professor Susan Paxton, of La Trobe University's School of Psychological Science, who was consulted on the report, said: 'It does highlight we do need to get more of our own Australian information and be providing more resources into this area.'

Ella Graham, 23, founder of Fed Up NSW Health lobby group, who has an eating disorder herself, welcomed the report.

'In NSW, two state-wide beds is atrocious, especially because by the time people get those beds, they're so acutely ill that they need months and months of care, rendering those beds inaccessible by anyone who needs care during that time,' she said.

Source: 'A surge in cases of eating disorders', by Caroline Marcus, *The Australian*, 11 December 2012.

ACTIVITIES

1 Identify how our modern society is contributing to the increase in eating disorders.

2 Discuss the implications of eating disorders for our society.

3 Visit the Virtual Medical Centre and search for 'an introduction to obesity treatments'. Watch the video 'The Management of Obesity' and provide a summary outlining suggested problems caused by and the possible treatments for this disease.

The Virtual Medical Centre – Introduction to Obesity Treatments

4 Visit the Butterfly Foundation website and select an eating disorder. Suggest two main meals for a person who is suffering from the disorder.

The Butterfly Foundation

UNIT REVIEW

LOOKING BACK

1 Explain why females and vegetarians may develop anaemia.
2 Compare the symptoms, causes and long-term effects of anorexia and bulimia.
3 Outline the risk factors for coronary heart disease. What dietary advice could you provide to a person who was at risk of developing heart disease?
4 List five strategies you could adopt to avoid dental cavities.
5 Some doctors suggest that one in every 100 girls in their late teens suffers from anorexia nervosa. What do you think may have led to the high incidence of this disorder among adolescents?

FOR YOU TO DO

6 Collect or design five iron-rich recipes. Prepare one of these at home or at school.
7 Design an advertisement promoting healthy and safe guidelines for reducing weight.
8 Using the internet or magazines, locate some examples of weight-loss diets. Evaluate the soundness and effectiveness of these diets.

TAKING IT FURTHER

9 Contact the diabetes association near you. Invite a representative to come and speak with your class about the different ways you can support a person with diabetes and their family.
10 Research some food service and catering establishments near you. What provisions do they make for customers with special dietary needs?
11 Design and prepare a PowerPoint presentation about one of the diet-related disorders. This presentation could be used to educate members of your community. Include information about the groups of people most at risk, causes, symptoms and treatment of the disorder, and strategies to stop people getting this disorder.

8.4 Lifestyle choices

Different lifestyles require different food and energy intakes to encourage and support a healthy and long life. People with very active daily lives need different foods from those who lead more sedentary lives, as they expend more energy.

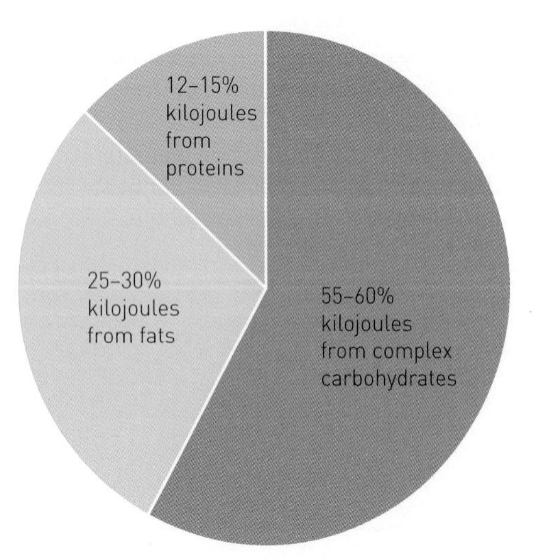

Figure 8.9 Energy sources from nutrients for athletes

Athletes

The quality and quantity of food eaten makes a great difference in the way an athlete can train and compete. Good nutrition is important to develop a strong body and energy has to be stored for endurance. Special attention to nutrition will ensure that you get the best from the activities of your choice, and maximum benefits over the long term. Here are some important principles of sports nutrition.

- Limit fat intake to 25–30 per cent of your kilojoules.
- Keep your muscles fuelled by eating plenty of complex carbohydrate foods, as they supply the body with sustained energy. 55–60 per cent of your total kilojoule intake should come from complex carbohydrates such as breads, cereals and grains.
- Consume 1 gram of protein per kilogram of body weight per day – 12–15 per cent of kilojoules should come from protein-rich food.

- Drink fluids before, during and after exercise sessions to prevent dehydration, overheating or heat stroke.
- Female athletes should make sure that their iron intake is high by including iron-rich foods in their diet.

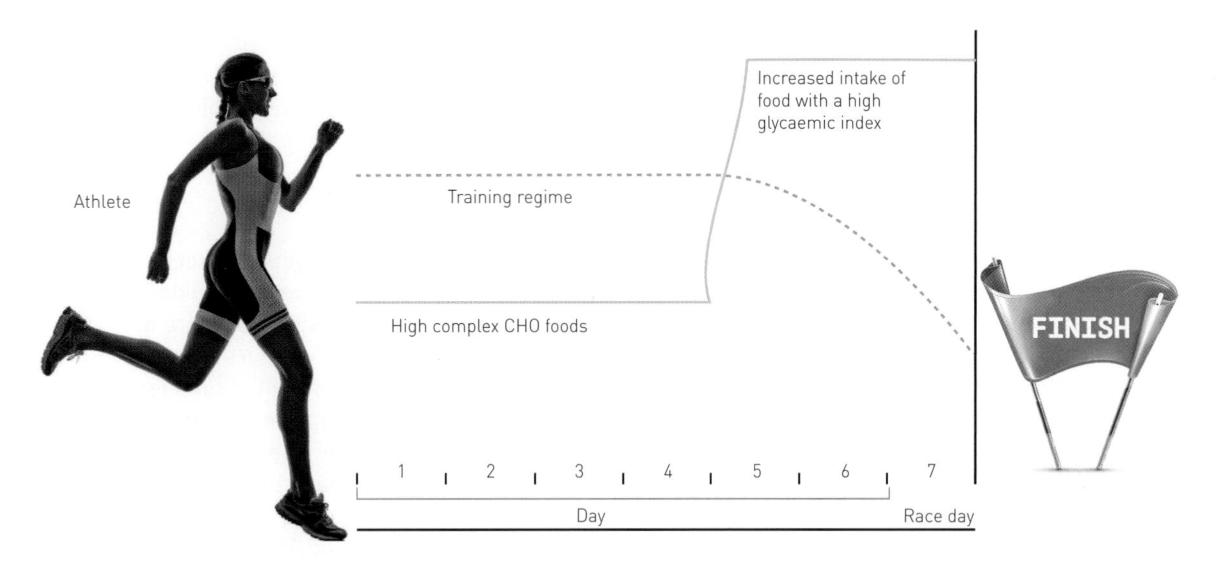

Figure 8.10 Carbohydrate consumption for endurance athletes

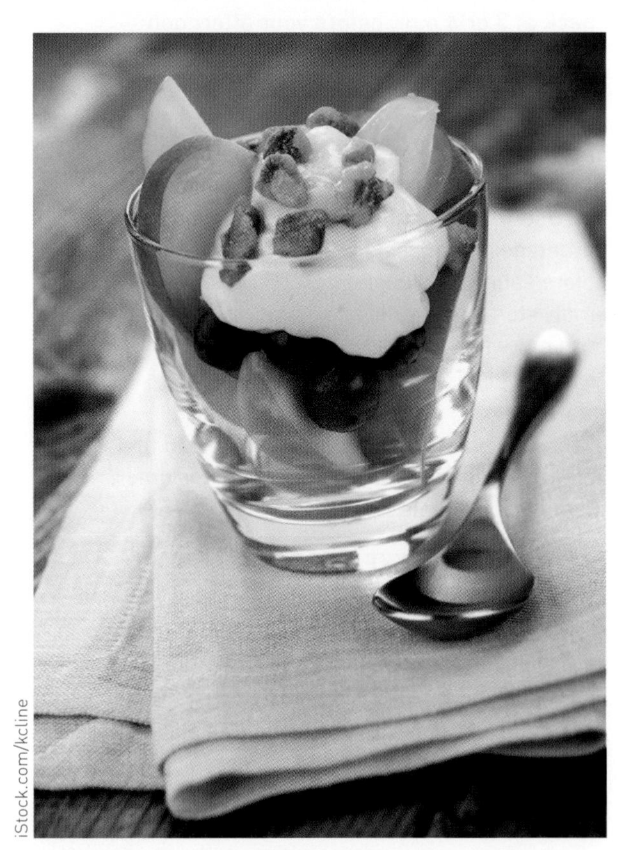

iStock.com/kcline

Figure 8.11 Fruit, yoghurt and nuts: the perfect training fuel

Vegetarians

Vegetarians do not consume meat. Many vegetarians have lower rates of heart disease, diabetes and bowel cancer, and are less likely to suffer from high blood pressure or high cholesterol. They also tend to be slimmer and have longer life expectancy than meat eaters. Many vegetarians choose this lifestyle for these health benefits. Other reasons for adopting a vegetarian diet include compassion for animals, environmental concerns and religious or spiritual beliefs.

There are three types of vegetarian diets:

- pure vegetarian diet or vegan diet, which consists entirely of plant foods, such as cereal products, legumes, nuts, seeds, fruits and vegetables, and therefore excludes all animal-based foods, including dairy, eggs and, in many cases, honey
- lacto-vegetarian diet, which excludes meat and eggs, but includes dairy products
- lacto-ovo-vegetarian diet, which excludes meat, fish and poultry, but includes dairy products and eggs.

FOOD IN FOCUS

THE BEST FUEL FOR TRAINING

Whether you are taking part in a half or full marathon, or just exercising to keep fit, training involves more than just pounding the pavement. What you eat plays a pivotal role. The question is: which foods are best? And when should you eat?

Everything you eat and drink has an effect on your body when you exercise. And it's not just what you eat weeks leading up to the race – you need a balanced diet every day. This includes the right meal before and after you train to help your body recover and repair.

THE TRAINING DIET

There's been a lot of debate around low-carb diets and distance running (aka Banting diets or LCHF eating). However, Sports Dietitians Australia warns that low-carb eating will not only deprive the body of important nutrients, but will also compromise performance.

Carbohydrate is a key fuel source for exercise and especially recommended for distance runners to support the high-energy demands of training, fast recovery, and optimal race performance.

Exactly how much carbohydrate you need is dependent on the frequency, duration and intensity of the run, and should reflect your daily training load. In other words, consume a higher proportion of carbohydrates for harder training days and less on easy/recovery days.

But it's not all about carbohydrates; moderate amounts of protein and smaller amounts of quality fats, such as those found in oily fish, avocado, olive oil and nuts are also needed to ensure sufficient energy, muscle repair, and an adequate supply of vitamins and minerals.

SO WHAT'S THE IDEAL TRAINING DIET?

As a general rule, opt for a combination of low GI carbohydrates to fuel the muscles with a small amount of protein to ensure a slow, gradual release of energy. Examples of pre-training food options that fit this nutrient balance include:

- wholegrain crackers with hummus or reduced fat cheese
- fruit salad with yoghurt
- fruit smoothie
- rolled oats with milk
- baked beans on toast
- lean meat and plenty of salad on sandwich or wrap
- pasta or rice with a sauce based on low-fat ingredients (e.g. tomato, vegetables, leaner cuts of meat)
- cereal or muesli bars.

TIMING IT RIGHT

When it comes to fuelling your workout, timing is everything. As a general guide, eating a meal roughly 3–4 hours before or a lighter snack roughly 1–2 hours before exercise is a sufficient amount of time for the stomach to empty and to have a positive effect on training, boost energy levels without an upset stomach. This means that if you have eaten your breakfast around 7 a.m., you may need a light snack around 11 a.m. before your lunch time workout. Or, if you have eaten your lunch at 12 or 1 p.m., you will need to eat snack at 3 or 4 p.m. before your afternoon training session.

If you train early in the morning and your workout is less than one hour, going without food or drink probably won't do you any harm. Just make sure you're staying hydrated. Similarly, exercisers with weight loss goals may find an advantage in exercising first thing in the morning before eating breakfast, as exercising on an empty stomach results in a greater potential to burn more fat.

However, if your goals are performance related (e.g. to improve speed or time), exercising on an empty stomach can lead to an early onset of fatigue, plus a tougher time meeting your goals.

EATING DURING EXERCISE

You shouldn't need to eat when you exercise less than 90 minutes. If you do feel tired, it may mean you haven't eaten enough carbohydrates, or your diet is not well-balanced, or you're dehydrated.

If you train longer than 90 minutes at a moderate to high intensity, consuming extra carbohydrates is recommended as carbohydrates supplies start to run low. This is where sports drinks or gels are handy and practical. Research with athletes shows that 30-60 g of carbohydrates needs to be consumed each hour of exercising to delay fatigue.

>

HYDRATION

Regardless of how long you exercise for, hydration is always key and training provides you with the opportunity to practice your fluid-replacement strategies. While there's no set recommendation for daily fluid intake, a good rule of thumb is to drink enough fluid so that your urine is pale in colour – this is your quickest guide to know if you're drinking enough or not. Water is the best thirst quencher in most cases, but a carbohydrate-containing beverages, such as sports drink is recommended for exercising longer than 90-minutes to replace electrolytes and top up glycogen stores.

OPTIMISING RECOVERY

For training and race recovery, stick with the three Rs of refuel, repair and rehydrate. To refuel, have a protein-rich snack to aid muscle repair, combined with carbohydrates to restock your spent energy stores, along with drinking enough fluids to help enhance recovery and rehydrate. Good examples include:

- stir-fry with lean red meat or chicken, vegetables and rice
- wholegrain rolls made with salad and lean meat
- flavoured milk or fruit smoothies
- jacket potato with reduced-fat cheese and legumes.

Source: 'The best fuel for training', Kathleen Alleaume, News.com.au, 4 July 2015.

ACTIVITIES

1. Outline the role of carbohydrates in a pre-event meal.
2. What is the main indicator that an athlete is dehydrated?
3. What are the 'three R's'?
4. Design a pre-event meal for an athlete planning to run a half marathon, running for approximately two hours.
5. What nutritional advice would you give an athlete for eating during the half marathon?

Figure 8.12 Vegetarians enjoy a wide variety of dishes.

Clockwise from top left: istock.com/ehaurylik; istock.com/Olha_Afansieva; iStock.com/GitteMoller; Shutterstock.com/Arayabandit; istock.com/OksanaKiian; istock.com/Lesyy

Vegetarians need to take special care when designing their diets to ensure that the correct proportions of all nutrients are included. Meat, which is an important source of protein and iron, can be replaced with soy beans or soy bean products, which are sources of complete protein. Lentils and beans are also sources of protein but, being incomplete, they lack some of the essential amino acids and need to be combined with other protein-rich foods to make them complete and more nutritious. This is often called the supplementary value of protein.

A vegetarian diet should be based on the following guidelines:

- Eat a wide variety of foods from the vegetarian food groups.
- Eat plenty of fresh fruit and vegetables and wholegrain cereal products.
- Ensure an adequate and reliable supply of Vitamin B12.
- Consume foods rich in Vitamin C with meals to increase iron absorption.
- Ensure regular exposure to sunlight (a few hours a week) in order to maintain Vitamin D status.
- Limit your intake of foods high in saturated fats and cholesterol. Vegetarian foods in this category are eggs, full-fat dairy products, and butter and coconut products.
- Limit your intake of added salt and salty foods.
- Limit your intake of sugary foods, such as lollies and soft drinks.
- Drink plenty of water.
- Avoid excessive alcohol consumption.

Cultural backgrounds and religious beliefs

Many religious and cultural groups have special food needs.

- Seventh-Day Adventists and Buddhists are vegetarian.
- Hindus do not eat beef, and many Hindus are vegetarians
- Judaism religion has many strict laws associated with the foods eaten and the way in which it is obtained and prepared; for example, meat from the pig (pork, ham or bacon) must not be consumed. The meats that are permitted can only come from an animal that chews its cud and has divided hooves, such as cows, sheep, ox and goats. In very observant Jewish homes, meat and dairy products are not served at the same meal. They have to be prepared and served on separate plates using different utensils. The dishes must be cleaned in separate sinks.
- Muslims do not eat pork or drink alcohol, and follow very strict guidelines when it comes to selecting foods to include in their diet. Animals must be slaughtered according to Islamic law. The word 'haram' means unlawful and the word 'halal' means lawful. Pig meat, blood, carnivorous animals and alcohol are haram. Milk, honey, fish, legumes, grains, fruits and vegetables are halal. Animals such as cows, sheep, goats, deer, moose, chickens, ducks and game birds are also halal, but they must be slaughtered according to Islamic rites in order to be suitable for consumption – this is known as 'zabihah'.
- Fasting is required by some religions during particular times, such as during Ramadan among Muslims. The sick, as well as children and pregnant women, are usually excused from fasting.

Fairfax Photos/Quentin Jones

Figure 8.13 Muslims choose halal food as part of their diet.

Table 8.1 Food preferences of major cultural and religious groups

FOOD	HINDUS/BUDDHISTS	MUSLIMS	JEWS
Eggs	Some*	Yes	Yes
Milk and yoghurt	Yes	Yes	Yes
Cottage/curd cheese	Yes	Yes	Yes
Chicken	Some*	Halal#	Kosher+
Mutton	Some*	Halal#	Kosher+
Beef	Hindus no, Buddhists some*	Halal#	Kosher+
Pork	Some*	No	No
Fish	Some*	Yes	Yes
Butter/ghee	Yes	Yes	Yes
Margarine/vegetable oils	Yes	Yes	Yes

* Very strict followers avoid this.
# Halal meat must be killed, dedicated and prepared in a special way.
+ Kosher meat for Jews requires special rituals and butchering procedures in preparation.

Logistical considerations

The supply of food for a large number of people presents **logistical** issues. This could include preparing food for people who are bushwalking or camping, in hospitals, in canteens, in nursing homes and for plane travel.

Enjoying outdoor leisure activities is something that many Australians do. When the decision is made to eat food outdoors it is important that some planning takes place. When undertaking outdoor activities it is important to remember to keep foods either very cold (refrigerator cold) or very hot (steaming hot) so that foods do not enter what is known as the danger zone. As soon as the temperature of **perishable** foods enters the danger zone, bacteria begin to multiply – and this can be very dangerous.

Bushwalking

There are many foods available that are suitable to take with you when you are planning to go on

HANDS-ON

INVESTIGATING HALAL FOOD

PURPOSE

To explore how cultural backgrounds and religious beliefs impact on your dietary requirements.

MATERIALS

Five everyday food products with and without halal certification labelling attached

STEPS

1 Divide into groups of four and review the definitions of 'haram' and 'halal'.
2 View the five foods presented and read each label.
3 Decide whether the food would be classified as halal or haram.
4 Copy the grid below into your notebook and list each item you have classified.

FOOD	HARAM	HALAL
1		
2		
3		
4		
5		

ACTIVITIES

1 Discuss the constraints Australian Muslims may face when shopping for halal foods.
2 Use the library or the internet to investigate meals that would meet the dietary needs of this particular group.
3 Review your completed worksheet and findings with the rest of the class.

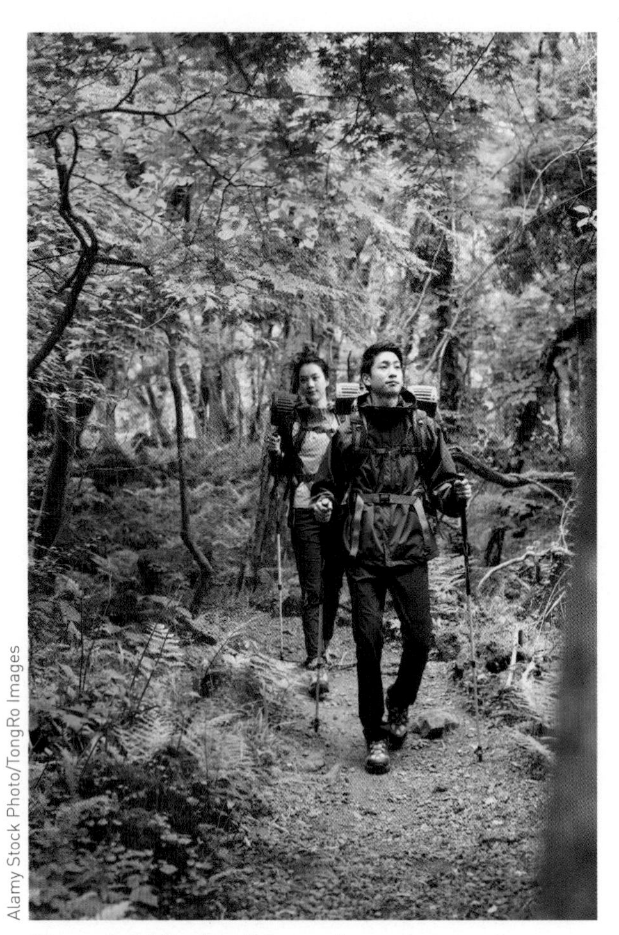
Alamy Stock Photo/TongRo Images

Figure 8.14 Bushwalking and camping require lightweight, easily transported nutritious food.

a bush walk. The selection really comes down to personal preferences. The basic requirement is to provide a balanced healthy diet to give you enough energy to complete the walk, depending on the distance, terrain and the load you are carrying.

Food selected for your backpack should be light in weight, high in energy value, easy and fast to prepare, appetising and not be at risk of spoiling in hot weather. With freeze-dried and dehydrated foods readily available, the bush walker has a wide range of foods to choose from that are lightweight, quick and tasty. All you really have to do is add water, wait, heat and eat!

Be sure to pack some high-energy food bars or sultanas or nuts to snack on as you are walking. These foods also come in handy if you are required to spend an unexpected night in the bush. You should eat small amounts often, so that your energy is high throughout the entire walk and your body functions effectively. Drinking water constantly is also essential to keep your body hydrated and to convert the food eaten into energy.

Mark Fergus

Figure 8.15 Foods for bushwalking and camping need to be convenient and lightweight.

Camping

Meals on camping holidays need not be complicated, boring or lacking in nutrition if you plan and pack with a little imagination.

Perishable foods are generally unsuitable for camping unless you have access to a refrigerator. It is best to use dry, UHT and canned products such as:

- pasta, rice and instant noodles
- long-life fruit juice, milk, cream and custard
- bottled mustard and pasta sauce
- canned fruit in natural juice, or meats such as tuna, salmon and crab
- dried peas, beans, herbs, soups and fruits.

Hospitals

Apart from special diets advised by doctors, in general, hospital food should be light in fat, presented attractively, small in portion and offered frequently. Dietary restrictions may make it impossible for some patients to eat hospital food. For instance, unless the hospital caters for special dietary arrangements, it may be logistically impossible for kitchens to provide halal or kosher food, and patients may have to organise the delivery of their own meals.

Canteens

The school environment has a special role to play in the education, health and wellbeing of students and the school communities that they serve. Given the current trend towards buying more meals away from home and the increasingly busy lifestyle of Australian parents, school children are purchasing more of their daily food intake from school.

There are many different food environments in a school. In a healthy school, the nutrition messages are consistent across all food settings and promote the benefits of a balanced diet and healthy food choices. Healthy eating should be modelled at every opportunity and carried through into the classrooms, with teachers also demonstrating appropriate eating behaviours.

Schools can help to reduce nutritional problems by offering and promoting a good selection of nutritious, tasty and attractive foods. Schools make their own decisions about what to sell in the school canteen and other settings. This can sometimes be difficult because of the large number of students, the variety of nutritional needs that have to be met and the great number and type of food products available.

At school, food may be consumed in any of the following:

- canteens and tuckshops
- lunchboxes
- vending machines
- incursions and excursions
- school sponsorship events
- healthy food fundraising
- classroom rewards.

Nursing homes

Meals are an extremely important factor in the quality of life in nursing homes. Meals satisfy physical requirements as well as psychological, emotional and social needs. Care needs to be taken when designing and preparing food for residents of nursing homes.

Menu plans should take into consideration the individual preferences and special dietary needs of all residents.

FOOD IN FOCUS

CLERMONT NURSING HOME

Clermont Aged Care takes pride in the quality of food it serves. Each day a menu is handed out and the residents have a choice of quality foods; three meals a day plus morning and afternoon tea. Extra fruit, tea and coffee are also available to the residents. A consultant dietitian reviews the menus and the meals regularly, especially when a new resident arrives at the nursing home.

ACTIVITIES

1 Visit the Clermont Aged Care website and click on the sample menu on the Life at Clermont page. Select one day's food for an elderly woman with weak bones.
2 The consultant dietitian reviews this menu regularly. Suggest appropriate changes that could be made to make it more appealing.

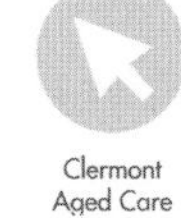

Clermont Aged Care

3 Identify five factors that the consultant dietitian would need to be aware of when reviewing the menu and residents' meals.

Courtesy of Clermont Nursing Home

Figure 8.16 Meals at Clermont Aged Care. Friday lunch: Fish and chips. Afternoon tea: Cake

Airline food

There are many constraints in the provision of food to airline passengers. However, airline food has improved dramatically over the years, with many airlines now treating their customers to a broad selection of international and regional foods in menus that fuse current trends with traditional flavours. Most airlines also cater for special dietary, medical, cultural or religious requirements, with the selection of special meal options varying depending on the airline. For example, Qantas offers diabetic, gluten-free, lacto-ovo-vegetarian, vegan, Hindu, halal and kosher alternatives if pre-ordered.

Menus are constantly being improved, particularly on long flights, and some airlines even enlist top chefs to help develop their menus. Qantas is one example of this – they have been working closely with leading Australian restaurateur Neil Perry of Rockpool to design a unique restaurant-style dining experience with the opportunity to indulge in an eight-course tasting menu, delicious canapés, light meals or even gourmet toasted sandwiches on selected services.

FOOD IN FOCUS

46 MEALS A MINUTE – AIRLINE CATERERS BRING NEW MEANING TO 'FAST FOOD'

Fast food giants have nothing on airline catering companies, that churn out an average of 46 meals a minute for as little as $3.50 each.

One of the world's largest is Brahim's Airline Catering in Kuala Lumpur that provides meals for around 30 international carriers, including Emirates, Etihad, Malaysia Airlines, Garuda, AirAsia and Cathay Pacific.

The airlines set the budget for each meal which can be as little as $3.50 (10 Malaysia Ringgit) for economy class passengers and more than $30 (100 Ringgit) for First Class travellers.

The 1200 staff adhere to strict health and safety guidelines to achieve their enormous output – which incredibly is still less than a quarter of that of Emirates' own facility in Dubai.

Each week, around 2.5 tonnes of seafood, 600 kg of beef and four tonnes of chicken are sliced up, cooked and blast chilled; poultry being the most popular protein 'because it's cheap'.

Computer screens display the exact portions required and the presentation of the dish, with staff expected to produce exact replicas thousands of times over.

All food is prepared from scratch including pastries, flame-grilled satay sticks and salads, with an emphasis on freshness and presentation.

Ten people are assigned to preparing the award-winning satay sticks served exclusively in Malaysia Airlines' First and Business Class cabins, with an individual output of 3000 sticks a day for each worker.

Alamy Stock Photo/Kevpix

Figure 8.17 Presentation: Getting food onto flights is no simple task.

>

So structured is the cooking procedure, the satay sticks can be served in the air, just 45-minutes after they are made.

Executive sous chef Fakhrul Aliff said the most challenging part of their work was keeping meals consistent and within budget, and creating new menus as required.

'We cater for around 160 flights a day, which can mean hundreds of different menus,' said Mr Aliff.

'New menus require months of planning and trials to get them just right.'

University of Queensland Centre for Nutrition and Food Sciences Senior Research Fellow Dr Eugeni Roura, said preparing food to be served at high altitudes was a science in itself.

'The caterer's challenge is undoubtedly being able to preserve the food 'as fresh as possible' to retain moisture and juiciness, and all at zero risk of bacterial growth or fermentation,' said Dr Roura.

'Obviously taste enhancers are a must, salt, sugars, glutamate, spices, chilli and garlic are always highly present in plane foods.'

Mr Aliff said Brahim's had perfected a process of 'blast chilling' that allowed cooked hot meals to be frozen at high speeds to minimise the risk of bacterial growth.

'It brings down the core temperature from above 65 degrees, to below 10 degrees in a matter of minutes,' he said.

Source: '46 meals a minute – airline caterers bring new meaning to 'fast food', by Robyn Ironside, News.com.au, 19 October 2015.

ACTIVITIES

1 Identify the constraints of preparing meals 9000 metres in the air compared to on land.
2 Explain why it is important for each assembly worker to be issued with a coloured photo of what their particular finished meal should look like.
3 Research Islamic dietary requirements and justify why a self-contained kitchen is required for food preparation.
4 Explain the process of cook-chill and what has replaced it.
5 Predict what the future of airline food might be.

HANDS-ON

CATERING FOR SPECIAL FOOD NEEDS

Visit the Nutrition Australia weblink. This website has links to private organisations and government bodies who provide support for all Australians to enjoy a healthy, balanced diet. Choose a special food need that interests you.

Nutrition Australia

1 Locate the groups and networks within Australia that support individuals.
2 Investigate how each provides support for individuals and families.
3 Contact each support network and ask them to send you some promotional material about their products and/or services. Display all brochures on a class notice board.
4 Imagine you were holding a special morning tea for the group you have researched. Plan the menu. Choose one dish, prepare it and share it with the rest of your class.

Support networks

There are many private organisations and government agencies that provide support for people with special dietary needs. Support, including information, is provided for:

- people who are on special diets for heart disease or cancer
- vegetarians
- elite sportspeople
- mothers who are breastfeeding
- people who have food allergies
- people who have diabetes
- people who have coeliac disease.

UNIT REVIEW

LOOKING BACK

1 Outline the role each major nutrient plays in sports performance.
2 When many athletes cease to participate in sports they tend to gain weight. Why do you think this happens?
3 Why is a vegetarian diet beneficial to health?
4 Identify three meals that combine protein sources to provide a complete protein.
5 Name and describe the different types of vegetarians. Explain any dietary difficulties that each may have.
6 Identify the factors that you would need to consider when developing food menus for a nursing home.
7 Identify the food safety issues that need to be considered when preparing and transporting food for bushwalking or camping.
8 Suggest four foods suitable to take on a camping holiday. Provide a reason for each choice.
9 Investigate the foods that are available from your school canteen. Suggest how the foods available could be modified to be more nutritious.

FOR YOU TO DO

10 Design a meal suitable for an athlete prior to a marathon run. Prepare this meal in class.
11 Explain why complex carbohydrates are important in an athlete's diet. List 10 examples of foods that contain complex carbohydrates.
12 Visit your local supermarket to investigate products targeted specifically at athletes.
13 Look through a variety of magazines dedicated to food, such as *Australian Table*, *Australian Good Taste* or *Super Food Ideas* and see how many recipes you can find that include grains, beans, peas, edible seeds or nuts. Which type of vegetarian diet does the recipe represent?
14 Design an entrée, main course and dessert for a dinner party that vegetarians will be attending.
15 List the nutrients that may be lacking in a vegan diet unless it is carefully planned.
16 Outline how your cultural or religious beliefs affect your food needs.
17 Design a poster promoting water consumption when bushwalking and camping.
18 Suggest some interesting sandwich fillings that can be easily and safely transported in a backpack.
19 Think of a healthy food you would like to see sold at your school canteen. How would you go about promoting this food? Put your ideas into practice (with the assistance of your school canteen manager).

TAKING IT FURTHER

20 Invite a sportsperson to your class to talk about their diet and training.
21 Consumption of sweets or sweet drinks in the hour before competing in sports is to be avoided. Use your library or the internet to research why this is so.
22 Predict the reaction of your family if you said that you wanted to become a vegetarian. If you are already a vegetarian, you may like to share with your class what your family's reaction was to your choice.
23 Investigate the social, religious and ethical reasons people have for being vegetarian.
24 Visit your local supermarket to compile a list of the range of convenience products that are designed to capture the growing vegetarian market.

8.5 Foods for special needs

Low-kilojoule foods

You eat food to fuel your body with energy and for growth and repair. The foods you eat provide kilojoules for the body. Carbohydrate, protein and fat provide the body with kilojoules. Fats and alcohol are by far the most energy-dense foods and should only be consumed in moderation, particularly if you are overweight or obese.

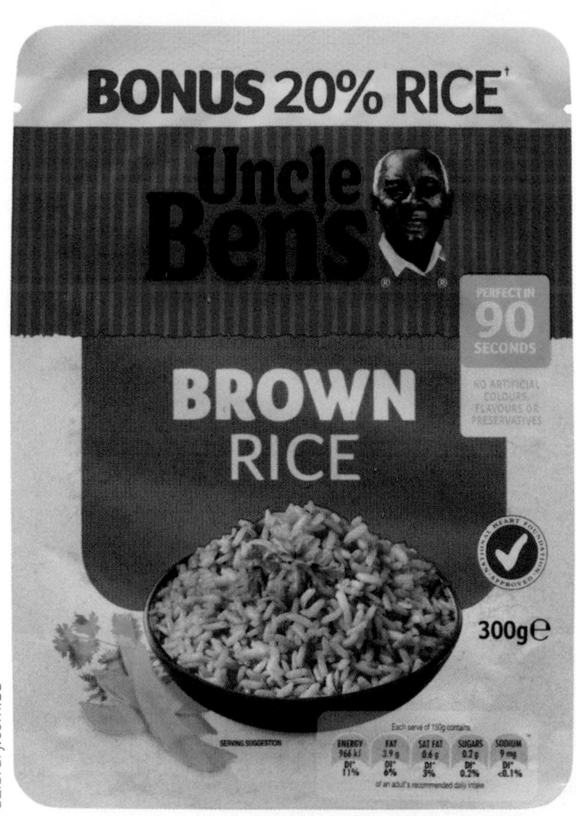

Figure 8.18 The National Heart Foundation works to improve heart disease prevention and care. Foods with the heart tick must meet strict levels of kilojoules.

During the past 25 years there has been a growing interest in low-kilojoule foods and beverages.

Low-salt foods

As a nation we tend to consume twice the recommended dietary intake for sodium. The salt in our diet comes not only from salt that we sprinkle on our food but also from processed foods. Salt consumption in large quantities has been linked to hypertension, high blood pressure, heart disease and stroke. It is possible to wean yourself off salt and begin to taste the natural flavours of food, rather than the salt.

Low-fat foods

Fats are vital for good nutrition; they act as a carrier for the fat-soluble Vitamins A, D, E and K, and they provide energy and the essential fatty acids needed for life and growth. Fat only becomes bad for you when you eat too much of it.

The key is to minimise the amount of fat consumed daily. There are now an increasing number of low-fat products available at supermarkets. Low-fat meals are the key to better health, as a diet low in fat and rich in fruit and vegetables, dairy products and wholegrains is recommended to avoid diet-related disorders.

High-fibre foods

It is important to include dietary fibre in your diet to help maintain regularity and reduce the risk of many illnesses. Foods that are high in fibre are high-fibre breakfast cereals, wholegrain and multigrain breads, fruit and vegetables.

High-protein foods

Protein is important for the growth of body cells and makes up virtually every part of the body. Children, adolescents, pregnant and breastfeeding women require more protein in relation to their body weight than adults. Accident and burn victims also require extra protein for rebuilding of damaged body tissue.

Examples of high-protein foods include:

- beans and legumes, such as lentils and chickpeas
- lean meat, poultry and fish
- eggs
- milk, yoghurt and cheese
- seeds and nuts
- soy products, such as tofu.

Planning considerations

Planning considerations for safe and nutritious foods for special needs include the following.

- Menus: special dietary needs may affect the menu, the ingredients needed and the overall health of the diners.
- Number of courses: the time of the day and place at which the meal will be consumed will affect the overall planning of a meal.
- Customer: the age, gender and preferences of the customer.
- Cost: it is important to be aware of your available budget.
- Time available: the time available for shopping, meal preparation and consumption affects the choice of foods.

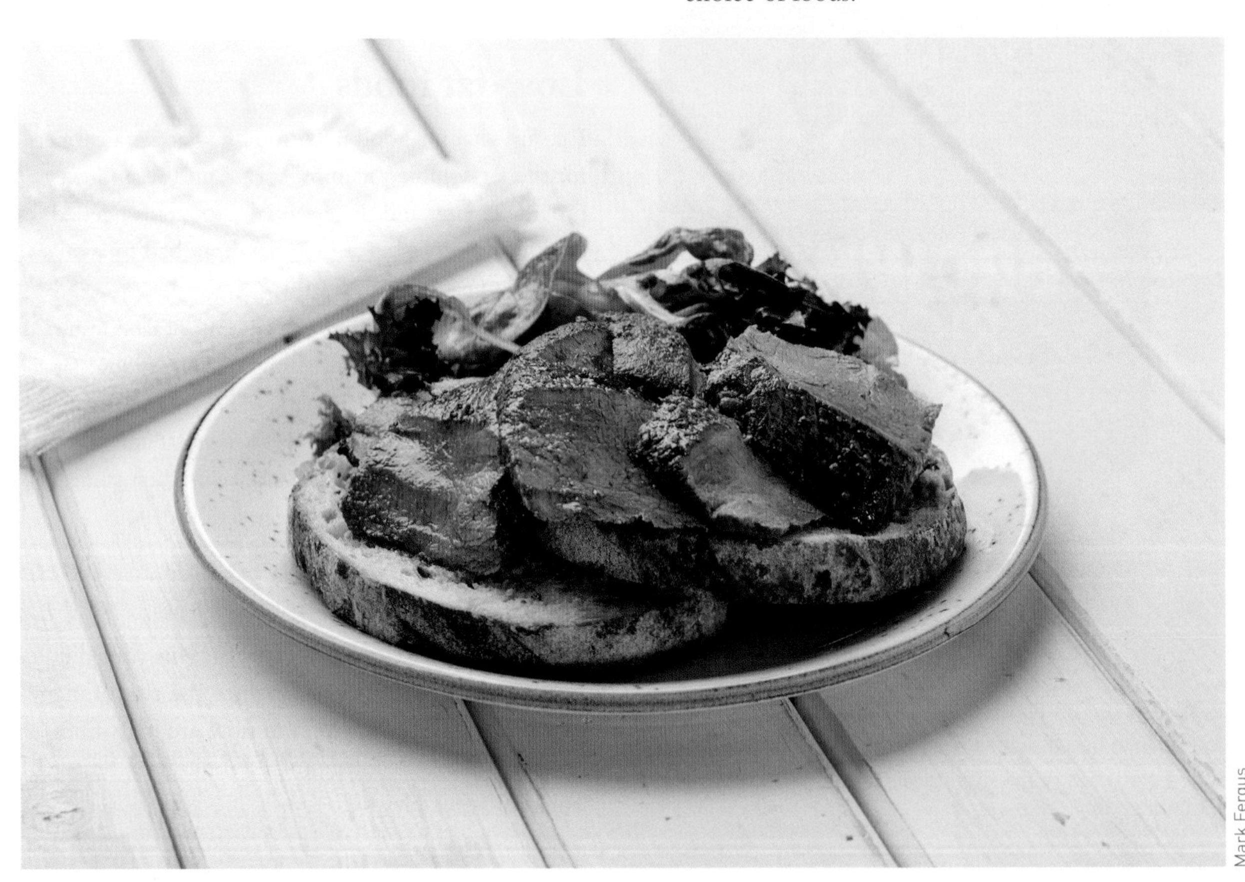

Mark Fergus

Figure 8.19 Lean meat is high in protein

KATHY USIC
NUTRITIONIST AND ACCREDITED PRACTISING DIETITIAN

Kathy Usic is an Accredited Practising Dietitian with over 30 years of experience in nutrition marketing and public health. She holds a Bachelor of Education and Master of Nutrition and Dietetics, and is currently Chief Executive Officer of the Glycemic Index Foundation (GIF), a non-profit health promotion agency committed to the education of scientifically backed low GI healthy-eating principles.

Prior to joining GIF, Kathy held senior leadership positions at several large companies, including George Weston Foods, Campbell Arnott's and Goodman Fielder. She has lectured in Nutrition and Health Studies at the University of Sydney, and also worked at the NSW School Canteen Association as a nutritionist and executive officer, where she was responsible for developing and implementing national nutritional guidelines for school canteens.

Prior to this, Kathy worked as a paediatric dietitian at the Children's Hospital at Westmead and Camperdown. In 1989, she joined the National Heart Foundation as a project officer where she helped implement various public health nutrition programs, including the successful Tick Food Program.

Before joining the Heart Foundation and undertaking further study in dietetics, Kathy worked as a secondary and tertiary teacher in Victoria and NSW, teaching physical education and home economics. Combining her interest in nutrition and good food, Kathy took a break from teaching and opened up a cafe known as Pelicans Fine Foods in Balmain in 1988. Kathy has also been the nutrition editor of *Practical Parenting* and a food writer for Time Inc. Publications (now Pacific Magazines).

A mother of four children, Kathy understands the demands of today's busy world and the complexity of food issues. She is a passionate nutrition marketer and believes that credible nutrition messages can help strike the right balance between good food and healthy lifestyle.

An accredited practising dietitian (APD) can provide you with personalised advice to give you the confidence to eat in a way that is best for you. APDs are university-qualified experts in nutrition and dietetics and are committed to the Dietitians Association of Australia's (DAA) Code of Professional Conduct, continuing professional development and providing quality services.

Accredited nutritionists (AN) are tertiary-qualified nutrition professionals who have expertise in a range of nutrition services, including public health nutrition, community health and tertiary education related to nutrition, but are different from APDs as they are not qualified to provide individual dietary counselling therapy and medical nutrition therapy. APDs can choose to use the APD and/or AN credential.

Source: With Permission of Kathy Usic. Photo: Kathy Usic.

QUESTIONS

1 Outline the difference between a dietitian and a nutritionist.
2 Investigate the university courses available in your state or territory to become a dietitian.
3 List 10 qualities required for Kathy to work in this industry.
4 Suggest the work Kathy would have been required to undertake when working on the 'national nutritional guidelines for school canteens' project.
5 Design a business card that highlights Kathy's role as an accredited practising dietitian.
6 Many people are referred to dietitians for nutritional advice. Explain why this is important for people at different stages of the life cycle.

UNIT REVIEW

LOOKING BACK

1 Identify a range of ways that people can reduce their intake of kilojoules, salt and fat.
2 Identify a range of ways that people could increase their intake of fibre and protein.
3 What considerations need to be taken into account to ensure food is safe to eat?
4 How will the location of the meal affect the type of food prepared?
5 Using examples, explain how the time available for shopping and meal preparation will affect choice of foods and cooking methods.

FOR YOU TO DO

6 Make a list of all the salty foods that you eat on a weekday and over the weekend. How could you cut down your salt intake?
7 Visit a supermarket or look in your food cupboards and find 10 foods that have 'salt' written on the label and 10 foods that have 'low-salt' written on the label.
8 Purchase two packets of chips, one low-salt and one not. Conduct a taste test and write a short report on your findings.
9 Herbs and spices can be used as alternatives to salt for enhancing food flavours. Develop a list of any other flavours you think of that can be added to food besides salt, herbs or spices.
10 A number of individuals have special dietary needs and require particular foods. Complete the table below by providing a description, the dietary needs and examples of suitable foods for each group.

GROUP	DESCRIPTION	DIETARY NEEDS	SUITABLE FOODS
Coeliac			
Vegan			
Lacto-vegetarian			
Muslim (religious)			

11 You have offered to cook a meal for a small group of friends that consists of a vegetarian, a person allergic to seafood and someone who is dieting. Design a two-course meal to meet the needs of all of your guests, including yourself. Justify your food choices. Provide recipes for all dishes.

TAKING IT FURTHER

12 Identify eight salt-reduced food products that are currently on the market. Outline the salt content in each product and compare these to products that are not salt-reduced.
13 Collect food labels from a range of food and beverage products. Read the nutrition information panel on each label and compare the level of kilojoules, sodium, fibre, protein and fat in each product. Display your findings in a table and analyse the results.
14 Research the amount of fat in common take-away foods; for example, hamburger, battered fish fillet and chicken nuggets.
15 Many traditional recipes, even though they may include ingredients that are not suitable for specific dietary requirements, can be altered to suit the diet of individuals with special needs.
   a Select a special food need that you have investigated in this chapter and choose a recipe for a meal or snack food.
   b Alter the ingredients according to the special food need you have chosen by either removing an ingredient, substituting a similar ingredient or cutting down on the quantity.
   c Prepare this food and evaluate it in terms of its taste, appearance and nutritive value.

Chapter review

LOOKING BACK

1. List three different types of dietary restrictions people may have and provide an example of each.
2. As a class, discuss the link between ageing, nutrition and health. List some important guidelines the aged should follow to maintain an active, healthy lifestyle.
3. Explain how the stages of the life cycle affect dietary needs.
4. List five examples of when a person may require a modified diet to meet a special need.
5. Describe the three types of vegetarian diets.
6. Identify foods that could be consumed as part of a daily diet of a person intolerant to gluten.
7. Suggest five eating habits that you and your family could change to improve your health and reduce the risk of coronary heart disease.
8. Discuss the implications of a diet high in sodium or fat on the health of our society.
9. Identify foods that could be added to recipes to replace salt.
10. Identify three activities or situations that present logistical issues when preparing and selecting foods. Share your results with your class and present your combined efforts on the whiteboard or smartboard.
11. Discuss how dietary modification can assist in managing people's special dietary needs.
12. Construct a table that summarises the circumstances that may lead to an individual having special needs.
13. Describe how a person's health status or lifestyle affects their food choices. Provide examples of when this may occur.
14. Explain how educating people with special needs in appropriate food selection influences the health system in your state or territory.

FOR YOU TO DO

15. Design a PowerPoint presentation naming the different stages of the life cycle and outlining specific age groups and nutritional requirements for each.
16. Design a humorous, informative comic strip about one stage of the life cycle.
17. Low-fat diets are not recommended for young children. Choose five products marketed to children and investigate the nutritional content of each. Rate them according to their suitability for a 3-year-old child.
18. Investigate the dietary requirements of one religion and discuss the implications for shopping.
19. Develop a one-day eating plan for a vegetarian of your choice. Give reasons for the foods selected.
20. Design and produce a brochure to promote one high-fibre product available at your local supermarket. Explain the benefits of purchasing and consuming this product.
21. Select one group with a diet-related special need. Use the internet to define the disorder, its symptoms, how it can be prevented and possible treatment options. Present this information as a poster. When completed, display it in your classroom.
22. Visit the Kids Health weblink and read the article titled 'A Guide to Eating for Sports'. Outline why it is important to eat the correct foods and quantities if you want to reach peak performance. As a class debate the following topic: 'Sports supplements really work'.

Weblink KidsHealth

23. Conduct a classroom forum to discuss the planning considerations for safe and nutritious foods for special needs. Consider whether it is more time-consuming and expensive to eat nutritious foods or processed fast foods.
24. Create a mind map to summarise this chapter. Use a highlighter pen to identify one new concept you have learned that could lead to changes in the choices you make when shopping for food products.

TAKING IT FURTHER

25. Plan a week's diet for a volunteer participant in your class. Take into consideration their health and physical status, their physical activity level and any special needs relating to food.
26. Conduct a search in your local telephone directory for restaurants that prepare and serve meals catering for people with special dietary requirements. Arrange to visit this restaurant to interview the chef and sample the food.

9780170383417

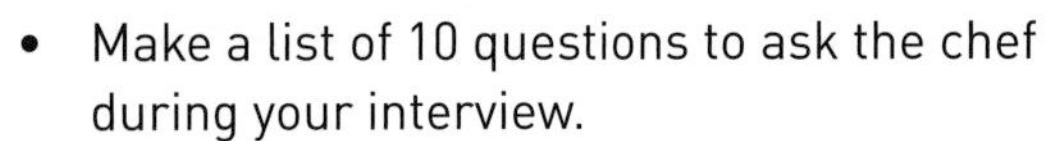

- Make a list of 10 questions to ask the chef during your interview.
- Prepare an oral report of your experience and predict any future trends in 'special need' restaurants. Support your predictions with specific reasons and examples.
- Share your experiences with your class and compare the different eating experiences you encountered.

27 Design and prepare a vegetable-based main meal. Calculate the cost of this meal and compare it to the cost of a meal that includes meat products. List the advantages and disadvantages of preparing and eating a meal that is vegetable-based rather than meat-based.

28 Form small teams and consider one diet-related disorder. Prepare an interactive presentation that highlights practical ways to educate people in your local area suffering from this disorder.

29 Design a menu for one week to show a 25-year-old male how to reduce his chances of developing diabetes.

30 Visit the Heart Foundation weblink and complete the following activities.

- Investigate the importance of foods gaining the Heart Foundation Approval Tick.
- Research what the Tick means, how companies earn the Tick and how the Tick standards are maintained.
- Download a full list of foods approved to use the Tick in the supermarket. Take this list to your local supermarket and highlight how many products you could easily locate from the list.
- Use the website's recipe finder and design a day's menu that meets the needs of following diets: low-kilojoule diet, high-fibre diet, low-salt diet, high-protein diet, low-fat diet.

Weblink
Heart Foundation

31 Invite a local professional athlete into your classroom to discuss how their food choices affect their athletic performance. Draft a series of questions before the visit so that you are prepared.

9780170383417

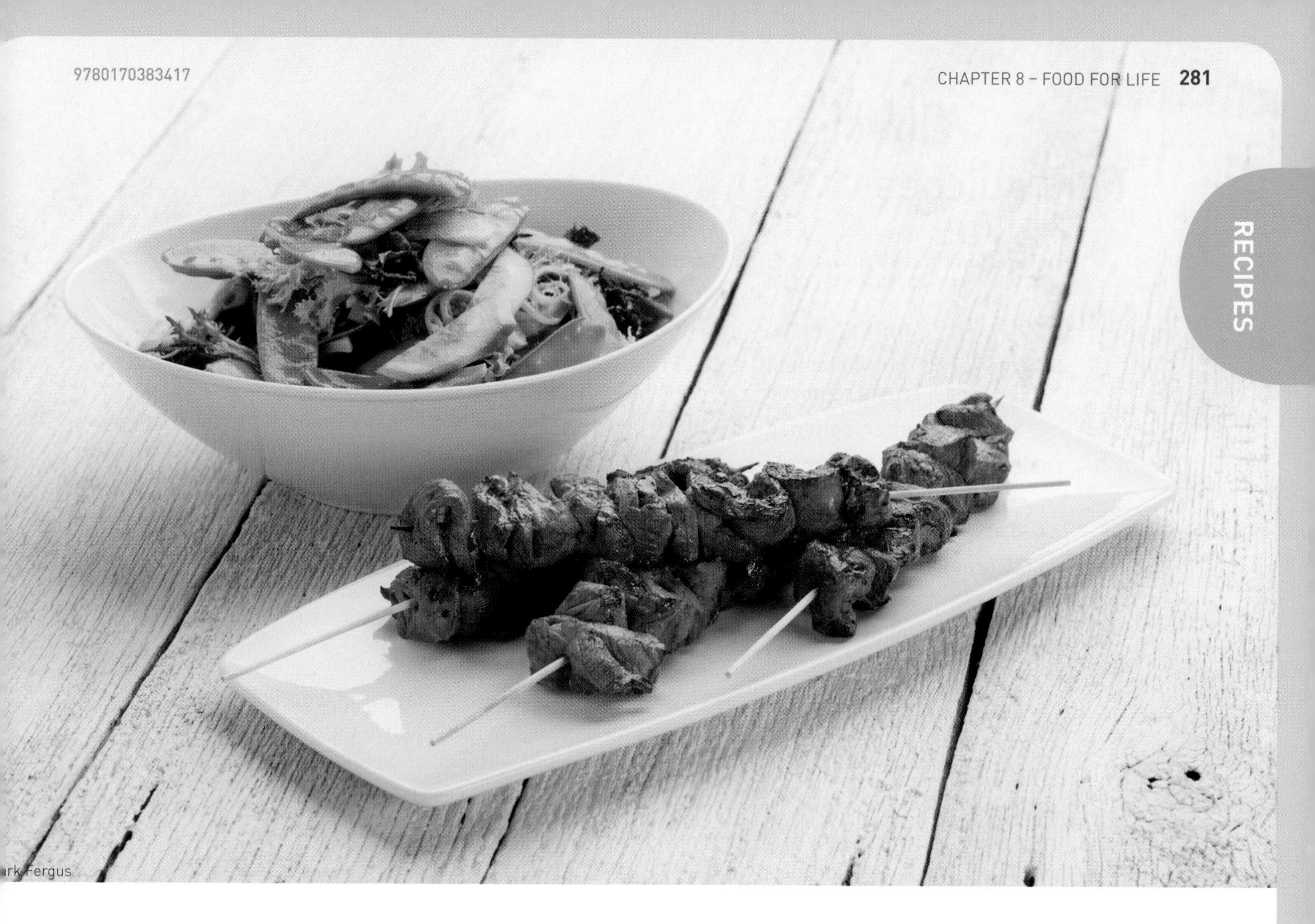

Teriyaki beef skewers with snow pea salad

 **Preparation time** 25 minutes  **Cooking time** 5 minutes  **Makes** 4

INGREDIENTS

650 grams rump steak, thinly sliced

⅓ cup (80 mL) teriyaki sauce

250 grams snow peas, halved lengthways

1 Lebanese cucumber, thinly sliced

50 grams mixed baby salad leaves (mesclun)

2 cups (400 grams) jasmine rice, cooked to packet instructions

DRESSING

1 teaspoon oil

1 tablespoon rice vinegar

1 tablespoon mirin (sweet Japanese rice wine)

1 tablespoon light soy sauce

1–2 teaspoons wasabi paste, to taste

1 teaspoon caster sugar

METHOD

1. Thread the beef on 8 metal or soaked bamboo skewers, folding each slice in half as you go. Place in a shallow dish and pour over the teriyaki sauce. Marinate for 5 minutes.
2. Meanwhile, cook the peas in lightly salted boiling water for 1 minute until just tender. Drain and refresh in cold water. Place in a bowl with the cucumber and leaves. Whisk the dressing ingredients together, then toss with the salad.
3. Heat a chargrill pan on medium-high. Remove the skewers from the marinade and cook for 2 minutes each side until charred and cooked through. Serve with salad and rice.

Recipe source: © News Life Media

QUESTIONS

1. Why do the skewers need to be soaked?
2. Evaluate your dish and teamwork in the making of this recipe. What worked well? What would you do differently?
3. Describe the experience of tasting wasabi.

9780170383417

RECIPES

Mini quiches

 **Preparation time** 30 minutes  **Cooking time** 15 minutes  **Makes** 8

INGREDIENTS

- 8 slices very large wholemeal bread, crusts removed
- 20 grams butter, softened
- 3–4 shallots, ends trimmed, thinly sliced
- 6–8 slices shaved ham, to fit
- ⅓ cup (35 grams) coarsely grated cheddar
- 6 eggs, lightly whisked

METHOD

1. Preheat oven to 160°C. Brush 8 x ⅓ cup capacity muffin holes with melted butter to lightly grease. Use a rolling pin to gently roll out the bread slices to 3 mm thick. Use a 11 cm diameter pastry cutter to cut bread into discs. Spread each slice lightly with butter. Line the prepared muffin holes with the bread slices.
2. Arrange the shallots, ham and cheese evenly among muffin holes, then top each up with egg. Bake in oven for 15 minutes or until golden brown and just set. Place in the fridge to chill. Place in airtight containers to store.

QUESTIONS

1. Evaluate the texture and flavour of your finished dish.
2. Provide a vegetarian alternative for the ham.
3. How did you judge the quiches were cooked? What were you looking for?

Mark Fergus

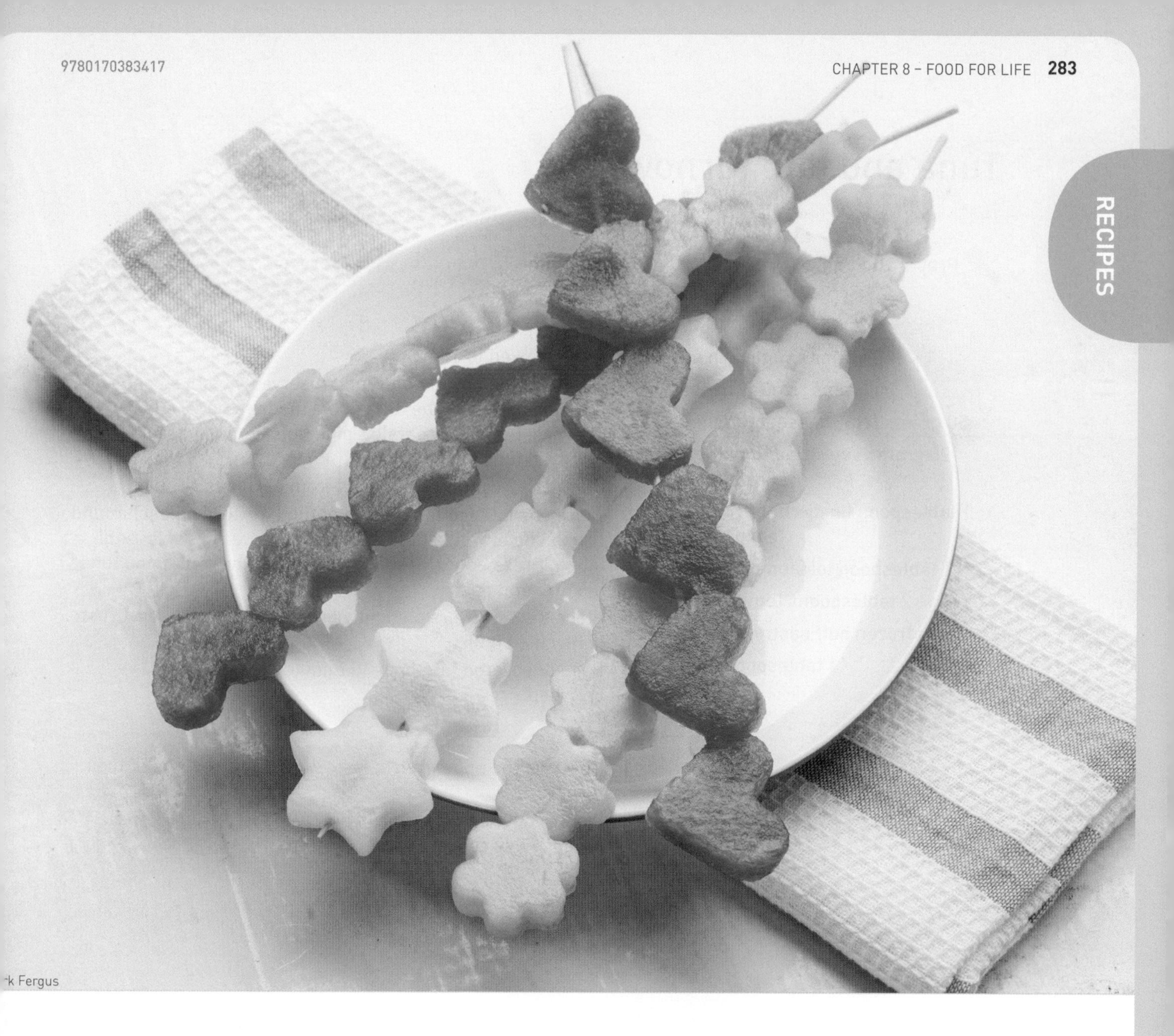

-k Fergus

Fruit wands

 **Preparation time** 30 minutes  **Cooking time nil**  **Makes** 24

INGREDIENTS

½ seedless watermelon, peeled, cut into 1 cm-thick slices

½ honeydew melon, seeded, peeled, cut into 1 cm-thick slices

½ rockmelon, seeded, peeled, cut into 1 cm-thick slices

Bamboo skewers

METHOD

1. Use 3 cm- and 4 cm-diameter flower, heart and star-shaped pastry cutters to cut shapes from the watermelon. Repeat with the honeydew melon and rockmelon.
2. Thread the shapes onto the skewers, using one of each melon type per skewer.
3. As a variation to the fruit shapes, cut carrots into very thin strips and cut out with number or alphabet shapes. Serve with celery battons and cheese spread.

QUESTIONS

1. This recipe can generate a lot of wasted fruit. Brainstorm uses for the off-cuts.
2. Suggest another alternative for presenting this selection of fruit.

9780170383417

RECIPES

Tuna and feta turnovers

These are a quick, convenient and nutritious alternative to take-away foods.

 **Preparation time** 15 minutes  **Cooking time** 15 minutes  **Serves** 2

INGREDIENTS

- 1 × 110 grams can tuna in oil
- 50 grams ricotta cheese
- 50 grams feta cheese, crumbled
- 25 grams sun-dried tomatoes, sliced thinly
- 1 tablespoon finely chopped flat-leaf parsley
- 1 tablespoon toasted pine nuts
- 2 tablespoons lemon juice
- 2 sheets frozen puff pastry, thawed
- 1 tablespoon milk

METHOD

1. Preheat oven to 180°C.
2. Drain the tuna over a bowl and reserve two tablespoons of the tuna oil. Flake the tuna into a medium bowl, add both cheeses, tomatoes, parsley, pine nuts, lemon juice and reserved tuna oil. Mix well.
3. Cut four 12 centimetre rounds from each pastry sheet.
4. Place one heaped tablespoon of tuna mixture on each round. Brush edges with a little milk and fold over to enclose the filling. Press edges to seal. Repeat with the remaining tuna mixture and pastry.
5. Place the turnovers on lightly greased oven trays and brush lightly with milk.
6. Bake uncovered for about 15 minutes or until lightly browned.

QUESTIONS

1. Describe the appearance, taste, texture and aroma of the turnovers.
2. Would you make any modifications to the recipe if you were to produce this recipe again?

Mark Fergus

9780170383417

rk Fergus

Zesty crab cakes

 **Preparation time** 25 minutes  **Cooking time** 20 minutes  **Makes** 12-15

INGREDIENTS

- 3 x 170 grams cans crab meat, drained
- 1½ cups fresh white breadcrumbs
- ½ cup coriander leaves, finely chopped
- ½ cup mint leaves, finely chopped
- 3 shallots, thinly sliced
- 2 limes, rind finely grated, juiced
- 1 egg
- 2 tablespoons sour cream
- 2 tablespoons olive oil
- 1 avocado
- lime wedges, to serve

METHOD

1. Line a baking tray with baking paper. Combine the crab meat, breadcrumbs, 2 tablespoons coriander, 2 tablespoons mint, shallots, lime rind and 2 tablespoons of lime juice in a large bowl. Whisk the egg and sour cream in a jug. Add to the crab mixture. Season with salt and pepper. Using clean hands, mix to combine.
2. Shape tablespoons of the mixture into patties. Place onto prepared tray. Cover. Refrigerate for 20 minutes.
3. Preheat oven to 140°C. Heat oil in a non-stick frypan over medium heat until hot. Cook the crab cakes, 6 at a time, for 2 minutes each side or until light golden. Transfer to a baking tray. Keep warm in oven while cooking the remaining crab cakes.
4. Mash the avocado until almost smooth. Add the remaining coriander, mint and 1 tablespoon of lime juice. Mix well. Spoon a dollop of mixture onto each crab cake. Top with lime wedges. Serve hot or cold.

QUESTIONS

1. Suggest a remedy if your mixture is too wet to form pattie. What about if it is too dry?
2. Outline some safety considerations when pan-frying.
3. What is the appropriate first aid for minor burns?
4. Suggest a suitable vegetable accompaniment for this dish.

RECIPES

Chicken risotto

There are wheat-free and gluten-free products on the market that make life much easier for people who are wheat- and gluten-intolerant. This is a tasty wheat, gluten and dairy-free dish that you can make at home.

 **Preparation time** 10 minutes  **Cooking time** 30 minutes  **Serves** 2

INGREDIENTS

3 tablespoons olive oil
200 grams arborio rice
1 onion, chopped
1 green capsicum, chopped
1 red capsicum, chopped
8 medium mushrooms
750 mL wheat-free/gluten-free vegetable stock
2 skinless chicken breasts, chopped
2 teaspoons fried oregano
salt and pepper to season

METHOD

1 Put 2 tablespoons olive oil in a heavy-bottomed pan and add the rice. Gently heat the rice for 2–3 minutes, until it looks translucent.
2 Add the capsicum, peppers and mushrooms and cook for another 5 minutes, stirring and being careful not to brown the rice.
3 Add the vegetable stock and bring to the boil. Turn down the heat and simmer the rice for 25 minutes, adding boiling water if necessary to ensure it does not dry out.
4 Put the chicken pieces in a frypan with 1 tablespoon oil and cook until lightly browned.
5 At the end of the risotto cooking time add the chicken, oregano and salt and pepper before mixing well.

QUESTIONS

1 List the nutrients provided by this dish.
2 Which ingredients in this recipe provide dietary fibre?
3 How could you modify this dish to make it completely balanced?
4 Outline which specific nutritional needs this dish would meet.

Mark Fergus

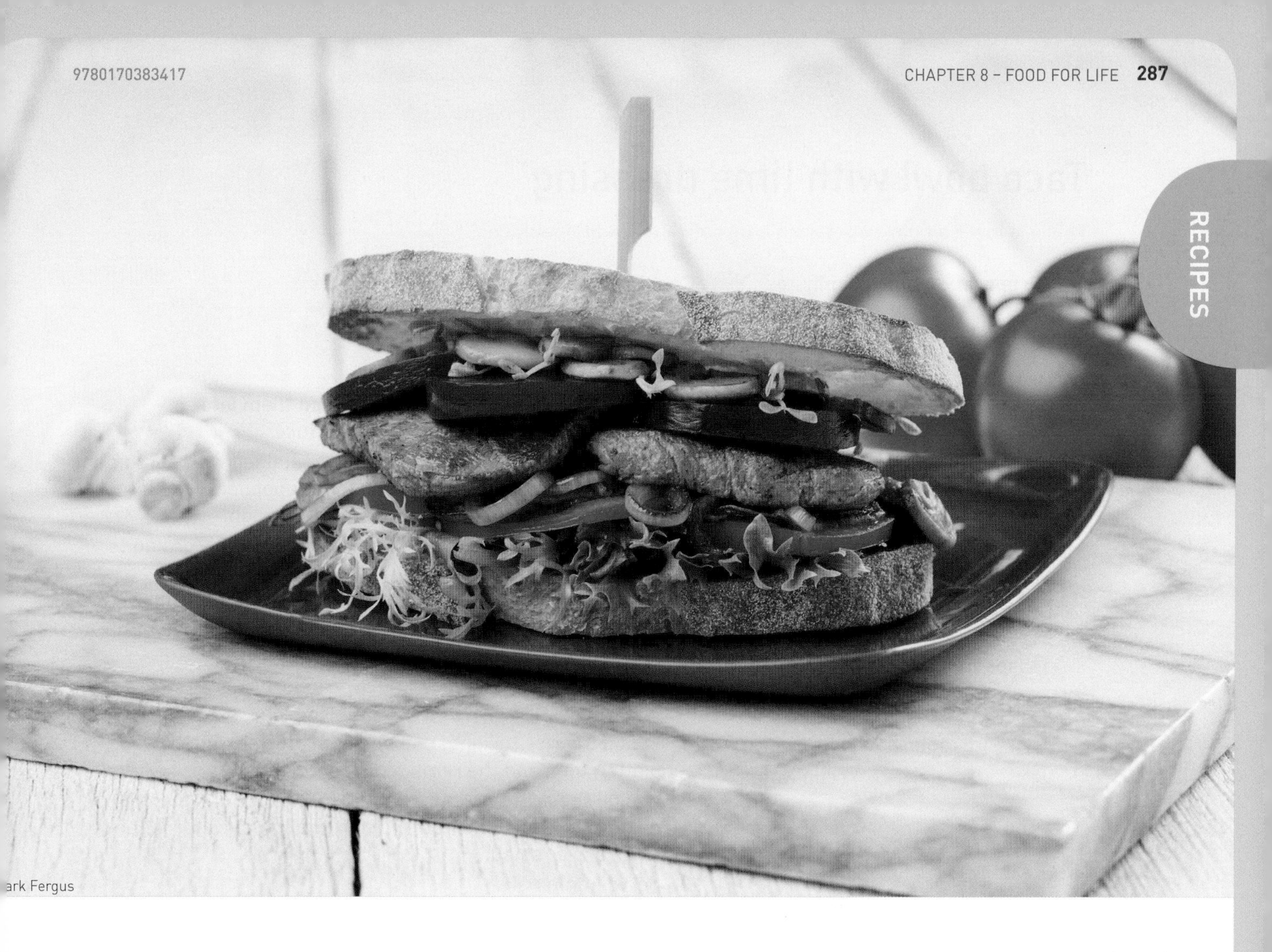
ark Fergus

Steak sandwich

This recipe uses a variety of ingredients that are both nutritious and appetising. Present this steak sandwich on a plate as though you were styling it for a magazine advertisement.

 **Preparation time** 10 minutes  **Cooking time** 30 minutes  **Serves** 2

INGREDIENTS

100 grams button mushrooms, sliced

1½ tablespoons balsamic vinegar

1 onion, sliced

½ teaspoon brown sugar

2 fillet steaks, flattened slightly

4 slices ciabbata or sourdough bread

25 grams mixed salad leaves

1 tomato, thinly sliced

150 grams canned beetroot slices, drained

25 grams snow pea sprouts

METHOD

1. Spray a non-stick frypan with oil and then heat the pan.
2. Add the mushrooms and 1 tablespoon balsamic vinegar and cook over high heat until browned and tender. Remove from pan.
3. Add the onion, remaining balsamic vinegar and sugar, and cook over low-medium heat for 10 minutes or until caramelised. Remove from pan. Clean pan, respray with oil and heat.
4. Cook the steaks over high heat for 3 minutes on each side, or to your liking. Toast bread until golden brown on both sides.
5. To assemble each sandwich, place salad leaves, tomato, onion, steak, beetroot, snow pea sprouts and mushrooms on a slice of bread and top with another slice of bread.

QUESTIONS

1. Outline the nutritional benefits of your sandwich.
2. What additional or alternative ingredients might you use to personalise your sandwich?

RECIPES

Taco bowl with lime dressing

This is a low-sodium, vegetarian alternative to traditional tacos.

 **Preparation time** 20 minutes  **Cooking time** 40 minutes  **Serves** 4

INGREDIENTS

1 cup brown rice
2 sweet potatoes
1 head lettuce
1/2 cup corn (frozen)
1 cup cherry tomatoes
1/2 onions
1 avocado
1 lime
1 teaspoon sea salt
1/8 teaspoon black pepper (ground)

LIME DRESSING

2 tablespoons lime juice
1/3 cup water
1 avocado
1 teaspoon garlic powder
1 teaspoon sea salt
1/2 teaspoon black pepper (ground)

METHOD

1. Preheat oven to 200°C.
2. Thoroughly wash the potatoes and cut them into 25-mm cubes. Shred or chop the lettuce.
3. Arrange the potatoes on a baking sheet lined with parchment paper; sprinkle with sea salt and black pepper. Bake for 35–40 minutes, turning once halfway through.
4. While the potatoes are cooking, cook the rice according to package instructions.
5. In a small pan bring 2 cups of water to a boil, then add the frozen corn; lower heat and cook for 5 minutes, then drain.
6. Fluff the rice with a fork and fold in the corn. Halve the tomatoes, chop the onion and avocado, and slice the lime into wedges.
7. Spoon the rice mix into one end of a large serving bowl or platter, then add the remaining ingredients.
8. Add all the dressing ingredients to a blender and blend until mixture is well combined and thin; add more water if needed. Drizzle over salad.
9. Serve immediately.

QUESTIONS

1. Calculate the serves of vegetables contained in their recipe.
2. What is the difference between brown and white rice? Justify the use of brown rice in this recipe.
3. Would you consider this recipe high in carbohydrates? Explain your answer.

TIP: Avocado is a great source of healthy fat, and is a yummy veggie to include that is really filling. If you aren't having meat with your dinner, include avocado!

Mark Fergus

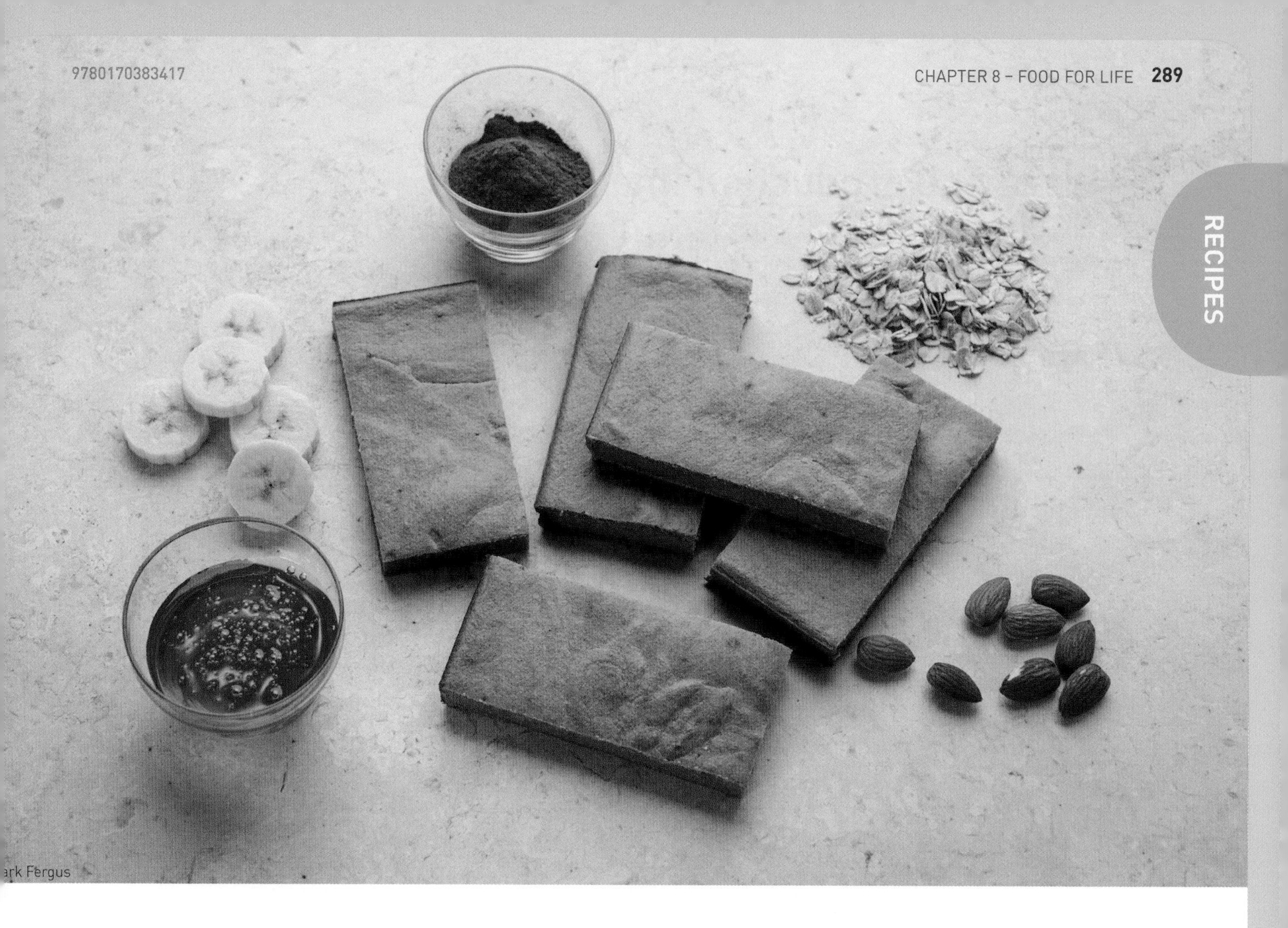

Protein bars

This recipe has been included in its original format. Your school may have a policy restricting the use of nuts, including almonds, in class. In your planning, consider an alternative ingredient of a similar texture to substitute for your lesson.

 **Preparation time** 5 minutes  **Cooking time** 20 minutes  **Makes** 8

INGREDIENTS

200 grams rolled oats
30 grams chocolate whey protein
1 teaspoon cinnamon
2 tablespoons almond butter
3 egg whites
2 medium bananas, mashed
1 tablespoon honey
100 mL skim milk

METHOD

1. Preheat oven to 180°C. Grease or line a 24 x 24 cm cake pan with parchment paper.
2. Using a food processor, process the rolled oats into a fine flour.
3. Mix the oats, whey protein powder and cinnamon. Add the almond butter and mix thoroughly. Add the egg whites, mashed bananas and honey.
4. Mix well and slowly add the skim milk. Once mixed, spoon or pour into prepared pan. Smooth with a knife. Bake for 15–20 minutes or until a toothpick comes out clean. Cut into 8 bars when cooled.

QUESTIONS

1. Explain reasons why the egg yolks have been omitted.
2. If you did not own a food processor, how could this recipe be made successfully?
3. Did you use an alternative to the nut butter? How successful was it?

9780170383417

Pizza sub production line

Your class has been asked to trial a new recipe for the school canteen. As a class you are required to work together to prepare and cook a pizza sub for each person in your class.

You will need to work collaboratively to reach the desired result.

 **Preparation time** 15 minutes  **Cooking time** 10 minutes  **Serves** 20

INGREDIENTS

- 10 bread rolls, cut in half and buttered lightly
- 2/3 cup tomato paste
- 160 grams canned crushed pineapple
- 250 grams cheese
- 5 slices ham, finely chopped by hand or in the food processor, or 1 tomato, sliced
- pinch of oregano
- aluminium foil

METHOD

1. Preheat oven to 160°C.
2. Spread the bread rolls lightly with tomato paste.
3. Top with drained pineapple, cheese, and ham or tomato slices.
4. Sprinkle with oregano.
5. Wrap base of each pizza sub with foil for easy handling and to prevent the roll from drying out.
6. Place in oven for 10 minutes to melt the cheese.

QUESTIONS

1. Develop a production plan for this recipe.
2. Develop a costing schedule for this recipe.
3. Invite your canteen manager or a representative to evaluate your end result.
4. Design and prepare a poster advertising your dish. Remember to put the price on your poster!

Mark Fergus

RECIPES

rk Fergus

Traditional sausage rolls

Sausage rolls are a convenience food commonly served at parties and available from bakeries and corner stores as a take-away food item. The basic composition of a traditional sausage roll is generally a sheet of puff pastry sliced into two, wrapped into tubes around sausage meat and brushed with egg before being baked.

 **Preparation time** 10 minutes  **Cooking time** 30 minutes  **Makes** 12 sausage rolls

INGREDIENTS

- 500 grams sausage mince
- 1 small onion, finely chopped
- 2 tablespoons tomato sauce
- 2 tablespoons parsley, finely chopped
- 2 sheets frozen puff pastry, thawed
- 1 egg yolk or milk, for glazing

METHOD

1. Preheat oven to 200°C. Line a baking tray with baking paper.
2. Place the sausage mince, onion, tomato sauce and parsley into a large bowl and mix with hands. Season to taste.
3. Cut each pastry sheet in half. Wet hands and divide the mince into four portions.
4. Take one portion and roll into a sausage shape, about 3.5 cm wide and as long as the length of the pastry (approximately 25 cm). Place it halfway down the pastry piece.
5. Fold one side of the pastry over the mince and continue rolling. Cut into three pieces and place on prepared tray. Cut slits on top of each piece and brush with egg yolk or milk.
6. Repeat with the remaining ingredients.
7. Bake the sausage rolls in oven for 30 minutes or until tops are golden.
8. Serve immediately.

9780170383417

Revamped sausage rolls

The traditional sausage roll recipe can be varied and revamped in many different ways. Some variations on the basic recipe include adding ingredients such as seasonings and sauces or using a different pastry. Sausage rolls can also be served in various lengths, including party or jumbo sausage rolls.

 **Preparation time** 15 minutes  **Cooking time** 10 minutes  **Makes** 16 sausage rolls

INGREDIENTS

- 1 small onion, finely chopped
- 400 grams lean beef mince
- ½ cup fresh breadcrumbs
- 1 egg
- ½ cup barbecue sauce
- 2 tablespoons finely chopped parsley
- 16 sheets filo pastry
- 1½ tablespoons light olive oil
- poppy or sesame seeds (optional)

METHOD

1. Preheat oven to 180°C. Line a baking tray with baking paper.
2. Spray a non-stick frypan with cooking spray and cook the onion on low heat for a few minutes until soft.
3. Place the onion in a bowl with the mince, breadcrumbs, egg, barbecue sauce and parsley. Mix with hands and season to taste.
4. Unfold the filo sheets and cover with a damp tea towel. Remove one sheet and use a pastry brush to lightly coat with oil. Place two filo sheets on top, lightly oil the top sheet, then place another sheet on top. Turn pastry so the length is facing you and paint the furthest edge away from you with water.
5. Divide the mince mixture into four portions. Take one portion and roll into a sausage shape about 2 cm wide, and place the mince about 2 cm from pastry edge closest to you. Carefully the roll up. Place roll joined-side down on work surface and gently pat to flatten slightly.
6. Brush the top with oil and sprinkle with poppy or sesame seeds, if desired.
7. Cut into four pieces and place on prepared tray. Repeat with the remaining ingredients.
8. Bake for 35 minutes.
9. Serve immediately.

QUESTIONS

1. Did you make both sausage roll recipes? Evaluate both in terms of taste, texture and presentation.
2. Suggest a further modification to the recipe to suit vegetarians.
3. What alternative mince could you substitute to reduce the fat content further?

Mark Fergus

CHAPTER 9

Mark Fergus

Great expectations

FOCUS AREA: FOOD FOR SPECIAL OCCASIONS

FOCUS AREA

Food plays an important role at special occasions. This chapter investigates the significance and reasons for celebrating, ways to prepare foods for small- and large-scale catering occasions, what to consider when planning food for special events, and the importance of presentation and display of foods. You will follow a workflow plan and use correct food-handling skills to organise and prepare foods for a variety of fun and interesting events.

CHAPTER OUTCOMES

In this chapter you will learn about:

- the role and significance of food
- celebrations
- foods for special occasions
- menu planning
- the presentation of food

In this chapter you will learn to:

- outline the historical importance of food
- explore a variety of special occasions
- create food for special occasions
- devise a workflow plan
- demonstrate appropriate food-handling and presentation skills.

RECIPES

- Double corn bread, p. 322
- Butterfly cakes, p. 323
- Quick Christmas fruit mince tarts, p. 324
- Spicy roasted chick peas, p. 325
- Chicken and ricotta sausage rolls, p. 326
- Potato and corn chowder with warm bread rolls, p. 327
- Basic butter cake, p. 328
- Vienna cream icing, p. 329

'To break bread' means to show hospitality, and it symbolises peaceful relations between people.

Commercial buffets and salad bars must have special protective covers called 'sneeze barriers'. This helps to prevent people from coughing or sneezing over foods that others are to consume, thus helping to prevent food contamination.

Many foods cannot be pre-prepared too far in advance. Bread becomes stale, cut or peeled potatoes go brown and whipped cream begins to liquefy.

Babies and young children have tastebuds on their tongues and on their cheeks. Flavours and temperatures are very strong to them as their mouths are highly sensitive. As children get older, the tastebuds on the cheeks begin to disappear.

Canned food such as peas remain fresh for up to 2 years when unopened in your cupboard. Frozen peas will last approximately 3–6 months and fresh peas 5–10 days.

Food stylists make food look appealing and mouth-watering in photographs. When setting up the photo, mashed potato may be used instead of ice-cream so that it doesn't melt. Roast meats may be painted with wood lacquer to enhance the shine and hot water is placed behind or below food to make it appear that it is steaming.

FOOD WORDS

canapé small piece of toast spread with a variety of toppings and served as an appetiser

commercial for profit

cross-contamination transfer of bacteria or other micro-organisms from one food or item of equipment to another

crudité raw vegetable cut into bite-sized strips and served with a dip

fast eat very little or nothing

julienne narrow strips that are approximately 3 mm wide

leavened bread bread that has risen because of the addition of a leavening agent such as yeast

portion to measure or divide foods into pre-determined quantities

9.1 Role and significance of food

Foods around the world

Food has always been and will always be important. It plays an essential part in people's lives, for without it we would die.

Food also has an important social role. At a children's birthday party, you are likely to find sweets, chips, soft drinks and, most importantly, a birthday cake. At a Christmas celebration, you could be served turkey, ham and plum pudding with custard.

People have always met together to relax, talk and share common experiences and beliefs. We like to share food at the same time because it makes the occasion even more enjoyable. It shows great hospitality and friendship when we share food with another person.

HANDS-ON

SENSORY EVALUATION OF BREAD

PURPOSE

This activity allows you to familiarise yourself with different types of breads. While you taste each sample, make notes on what you taste. How is each type of bread similar or different? Identify any unique features for each type of bread. Think about the cultural influences.

MATERIALS

One portion of the following breads: naan, bagel, croissant, crumpet, lavash, focaccia, pitta and damper

STEPS

1. Sample each of the breads listed.
2. Make brief notes about each bread. Comment on appearance, texture, taste and colour. Discuss your findings with another student or as a class.

ACTIVITIES

1. Which breads were similar? Explain.
2. Select one bread and explain how it is served and what it is served with.
3. Categorise each bread as either leavened or unleavened.
4. Select one type of bread and explain in 100–150 words how you think it reflects the culture or history of the country from which it originated.

TIP

Some bread, such as crumpets, require cooking or heating before serving. You may wish to serve these breads with butter or another plain spread to avoid altering the taste.

FOOD IN FOCUS

BREAD ALL OVER THE WORLD

Bread is broken and shared during special celebrations in many countries. Many types of bread have great historical and religious meaning. Bread comes in different forms, is a staple of many people's diets, provides good nutrition, is relatively cheap and may be eaten with most meals.

The type or form of bread eaten reflects the customs, culture, religious beliefs, climate and history of people.

- Australians ate damper in the early days of European settlement because the early settlers did not have yeast to help the bread to rise.
- Indian people eat naan bread because it is soft and absorbent, and when eaten with curries helps to soak up the liquid.

Bread has been written about throughout history. Look at this brief timeline.

>

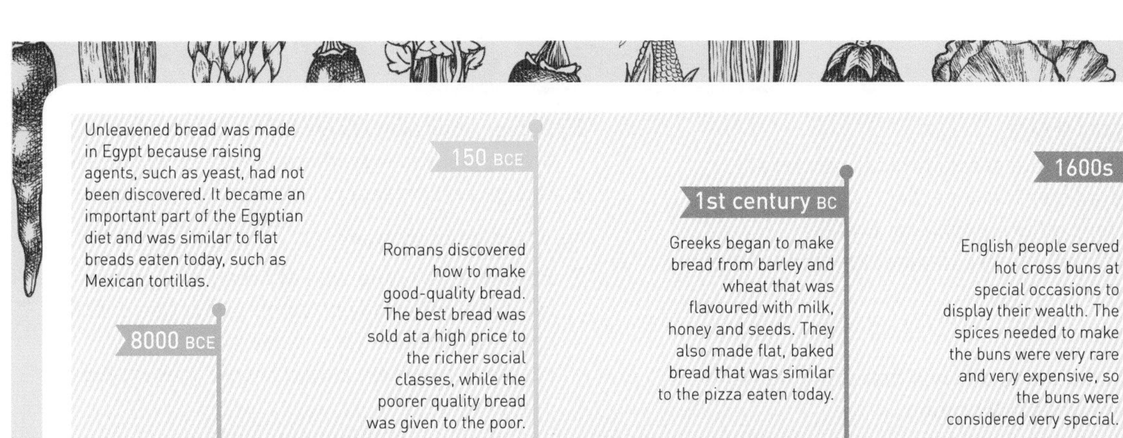

8000 BCE
Unleavened bread was made in Egypt because raising agents, such as yeast, had not been discovered. It became an important part of the Egyptian diet and was similar to flat breads eaten today, such as Mexican tortillas.

3000 BCE
Ancient Egyptians discovered how to make leavened bread. This bread was considered to be very special, so was offered to the god Osiris (the god of grain) in ceremonies. Bread became so valuable that it was sometimes used instead of money.

150 BCE
Romans discovered how to make good-quality bread. The best bread was sold at a high price to the richer social classes, while the poorer quality bread was given to the poor.

40 BCE
Roman authorities saw that bread was such an important part of the people's diet that they allowed bread to be given without charge to adult males.

1st century BC
Greeks began to make bread from barley and wheat that was flavoured with milk, honey and seeds. They also made flat, baked bread that was similar to the pizza eaten today.

1202 CE
King John in England introduced laws that governed the price of bread. People had to be able to afford it in order to survive.

1600s
English people served hot cross buns at special occasions to display their wealth. The spices needed to make the buns were very rare and very expensive, so the buns were considered very special.

Bread may:

- remind people of a religious occasion. For example, matzo bread helps Jewish people to recall when their ancestors fled from slavery in Egypt; they had to leave quickly and could not wait for the bread to rise.
- be eaten to mark a special event or season. For example, Italians like to eat rich, sweet and fruity panettone during the Christmas period.
- be shared among family and friends as a symbol of love and affection. For example, Jesus Christ broke bread and shared it with his disciples to symbolise the sacrifice of his life.

Alamy Stock Photo/Bon Appetit

Figure 9.1 How Indian naan bread is traditionally made

ACTIVITIES

1. What is unleavened bread?
2. Why were hot cross buns served only at special occasions?
3. Name two unleavened breads.
4. Imagine that bread suddenly became unavailable or very expensive in Australia. Describe what might happen. What would people eat instead?

iStock.com/Denira777

Figure 9.2 Breads from different countries of the world

UNIT REVIEW

LOOKING BACK

1 List foods that are commonly served at children's birthday parties.
2 List four different types of bread.
3 Why do people like to meet together and share food?

FOR YOU TO DO

4 Design a menu of foods that could be served at a Christmas celebration.
5 Design your own recipe for an interesting type of bread. List the ingredients that you would use.
6 Design and make a recipe card for your bread.

TAKING IT FURTHER

7 Research how one type of bread is used in a religious ceremony or celebration. Use a KWL chart to organise your research. Use your research to create a multimedia presentation.

9.2 Reasons for celebration

9.2a Reasons for celebration

There are many reasons why people celebrate special occasions. Think about the celebrations in which your family regularly participates. What are the reasons for these events? Why do you celebrate them?

Social reasons

Social celebrations are those that commonly involve friends and relatives. They may be planned events, like a school dinner for Year 10 parents, or they may be celebrations that are organised for no particular reason, such as morning tea for old friends.

Social celebrations can be either formal or informal. In Australia they are often informal, or casual – think of the popularity of barbecues, picnics and buffets.

Certain social events are linked with particular foods that are commonly served and eaten. For example:

- foods served at a barbecue or picnic often include steak, sausages, barbecue chicken, bread rolls, salads, cheeses and fruits
- foods and beverages served at a cocktail party often include **canapés**, **crudités**, crackers, cheese and fruit, wine, cocktails and champagne
- foods and beverages served at a formal dinner often include soup, terrine, seafood cocktails, roast meat and vegetables, cheesecake, alcohol and coffee
- foods and beverages served at a morning or afternoon tea often include tea, coffee, scones, muffins, cakes and tarts

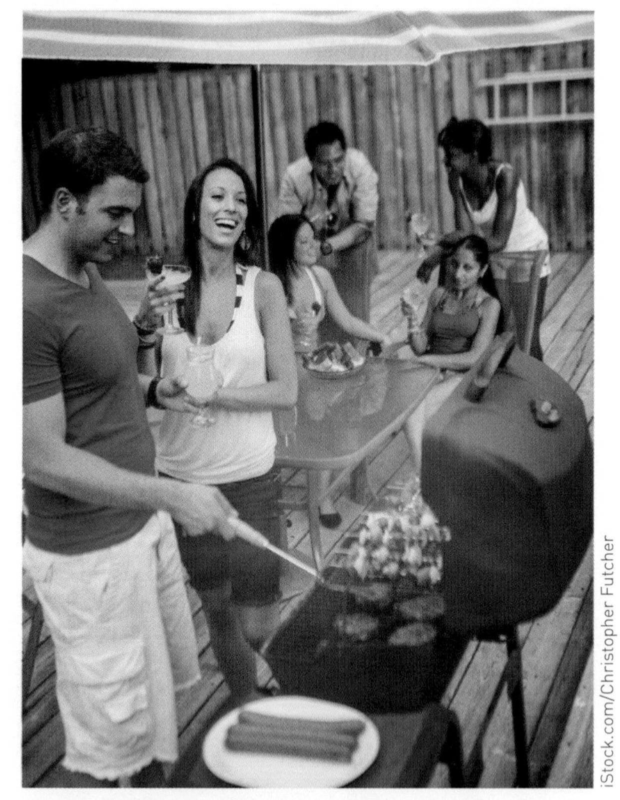

iStock.com/Christopher Futcher

Figure 9.3 Eating outdoors has become very popular.

- foods and beverages served at a buffet often include cold meats, salads, pasta, potato bakes, puddings, soft serve ice-creams and toppings.

Cultural reasons

Culture is understood as the values, knowledge, beliefs, behaviours, dress and foods that are passed on through each generation of a particular society. We usually gain an understanding of our culture from our parents and those around us. Cultural celebrations often have historical foundations. For example:

- in Australia, many people exchange gifts at Christmas and give chocolate eggs at Easter, even though they may not necessarily be Christian
- many Australians eat a cold Christmas dinner, while others still follow the English cultural tradition and consume a hot turkey roast.

There are many different types of cultural celebrations in multicultural Australia, many of which involve the preparation of special foods.

Australia Day

This celebration is held on 26 January and signifies the founding of a modern Australian nation and culture. Australia Day marks the landing of Captain Arthur Phillip at Botany Bay in 1788 and the beginning of European settlement. People celebrate with flag-raising ceremonies, fireworks, regattas and barbecues. Foods eaten on this day include green jellies, red cordial, lamingtons, meat pies, Vegemite sandwiches and Pavlova.

For Aboriginal Australians, Australia Day can be a day of sadness and mourning, recalling a time when their traditional lifestyle was destroyed and diseases, death, foreign laws and restrictions were introduced.

Aboriginal Australian initiation ceremonies

In most traditional Aboriginal Australian communities, boys who are 11–13 years old are initiated into manhood in a series of ceremonies that teach them their traditional songs and dances and the Dreaming. Boys leave behind their world of freedom and immaturity and emerge as responsible adults who are educated in Aboriginal Australian history and culture. People bring gifts to the ceremonies and the finest traditional foods are prepared. Traditional dance, dress and face painting also feature in this special event.

Chinese New Year and the Lantern Festival

This event is celebrated by the Chinese community in January and February when the moon is in a certain position. Families and friends get together to celebrate the New Year by wishing each other good luck and good fortune. Foods eaten during Chinese New Year include pork dumplings, pork crackling, roast suckling pig, sweets, lychees and Chinese tea.

After the New Year, the Lantern Festival begins. Paper lanterns are hung in houses and along streets. Children parade through the streets with paper lanterns and then return home for a feast of special foods such as fruit, sweets, mooncakes and rice cakes.

Alamy Stock Photo/MediaServicesAP

Figure 9.4 Chinatown during Chinese New Year celebrations

FOOD IN FOCUS

JAPANESE TEA CEREMONY

The Japanese Tea Ceremony (*cha-no-yu, chado* or *sado*) is a traditional Japanese activity that was influenced by Zen Buddhism. A green powdered tea is ceremonially prepared and served by a skilled host.

Cha-no-yu means 'hot water for tea' and refers to a single tea ceremony, while *sado* or *chado* means 'the way of tea' and refers to the study of the tea ceremony. The host must be experienced in types of tea, tea-making techniques, the kimono, calligraphy, flower arranging, ceramics and incense, to name just a few.

It is thought that the drinking of tea in Japan commenced in the ninth century. A Buddhist monk returned to Japan from a visit to China in 815; soon after, a type of simmered tea was served to the emperor. In 816, tea plantations were established in the Kinki region of Japan.

In the 12th century, powdered green tea was introduced; by the 13th century, Samurai warriors had begun preparing and drinking this powdered green tea. It is thought that the foundations of the tea ceremony were established at this stage.

The Japanese tea ceremony evolved and transformed over time. By the 16th century, tea drinking had spread to all levels of Japanese society. Today tea schools, community centres and private homes and institutions offer classes on traditional tea ceremony technique and know-how.

ACTIVITIES

1 When was tea thought to have been introduced into Japan?
2 Define the Japanese word *cha-no-yu*.
3 Investigate and list equipment required to conduct a tea ceremony. Hint – use the search word 'chadogu'.
4 Use a computer program such as OneNote to make a summary of the usual sequence of a traditional tea ceremony.

Alamy Stock Photo/Steve Vidler

Figure 9.5 A kimono is traditionally worn when serving tea at a tea ceremony.

Religious reasons

Religions celebrate their beliefs at different times of the year, and even the same religious event may vary from country to country in the food that is eaten and the rituals that are performed.

People may eat special foods on certain religious days to remind themselves of their god or gods, or their beliefs. People may go without food because it allows them to think more about their god or gods and others that may be suffering or have suffered in the past.

Christianity

Christmas

Christmas marks the birth of Jesus Christ, and Christmas Eve or Christmas Day services are held to celebrate the occasion. Christians believe that three wise men arrived at Jesus' birthplace with gifts for the new king. This tradition remains today when people exchange gifts during the Christmas period. People set up Christmas trees and send cards to help celebrate the occasion. Foods commonly eaten in western Europe include roast turkey, gravy and vegetables, shortbread, mince pies, plum pudding and custard. In Australia, a cold Christmas lunch or dinner – such as cold meats, seafood, trifle and salad – is often served because of the warm climate.

Easter

Easter marks the death and resurrection of Jesus Christ, and Lent is the period of forty days before Easter. It is a time of preparation and repentance. Some Christians deny themselves something during this period, particularly luxuries or pleasurable foods. Meat is not permitted on Fridays during the

Lent period, but fish is. Christians celebrate Easter Sunday, when they believe that Jesus rose from the dead. Foods commonly eaten during the Easter celebration include hot cross buns, pancakes and Easter eggs. Eggs symbolise the rebirth and new life that is associated with Jesus' death, and hot cross buns remind people that Jesus died on the cross.

Judaism

The Sabbath

Jewish people believe that when God made the universe, he left one day for rest. The Sabbath period begins at sunset on Friday and ends after sunset on Saturday. On the Saturday afternoon there is a special family meal. All foods are prepared beforehand to allow followers to rest on the Sabbath day. Wine that has been blessed is served, and sweet bread called 'challah' is made to celebrate the occasion. Jewish food is prepared following particular rules so that it is considered 'kosher' or appropriate. Animals must be killed in a special way, meat and milk products cannot be eaten at the same meal, and a large number of animal products, including pork, rabbit and shellfish, are not allowed.

Passover

Passover lasts for eight days and marks the time when God freed the Israelites from slavery in Egypt. It is a time of rebirth and renewal. At the start of Passover, special foods such as eggs, fruit, nuts, salt water, horseradish and a flat, unleavened bread called matzo are prepared.

Shutterstock.com/ajt

Figure 9.6 A Jewish dinner on the Sabbath includes challah.

Islam

Ramadan

Ramadan marks the occasion when the prophet Mohammad received his first revelation from the angel Jibril (also known as Gabriel) for the Muslim holy book. For one month every year, followers **fast** from sunrise to sunset to grow closer to God and reflect on their own lives. No food or water is consumed during daylight hours during this month. Fasting reminds Muslims what it is like to go without, and promotes discipline. A light meal is eaten in the evening, including dates, rice, yoghurt and meat. A full meal is eaten before sunrise each day. Elderly people, those who are ill, pregnant women and young children do not have to fast.

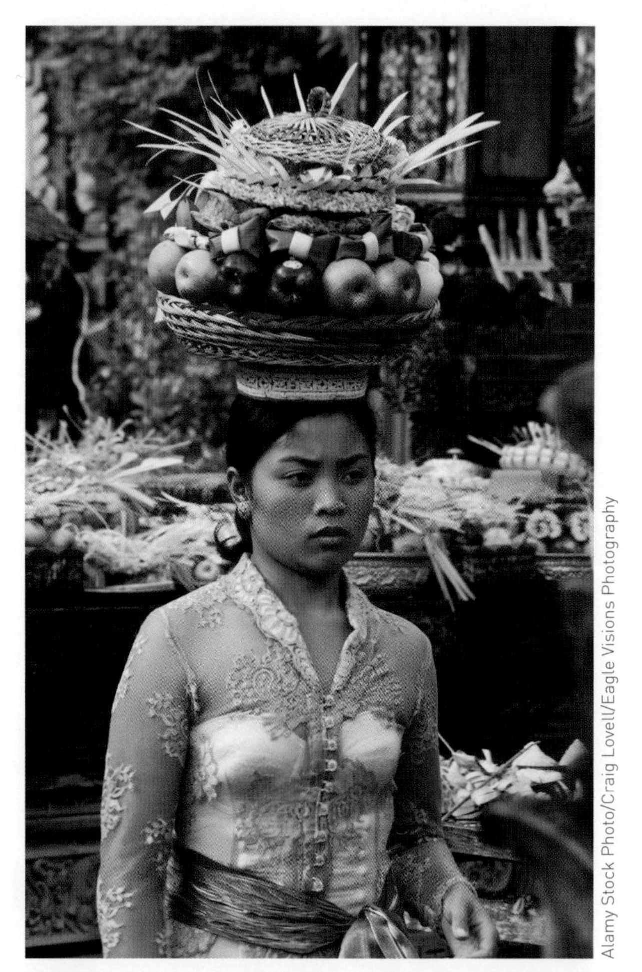

Alamy Stock Photo/Craig Lovell/Eagle Visions Photography

Figure 9.7 Hindu people offer food to the gods as a sign of respect and thanks.

Hinduism

Rama Navami

Rama Navami celebrates the birthday of the god Rama. Followers worship Rama and have a day of feasting on exotic, expensive and unusual delicacies, offering food such as bananas and coconut to the god. Hindu people do not eat beef because the cow is considered sacred and relatives may be reborn as cows in another life. Many Hindus are vegetarian, and those who do eat meat may require it to be first sacrificed to the gods. Hindus eat with their right hand only as the left hand is considered unclean.

Historical reasons

Events in history can influence when and how a special occasion is celebrated. Think back to the celebrations that you have already investigated. Can you identify a historical reason for any of these occasions?

Anzac Day

Anzac Day is celebrated on 25 April and reminds people of the courage, unity, sacrifice and loyalty of the Australian and New Zealand Army Corps, who fought in a place in Turkey called Gallipoli in 1915. Women baked biscuits to raise funds for the war effort, and today, people still bake Anzac biscuits.

Valentine's Day

Valentine's Day is celebrated on 14 February. Its origins are unclear. One story that has survived since the time of ancient Rome tells of a Christian priest who was sentenced to death by the Roman Emperor because he performed secret marriage services for Christian soldiers. The priest was sent to jail, where he fell in love with the jailer's daughter. Before he was killed, he left her a love letter that he signed 'Your Valentine'.

On Valentine's Day people send anonymous cards and buy chocolates, flowers and sweets for their loved ones.

9.2b Reasons for celebration

Family reasons

The family is an important and influential unit. Your family helps to determine your values and ideas. Every family differs in terms of structure. Some common family structures are the nuclear family, the single-parent family and the blended or adopted family. Families all over the world choose to celebrate occasions that are important to them.

- Birthdays celebrate a person's age. Foods commonly served at children's parties include chips, sweets, fairy bread, chocolate crackles, sausage rolls and soft drinks.
- Weddings celebrate the marriage of two people. Foods commonly served include wedding cake (either fruit, chocolate mud or profiterole), champagne and sugar-coated almonds.
- Family reunions celebrate the bringing together of families that may have become separated over a period of time. A buffet style of food service such as a spit roast may be appropriate, as usually large numbers of family members have to be served.

Alamy Stock Photo/Bill Cheyrou

Figure 9.8 During a wedding celebration among family and friends, many elaborate foods may be served.

UNIT REVIEW

LOOKING BACK

1 Define the term 'social celebration'.
2 List foods commonly served at a barbecue or picnic.
3 What foods are eaten during Chinese New Year?
4 How did Anzac biscuits originate?
5 Explain in your own words what an informal occasion is and give examples.
6 Why do you think that buffets are suitable for large numbers of people?

FOR YOU TO DO

7 How would catering for a children's birthday party be different from catering for an adult's birthday celebration? Design a menu of foods to be served at an adult's birthday and present this in the form of a 'table menu'. Use a computer to make your menu look as professional as possible. Pictures and clip art help to add interest and appeal.
8 Interview a family friend or relative who has participated in one of the cultural celebrations listed in the text. Ask them about their experiences, document your findings and report these to the class.
9 In pairs, create a collage of pictures and illustrations for a particular festival celebrated in Australia. On the back of the page, describe the history of the festival and list the foods associated with celebrating the event. Display these in the classroom so that others can see your work.

TAKING IT FURTHER

10 Research Aboriginal Australian celebrations. Find out what happens and what foods are served. Use a variety of sources such as the internet, books, journals and magazines.
11 Consider various locations that may be suitable to hold social celebrations, such as an informal barbecue or picnic, and those locations suitable for holding a formal celebration, such as a Year 12 graduation.
12 Using library resources and the internet, research and outline the historical foundations and associated foods linked to a festival such as Thanksgiving or Diwali. Provide an oral presentation and use digital media, pictures and images to add interest to your presentation.

Production and preparation

Foods, techniques and equipment

Producing food for any special occasion requires a great deal of thought and preparation. Food ingredients, equipment and facilities must be of a high standard and preparation techniques must follow hygiene and safety regulations.

Foods

Below are a set of guidelines for selecting and preparing food for special occasions. Foods should be:

- appropriate to the occasion and appeal to the ages, cultures and likes of the group
- in season so that they are of the best quality and price
- varied in colour, flavour and texture
- in top condition – spoiled, contaminated, damaged, rotten or out-of-date foods are not acceptable
- ordered in suitable quantities to avoid wastage
- prepared at a suitable time to ensure the best quality – some foods can be prepared the day before, while others must be prepared just before the event.

Techniques

The host or caterer of any event must work in an organised and professional manner. Most caterers use a special time plan to ensure that things run smoothly. There are a number of practices that a good host should follow to ensure that food is safe, hygienic and appealing. These include:

- tying back or covering hair, having clean, short nails and wearing proper protective clothing such as aprons and closed-in shoes
- washing hands regularly with hot soapy water, especially after using the toilet
- using clean utensils and not licking fingers or cooking spoons when preparing food

- using clean chopping boards to prepare meat and vegetables to avoid **cross-contamination**
- using clean tongs to handle or transfer food
- employing skilled workers who are trained to operate special equipment, such as a spit roast
- preparing food using a production-line approach with a set time plan
- dividing the kitchen into four main kitchen areas for food preparation, food service, cooking, and cleaning so that workers do not get in each other's way and food is not accidentally contaminated
- following recipes accurately
- remembering any religious rules for special religious occasions.

Equipment

Kitchen equipment must be safe and reliable if it is to be used to prepare and serve food for special occasions. Remember that:

- facilities must be appropriate for the number of guests being served
- if the celebration is to be held at a venue away from home, the host or caterer must consider what equipment and facilities are available
- cutlery, plates and storage space in refrigerators, freezers, cupboards and benches must be organised before the event to avoid any last-minute shortages
- kitchen equipment such as hand mixers, microwave ovens, blenders and stoves must be tested and checked before the event to detect any faulty equipment

Figure 9.9 A chef is required to be properly attired, in keeping with hygiene and safety practices.

- only equipment that saves time should be used; the preparation and cleaning of some equipment outweighs its value
- oven, refrigerator and freezer temperatures must be tested to see that they are suitable for keeping food safe and so preventing any risk of food poisoning
- equipment should be cleaned before and after use to ensure proper hygiene; stored equipment can often attract insects, rodents and cockroaches
- there must be enough serving equipment such as spoons, forks and tongs; when people swap serving utensils, cross-contamination can easily occur.

Small- and large-scale catering

The size of an event can have a great impact on the preparation involved. Planning an event for five or 10 people is very different from organising an event for hundreds of guests.

Planning ahead allows the host to organise an appropriate menu according to the information they have on the number of guests attending. This planning is necessary to:

- avoid excessive waste of food and money
- avoid the embarrassment of running short of food
- ensure that the food is the best choice for the event.

Consider the following: A sit-down awards dinner for 250 business people is to be held. If the caterer serves steak with pepper sauce to the guests, it would be hard to prepare and keep the steaks warm and prevent them from going dry and tough before serving. A suitable alternative would be a beef casserole, lamb curry or chicken cacciatore that could remain simmering at the correct temperature and still maintain its high quality.

When catering for small numbers of guests, it is relatively easy to estimate rough quantities of food.

When catering for large numbers of guests, a rough guess or estimate is not suitable. Every caterer follows a set of guidelines similar to these shown below. A host must consider the factors listed when planning a menu.

WS 9.3 Production and preparation

Shutterstock.com/Monkey Business Images

Shutterstock.com/Rawpixel.com

123RF/Cathy Yeulet

Figure 9.10 Catering for small numbers of guests is very different from catering for hundreds.

Table 9.1 Caterer's quantity guide

TYPE OF EVENT	APPROXIMATE NUMBER OR ITEMS OR PORTIONS PER PERSON	FOODS COMMONLY SERVED
Evening cocktail party	8–10 items of food per person (sweet and savoury) 2 or 3 serviettes a small handful of nuts (20–30 grams) a small handful of chips (40–50 grams) 10 mL serving of dipping sauce	Finger food such as hors d'oeuvres, canapés, nuts, chips, cocktail frankfurts, champagne and mineral water
Lunch or dinner party	100 grams meat per person 1 medium potato 1 cup cooked rice or pasta ¼ cup peas 1/10 of a lettuce ¼ medium tomato ½ medium carrot 1 cup of fruit salad ½ cup whipped cream 3 cold beverages per person 1 hot beverage per person	Roast meats, salads, baked vegetables, pasta, rice, fruit salad, cakes, ice-cream, soft drinks, juice, tea and coffee
Morning or afternoon teas	4 items of food per person (sweet and savoury) 2 or 3 hot beverages per person 1 cold beverage per person	Light foods such as slices, cakes, muffins, biscuits, flans, juice, tea and coffee

HANDS-ON

CLASS BUFFET ACTIVITY

PURPOSE

As a class, organise a lunch buffet that your class would enjoy. You will need to know the number of students in your class.

MATERIALS

Recipes for lunch buffet food, pen and paper, a food order and your notebook

STEPS

1 Select a theme and decide on a suitable menu. Every student should suggest a possible recipe and then the class should vote on which would be best.
2 Establish suitable quantities for each recipe selected. Try not to prepare too much food. Your teacher will be able to guide you.
3 Construct a food order and include the date on which the event will be held.
4 Allocate tasks for the event. Remember, workflow is important. Try to organise your school kitchen into the four main work areas. Set out a time plan.
5 Organise the layout of food and the decorations needed for the day. Each person can bring something to make the event special.
6 Design and produce a menu to paste into your notebook.

ACTIVITIES

1 Did you encounter any problems while organising and planning your buffet?
2 Was the workflow plan helpful? Explain.
3 Was there food left over?
4 Should you have had more of any foods?
5 Were all of the recipes successful?
6 How could you improve the buffet if you were to do it again?

TIP

Use the caterers' quantity guide to help you when ordering the foods.

UNIT REVIEW

LOOKING BACK

1 List one guideline for either selecting or preparing foods for special occasions.
2 Why is food prepared using a production-line approach?
3 Why should the temperatures of refrigerators and warming ovens be tested before a special occasion?
4 What type of food would not be appropriate to serve at a function for 250 people?
5 What quantity of nuts should be ordered per person for a cocktail party?
6 How might cross-contamination occur when preparing a buffet?
7 Imagine that you are holding a dinner party for fifteen people. How much minced meat would you need to order?

FOR YOU TO DO

8 Read through the following situations, then select one event from the list and prepare one part of your menu in class with a partner. Have your idea approved by your teacher and fill out a food order for the practical lesson.
   a Australia Day celebration: This is to be an informal celebration held in a local recreation area, such as a park overlooking Sydney Harbour. Eighteen guests of all ages will be attending.
      i Design a suitable invitation.
      ii Plan an appropriate menu.
      iii Justify your menu choices.
   b Valentine's Day dinner: This is to be a formal celebration for two adults to be held in an exclusive restaurant in the city.
      i Design a suitable room and table decorations (draw these).
      ii Plan an appropriate menu.
      iii Justify your menu choices.
   c Christmas lunch: This is to be an informal celebration for thirty guests of all ages (some with dietary disorders such as high cholesterol and obesity). The lunch will be held at home. As the host, you wish to spend as much time with your guests as possible, so some convenience foods may be required.
      i Describe the location.
      ii Plan an appropriate menu.
      iii Outline menu items that may use a convenience food and give details.

>

iv Justify your menu choices for those with special dietary needs.

d Wedding breakfast: This may be either a formal or an informal celebration. The bride and groom are an older couple and have invited 100 guests, most of whom are over the age of 55.

i Describe the location and type of celebration.

ii Design and display an appropriate menu.

iii Justify your menu choices.

e Camping picnic lunch: This is to be an informal occasion for eight people, held in a national park without powered cooking facilities. Foods will be served cold and may be prepared beforehand if required. Camp-fire cooking facilities may be used if needed.

i Describe the location.

ii Plan an appropriate menu.

iii Justify your menu choices.

iv List problems that may occur when serving food in this situation.

TAKING IT FURTHER

9 Design your own guidelines for selecting and preparing food for special occasions

10 Find two other recipes that may be suitable to use at an awards dinner for 250 people. Display these for the class to see and justify why they are suitable.

9.4 Menu planning considerations

Nutritional value

All menus for special occasions should be as nutritious as possible. Special occasions do require special foods, and this can often be at the expense of nutrition. Foods that fit into the 'eat least' category of the Healthy Diet Pyramid can be eaten in moderation on special occasions, but should not be a part of a regular healthy diet.

With any menu, it is always a good idea to offer an assortment of healthier alternatives, particularly if guests have special dietary requirements. Foods served at any occasion should consist of a balance of foods from the five food groups.

Imagine an adult who was overweight and has high cholesterol attended a dinner where the following was served:

- entrée – deep-fried spring rolls
- main – battered fish and chips
- dessert – chocolate fudge cake with cream.

Can you suggest two alternatives for each course that may be more nutritionally balanced?

Food appeal

Foods served at any meal must appeal to the senses. A variety of foods should feature in every menu.

Shutterstock.com/JeniFoto

Figure 9.11 A well-presented meal has a balance of colour, flavour and texture.

Colour

Different-coloured foods make a meal more appealing. This can be achieved by using fruits, vegetables, sauces, garnishes and table decorations. Imagine an uninteresting plate of boiled rice, cauliflower, butter beans and poached chicken. The lack of colour certainly would not stimulate the senses.

Aroma

Appealing food smells can stimulate the tastebuds. The aroma of a food has a great impact on its actual flavour, because flavour is a combination of smell and taste. Hot foods have a more potent aroma than cold foods; for example, compare the smell of a roast dinner to that of ice-cream. Think back to the last time you had a cold. Could you smell the food? Did it taste bland and uninteresting? Try to recollect the aroma of:

- freshly brewed coffee
- hot baked bread
- sizzling bacon, eggs and sausages.

Flavour

Flavours in any menu must be varied. The tongue can detect five main flavours and many different sensations. The five flavours that the highly sensitive taste buds can detect are sweet, salty, sour, bitter and umami. Guests at a dinner party would be unimpressed if all foods served were hot and spicy. How would you feel sitting down to the following menu?

- Entrée – curry puffs and hot chilli dipping sauce
- Main – peppered steak or chilli chicken with spicy seasoned vegetables
- Dessert – spicy fruit mince tarts.

Texture

Texture is also known as 'mouthfeel', which basically refers to how the food feels in the mouth. A well-planned menu will consist of differently textured foods in order to keep the guest interested. Textures can be smooth, grainy, rough, sharp, gritty or creamy. Imagine sitting down to a meal of foods that were all soft and creamy; for example:

- entrée – cream of pumpkin soup
- main – macaroni cheese, puréed beetroot and mashed potatoes
- dessert – cheesecake and ice-cream.

Occasion and setting

A menu must not only be suitable for the type of celebration, but also be appropriate for the setting or location of the event. For example, you could not successfully serve pumpkin soup at a stand-up cocktail party, on a boat during a harbour cruise or at a summer pool party.

If guests are at a picnic and sitting on grass, then finger foods would be suitable. Imagine having to provide a meal for friends in the middle of the bush after a day of hiking. What would you serve them?

FOOD IN FOCUS

GOURMET DINNER SERVICE

Gourmet Dinner Service (GDS) was founded by Janel Horton on Sydney's Northern Beaches, and offers delicious and nutritious prepared meals with the convenience of home delivery. Meals are either frozen or chilled and all you need to do is add salad or vegetables.

GDS employs a team of chefs and kitchen hands who work in a custom-built commercial kitchen with a walk-in freezer room in Brookvale, Sydney. They also have a consulting nutritionist who helps to plan balanced and nutritious meals for customers.

Janel has campaigned to dispel the '80s concept of a frozen dinner. She says that 'educating the market that frozen dinners are good for you has been a major challenge. The era of the '80s supermarket frozen dinner has given food a bad name'.

Most orders are placed online via their website and the menu offers an extensive range of meals. Customers can sign up for monthly newsletters and e-news that outline monthly specials or cost-effective seasonal options. This type of marketing has been highly effective and means GDS can be in direct contact with their customers on a regular basis.

Customers with special dietary needs can search meals online that are gluten-free, vegetarian, dairy-free or petite meals (35% smaller than GDS regular-sized meals for those watching their weight or those who have a smaller appetite).

The meal service is marketed towards a large sector of the community who are time-poor. This includes DINKS (dual income no kids) and singles, families with babies and toddlers (including those on a budget) and the elderly. Most of the meals are low-carbohydrate and reduced-fat to meet modern society's expectations. GDS takes advantage of

>

Fairfax Photos/John Reid

Figure 9.12 Janel in the Gourmet Dinner Service kitchen

seasonal foods in order to make meals fresher and more affordable.

The menu consists of meals such as:

- Arabian beef with lemon, almonds and tahini
- shepherd's pie
- san choy bau with rice
- lamb ratatouille with Italian vegetables
- cinnamon-scented lamb with raisons and eggplant
- chicken tagine in Middle Eastern spices
- spinach filo with fetta, zucchini and dill
- vegetable stacks with quinoa and tomato.

GDS offers meals for special occasions such as cocktail parties, Easter and Christmas functions. The cocktail menu includes foods such as goat's cheese tarts, Thai chicken balls with noodles and lemongrass, Italian risotto balls with bacon, mozzarella and herbs and chocolate praline petit fours.

The Christmas menu is a mix of traditional and modern foods suitable for either a formal dinner, casual buffet or family BBQ or picnic.

GDS provides families with a viable, cost-effective alternative to take-away food. They believe that if their customers have convenient, healthy meals at home on hand then it's easy to make the right choice and eat well.

ACTIVITIES

1. What type of meals does Gourmet Dinner Service offer?
2. Explain why the '80s frozen dinner has a bad name.
3. What niche market does Gourmet Dinner Service cater for?
4. Describe why this service may be useful for someone hosting a special occasion such as a Christmas picnic or barbecue.
5. Research recipe books of your own, or recipes on the internet, and give examples of other meals that would be suitable for a home delivery dinner service to sell chilled or frozen.

9.5 Characteristics of diners

Age

Foods on a menu must be appropriate to the age of the guests invited. Adults at a formal dinner party would be shocked to receive 'pigs in blankets', while children at a birthday party would be unimpressed by pâté and caviar. Elderly people cannot eat certain foods such as toffee, pork crackling, tough meats or sticky lollies, especially if they have dentures or false teeth.

Considering that some functions include a range of age groups, foods should be suitable for the full range of guests. Sometimes a separate children's menu is provided.

Meal **portioning** is another factor that requires consideration when age is concerned. Children consume a much smaller quantity of food compared to an adult male.

Health

Some people suffer from conditions such as obesity, high cholesterol, diabetes, food allergies and intolerances. It is very important to cater for people with special dietary needs when planning a menu. The provision of alternatives enables a person with a health problem to avoid the foods that they should not eat. Imagine that you suffer from a wheat allergy and cannot eat any foods that contain bread, flour, breadcrumbs or pasta. If you go to a pizza shop with your friends, what would you order?

Energy levels

Highly active workers, such as builders, carpenters, plumbers and gardeners who require a lot of energy, may tend to select more filling or energy-rich carbohydrate foods such as pasta, bread and rice from a menu. Less active workers, such as business men and women who work in an office, do not need as much energy from food because their jobs are more sedentary. These workers are more likely to select low-fat or low-kilojoule alternatives such as lean meats and salads.

Cultures

People from different cultures have different expectations for how food is planned and prepared for a special occasion. When menu planning, you must consider the strict dietary and food preparation requirements for a particular cultural group. There would be a lot of food left over if you served tomato and bacon quiche at a traditional Jewish function (considering that many Jews do not eat any pork products). A prior knowledge of cultural needs will ensure that problems do not occur.

Figure 9.13 Foods served at a children's birthday party reflect the ages of the guests.

FOOD IN FOCUS

THE KOREAN CULTURE AND KIMCHI

Kimchi is a fermented vegetable dish that is made in Korea and is similar to German sauerkraut. It comprises 70 per cent cabbage, 20 per cent radish and 10 per cent other vegetables.

In the Korean culture, kimchi is traditionally served with almost every meal. It makes up approximately 13 per cent of the total daily food intake!

Kimchi is made first by salting the vegetables. They are then washed, seasoned with spices and stored in a cool place for a few days in order for the natural process of fermentation to take place.

The table to the right outlines the nutritional composition of kimchi.

NUTRIENT	PER 100-GRAM EDIBLE PORTION
Protein	2 grams
Fat	0.6 grams
Total sugar	1.3 grams
Dietary fibre	1.2 grams
Calcium	45 milligrams
Vitamin B1	0.03 milligrams
Vitamin C	21 milligrams

Alamy Stock Photo/Julie Woodhouse

ACTIVITIES

1 What vegetables are used to make kimchi?
2 Use an Excel spreadsheet with appropriate formulas to calculate the quantity of protein, fat, dietary fibre and Vitamin C in 300 grams, 500 grams and 1 kg of kimchi.
3 What is sauerkraut? Use the internet to research why Germans value this food.

Tastes

Everyone has different food tastes. Some people adore the taste of caviar and blue-vein cheeses, while others cringe at the sight. When planning menus for special occasions it is important to remember that not everyone will like to eat all types of foods (particularly those foods that are quite exotic).

Number

The number of guests attending a special occasion has a direct impact on the menu planning. If large quantities of people are expected, then foods that can be prepared in bulk quantities are essential. There may be a long waiting time between when the meal is ready and when the guests sit down to eat, so foods that may be reheated easily are vital. There will also be people with special dietary needs, such as those who are vegetarians, or have coeliac disease or lactose intolerance. Foods must be provided for these people in order to save embarrassment and confusion.

If only a small number of guests are expected, then you should be able to cater for each individual's needs. If you know that one guest is vegetarian and another is allergic to peanuts, then you can easily plan your menu around these factors.

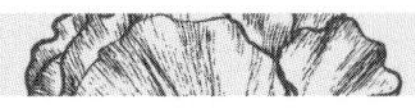
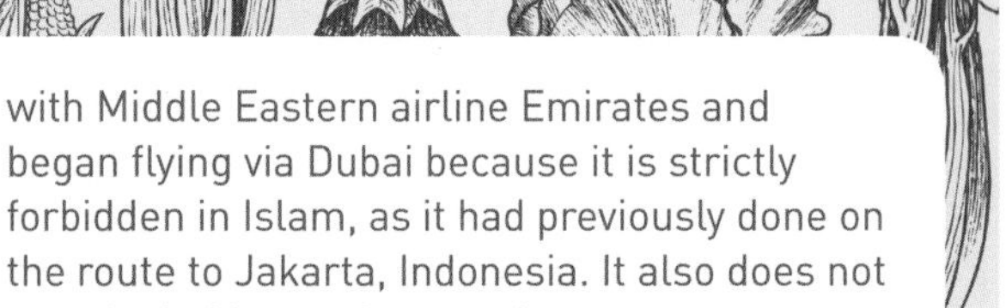

FOOD IN FOCUS

AIRLINES OFFER INCREASINGLY EXTRAVAGANT FOOD AS COMPETITION TO PROVIDE THE BEST HEATS UP

Food in planes has always had a bad rap. Soggy, reheated frozen meals were the norm for way too long. But with competition in the skies fiercer than ever, airlines are increasingly upping the ante when it comes to food served on-board with a new focus on using fresh, seasonal local produce, showcasing local ingredients before a traveller even arrives at their destination.

Many recruit celebrity chefs to design their business class menus. Virgin Australia has Luke Mangan, Qantas has Neil Perry.

There's a lot to consider when preparing meals for planes, and airlines have entire teams working year-round to develop dishes and keep menus fresh.

They also have to be culturally sensitive. For example, Qantas removed pork on flights to Europe when it inked a codeshare agreement with Middle Eastern airline Emirates and began flying via Dubai because it is strictly forbidden in Islam, as it had previously done on the route to Jakarta, Indonesia. It also does not use alcohol in meals served en route.

Some airlines spend a lot of time doing research about how food tastes different at 35 000ft.

'You lose about 30 percent of your tastebuds in the air', Mangan says. 'I try to make up for that, not through more salt, but through more vinegar in salad dressings, for example, or more fragrant spices with fish or chicken dishes.'

Perry says passengers expect fresh, contemporary dishes, with the chicken schnitzel sandwich the most popular in business class and a steak sandwich preferred in first class.

'Customers want flexibility to dine when they want and mix and match menu options,' he says.

Fairfax Photos/John Woudstra

Figure 9.14 Modern airlines meals are carefully planned to ensure fresh, seasonal and local produce is served.

ACTIVITIES

1 Why have airline meals historically had a bad reputation?
2 What is the new focus on modern airlines meals?
3 Name two celebrity chefs who design meals for airlines.
4 Why did Qantas remove pork from flights that were codeshared with Emirates?
5 Provide an example of a local and or seasonal food that may be used in an airline meal.

Resources

Ingredients

When planning a menu, the host must consider the ingredients to be used. Ingredients:

- should be in season to ensure that they are at the right price
- should be of high quality so must be purchased from a reliable supplier
- may be purchased in 'bulk' quantities.

Note that convenience food ingredients may be used to assist the host when planning for a big event, such as pre-prepared sauce mixes, frozen cheesecakes and frozen chips.

Equipment

When the host plans a menu, it is necessary to determine that the right equipment is available to prepare and serve the food. If the host is preparing food for 100 or more people, large-scale equipment may be necessary, such as a large soup vat for chicken noodle soup, or a food processor for grating large quantities of carrot. The host must be familiar with how to use each piece of equipment safely.

Skills

The skills of the host and kitchen staff must be appropriate to the menu. When planning foods for a special occasion, always remember to try the recipes beforehand to determine the degree of difficulty required. It would be a disaster to find out at the event that you misunderstood the recipe! All restaurants have a range of apprentice and fully qualified chefs to ensure that the skills required are always available.

Money

When planning a menu, the host should have a budget and select affordable ingredients that fit within that budget. For example, if the budget is limited, it would be silly to plan lobster as an entrée. Professional caterers must buy the raw ingredients and pay employees for the time and effort involved in preparing and serving the food. If ingredients are too expensive, and too much work is involved in preparing foods beforehand, the caterer has to pay more money for employees to stay overtime, and their business will certainly not succeed.

Time

The host will always want to spend as much time as possible with guests. Thus, planning a menu that allows for preparation prior to the event is helpful. Many people start by shopping for ingredients via the internet to save time. Convenience foods are also an ideal way to save time and effort in the kitchen. Instead of using fresh, steamed baby potatoes for a potato salad, the host may choose to select a can of 'tiny taters' and add dressing and seasoning. Lots of small convenience food alternatives save a great deal of time in the end.

iStock.com/atm2003

Figure 9.15 Large-scale kitchen equipment is needed to prepare food commercially.

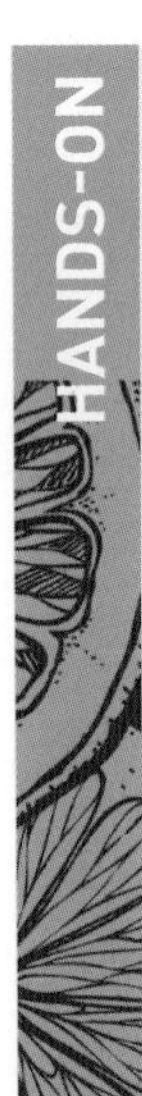

FRESH AND CONVENIENCE FOODS

PURPOSE

This activity allows you to plan, prepare and evaluate a menu that may be used for a special occasion of your choice. You must use a balance of both fresh and convenience foods.

STEPS

1 Select a special occasion with a partner or a team.
2 Plan an appropriate menu for the occasion.
3 Select one item from the menu that uses fresh and convenience foods. Show your teacher and have it approved.
4 Complete a food order and hand it to your teacher.
5 Prepare the food item in a practical lesson.

ACTIVITIES

1 What was the special occasion that you selected and why did you choose it?
2 Draw up a table in your notebook and list both the fresh and convenience food ingredients you used.
3 Comment on the final result. Was there anything that you would change? Discuss your finding with your partner or team.
4 Were the convenience foods helpful when you prepared the meal? Explain.

UNIT REVIEW

LOOKING BACK

1 What four things make up food appeal?
2 What types of foods have a more potent aroma?
3 List the five flavours detected by the tongue.
4 What types of foods do elderly people often avoid?
5 Name one dietary/health condition that an individual might suffer from.
6 When planning menus, list the five resources that must be considered.
7 In your own words, define the term 'sedentary lifestyle'.

FOR YOU TO DO

8 Consider the guidelines that should be followed when preparing a nutritious menu. Design your own guidelines that would guarantee that a healthy meal is planned.
9 Imagine that you are to host a children's lunchtime birthday party. You have been allocated a budget of $120. Eight 10-year-olds will be attending the event. What foods would you order? Start by planning a menu. Visit an online supermarket such as Coles or Woolworths and record prices. Prepare a list of all ingredients and prices. Can you come in under budget?
10 Make a list or collage of convenience party foods that a host could use in order to save time.

TAKING IT FURTHER

11 As a class, discuss the importance of using convenience foods as a time-saving device when preparing food for special occasions. Analyse the advantages and disadvantages.

9.6 Workflow plan

When food is prepared in a commercial setting, there must be a logical workflow. Workers must not waste time by getting stuck in areas of the kitchen in which others are trying to work. Cluttered kitchens where some staff are cleaning in the same place that others are trying to prepare raw ingredients are a disaster! Work does not flow freely and food can easily be contaminated.

Sequencing tasks and allocating time

Chefs must arrange a particular order of tasks to be completed. This is known as 'task sequencing'. In a well-organised kitchen, workers are allocated one task and a set period of time to complete it, and they must all work to schedule to get the job done.

Consider the preparation of marinated steak with creamy mushroom sauce and garden salad. The early steps of preparation in sequence would include trimming the meat and preparing a marinade, marinating the meat and setting it aside in the refrigerator, washing and slicing raw ingredients such as mushrooms for the sauce and salad ingredients, and so on. Each task would be performed in an exact timeframe. It would be pointless to prepare the sauce at an early stage because it is not required until just before serving. Large vats of sauce on the stove leave little room for the preparation of other foods.

Work areas in the kitchen

Dividing up the kitchen into four work areas ensures that workflow is smooth, time is not wasted and cross-contamination does not occur. Remember that any equipment needed at each stage or area of the kitchen must be located quickly.

1 The preparation area requires equipment to chop, slice, wash and peel raw foods.
2 The cooking area requires equipment to mix, bake, steam, roast, poach and simmer foods.
3 The service area requires equipment such as tongs, ladles and spatulas that are needed to serve food.
4 The cleaning and washing-up area requires equipment such as dishwashers, detergent, sponges and drying racks to clean anything used in the preparation of food.

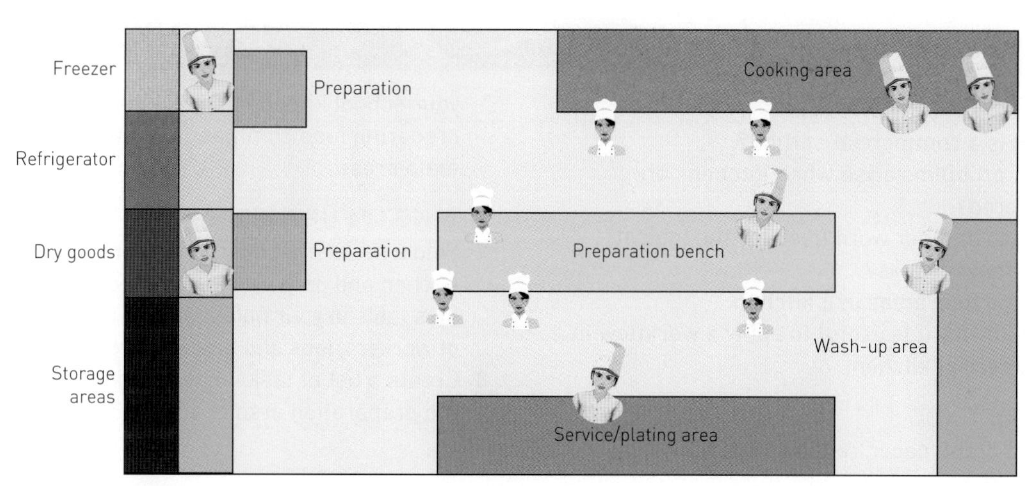

Figure 9.16 Each chef is allocated an area in which to perform their tasks to avoid problems such as cross-contamination.

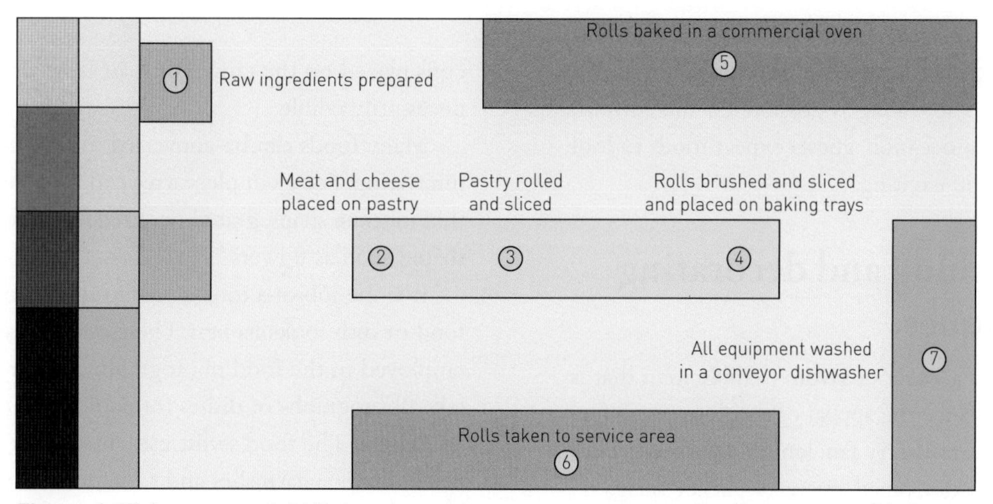

Figure 9.17 A commercial kitchen layout to prepare sausage rolls. You can use this layout when making the chicken and ricotta sausage rolls at the end of this chapter.

POTATO AND CORN CHOWDER WITH WARM BREAD ROLLS

PURPOSE

To devise your own workflow plan for a recipe

STEPS

1 Find the recipe for potato and corn chowder with warm bread rolls on page 327 and read the ingredients and method.
2 As a class, design a workflow plan that is appropriate for the layout of your school kitchen.
3 Allocate a task to each class member.
4 Count the number of students in your class. Work out how many batches of soup and bread you will require.
5 Fill out a food order and select a practical lesson.
6 Complete the practical class and answer the questions.

ACTIVITIES

1 What was your opinion of the workflow plan that your class designed? Were there any problems? Explain.
2 Did you encounter any problems with the size or scale of equipment available? Outline the problems associated with the size of your saucepans in a domestic setting compared to what would be required in a commercial setting.
3 Do you think that the time taken to prepare the soup as a production line was quicker than if you had all prepared it yourselves in pairs or groups of three or four? Explain.

UNIT REVIEW

LOOKING BACK

1 What is a commercial setting?
2 What problems arise when kitchens are cluttered?
3 What is another word used for the logical workflow of tasks?
4 List the four areas in a kitchen.
5 Explain why it is useful to apply a workflow in a commercial kitchen.

FOR YOU TO DO

6 Using graph paper, redesign the layout of your school kitchen to make it appropriate for preparing food commercially. Label the four main areas.

TAKING IT FURTHER

7 Find a recipe that may be used in a commercial kitchen and prepared in a production line. Draw up a table in your notebook that lists the number of workers, jobs and time required.
8 Create a list of tasks that would be performed in the preparation area of a kitchen.

9.7 Presenting and serving food

Every good chef knows the importance of making food look as appealing as possible for the consumer. For a special occasion, guests expect foods to look colourful and exciting.

Garnishing and decorating techniques

A garnish is a sweet or savoury edible item that is used to enhance the appeal of a dish or meal; for example, a strawberry fan, lemon twist, celery curl, cheese sprinkles, piped cream or parsley sprig.

Decorations are items that add appeal to a table or food; for example, an ice sculpture on a buffet table, a posy of flowers around food plates or white frilly cuffs placed on the crown of roast lamb. They are not necessarily edible.

Many foods can be converted into something interesting. For example, carrot can be sliced into thin **julienne** strips, grated or shredded, or cut into shapes such as flowers.

It is the job of a food stylist to make sure that a food or dish looks its best. These people are often employed in the food photography industry to take photographs of dishes for cookbooks or food magazines. The food stylist uses interesting backdrops, plates, cutlery, garnishes and table decorations to capture the desired image of the food. The consumer must almost be able to taste the food when they look at the picture. Look at the photograph in Figure 9.18, which was taken for a food magazine. What techniques has the food stylist used to capture the theme of the foods?

Presenting foods does not have to be time-consuming. If time is limited, there are many shortcuts that can be taken. For example, instead of piping icing in a decorative design on a cake, you can place a paper doily over the cake, dust it with icing sugar, remove the doily to uncover the pattern and then add some fanned strawberries.

When presenting foods for a special event you should make sure:

- table cloths are clean; white cloths are more attractive than coloured or patterned cloths
- decorations are simple yet appealing
- similar colours and themes are used
- a variety of garnishes are used
- plates are adequate in size
- serving plates have no messy spills and drips
- foods are presented at different heights on a buffet table and arranged around a central area
- serviettes are folded decoratively and displayed on plates or tables.

iStock.com/al62

Figure 9.18 A mouth-watering photo is the result of a food stylist's good work.

HANDS-ON

DECORATING BIRTHDAY CAKES

PURPOSE

To decorate a birthday cake and express your creativity.

MATERIALS

The recipes for a basic butter cake on page 328 and Vienna cream icing on page 329; pen and paper for planning your design, icing and decoration ingredients

STEPS

1. Design your birthday cake. Select a theme, appropriate colours and ingredients, draw your design and show your teacher.
2. Once your design has been approved, draw up a food order. Remember to order all the ingredients that you will need to decorate your cake.
3. Collect materials that will add to your cake display, such as a tablecloth, party hats and streamers.
4. Make the cake in one practical lesson and decorate it in the next practical class. Display your cake. Your teacher may be able to photograph it.
5. Look at the other cakes that are presented and complete the activities below.

TIP

Remember to use appropriate and hygienic handling techniques when preparing your design.

ACTIVITIES

1. What did you use to display your birthday cake?
2. What other methods and materials did students use to display their cakes?
3. List the decorations that were used.

iStock.com/Greg Nicholas

Shutterstock.com/Agnes Kantaruk

Shutterstock.com/Sergio Stakhnyk

Figure 9.19 Birthday cakes for children are made using a variety of designs, techniques, ingredients and flavours.

Sarah Carey

SARAH CAREY

THE FOOD FIGHTERS – CATERING

After 12 years of hard work and a great deal of experience in catering venues and á la carte kitchens, Sarah Carey established the Food Fighters and Mex on the Beach on Sydney's Northern Beaches. She wanted a fulfilling career that allowed her the flexibility to do what she loves and still find time for her family.

With a passion for Mexican food menus, weddings and cocktail functions, Sarah specialises in creating an exquisite dining experience for a range of catered events.

Sarah works either on-site, for example at a wedding function centre, or will work from customers' homes for a more personal in-home catering experience. She comments that it is often more affordable to have a caterer come to your home when having a party these days. The Food Fighters catering starts from just $10 per head depending on the event and menu involved.

Establishing a career in catering can begin with a variety of training courses. Sarah obtained a Certificate 4 in cooking via the Restaurant and Catering Industry Association of NSW. Caterers might have a variety of different backgrounds and experience that are sourced from an array of courses offered in different sectors such as culinary colleges, TAFE, universities and other private colleges. These courses offer training, practical advice, knowledge-base, techniques and skills that would apply to catering careers.

TAFE, local councils and other institutions offer courses on food safety, handling and hygiene for anyone working in the food industry. These sorts of courses are a valuable foundation for anyone wanting to start a career in the food industry. They cover aspects such as personal hygiene, food poisoning, food spoilage, legal procedures and regulations, cleaning and maintenance, pest control, rubbish storage and disposal and food safety plans.

A typical day for Sarah would be an early start. She drops the kids at kindergarten, does the daily ordering then goes shopping to pick up the orders for that day. Then Sarah arrives home and starts preparing the food ingredients. Starting the cooked foods first is critical, and salads must be made fresh on the morning of each catering job. Sarah's husband assists her by completing the customer invoices, which he then emails to Sarah for forwarding to her customers.

Sarah's success in catering is attributed to a variety of essential skills such as her great people skills for daily communication, attention to detail, good knife and fine-motor skills, a passion for food, sound business management skills, punctuality, reliability, perseverance and determination.

ACTIVITIES

1. What types of menus does Sarah specialise in?
2. Provide two examples of an event that might be catered for at a function centre.
3. Where did Sarah obtain her catering Certificate?
4. What are some of the skills required to work in the catering industry?
5. Summarise a typical daily routine of a caterer.
6. Imagine you were assisting Sarah to develop a new Mexican menu. List some entrees, mains and desserts you could suggest.

UNIT REVIEW

LOOKING BACK

1 Define the term 'garnish'.
2 Give an example of a decoration.
3 What does a food stylist do?
4 List two tips or guidelines that should be followed when presenting foods.

FOR YOU TO DO

5 Draw three other examples of garnishes.
6 Make up a tip of your own to follow when presenting foods.
7 In pairs, design a platter of garnishes for finger foods such as Indian samosas, crispy pork wontons, falafel, Vietnamese mini spring rolls and Indonesian beef satay sticks. First, decide on the types of garnishes to be used (for example, tomato roses), then fill out a food order for the appropriate ingredients.

TAKING IT FURTHER

8 Who might a food stylist work with in order to get the atmosphere and lighting of a food correct?
9 Using the internet, research the role of a food stylist. Include food photographs that show good presentation.
10 Find four pictures of food for special occasions in a magazine or online. Cut out or print the picture and paste it in your book. For each picture, comment on the decorations, garnishes, atmosphere or setting, appeal (use of colour, flavour and texture) and other techniques used to enhance the presentation of the food.

9.8 Using convenience foods for special occasions

Pre-prepared and partly prepared convenience foods can help save time and effort when preparing foods for a special event. There are many alternatives to using fresh, unprocessed foods.

Pre-prepared foods may be completely prepared and require only heating and/or serving; for example:

- simmer sauces, ready-made sauces and gravies in the pouch
- pizza bases/frozen pizzas/frozen pies and pastries
- canned vegetables and vegetable mixes in a can
- salad dressings
- whipped cream
- partly baked heat-and-serve bread rolls
- ready-made pastas such as ravioli or tortellini.

Partly prepared foods include foods that have had some treatment to make them easier to prepare, such as:

- boned and skinned meats
- peeled and chopped vegetables
- fresh fruit salad or soup mixes
- packet cake or bread mixes
- sliced and shredded cheeses
- powdered gravy mix.

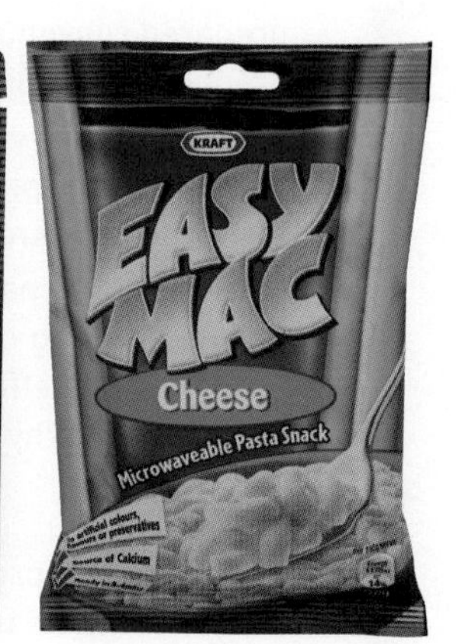

Figure 9.20 Pre-prepared convenience foods often only require heating and/or serving.

Many convenience foods are available to assist people with special dietary needs, such as:

- soy yoghurts for people with a milk allergy
- artificially sweetened soft drinks for people who suffer from diabetes
- non-animal-based cheeses for people who are strict vegetarians.

Time is critical when a caterer must prepare large quantities of different foods. For example, it is acceptable for a caterer to use a variety of prepared and pre-prepared convenience foods as well as fresh foods when presenting a smorgasbord.

Mark Fergus

Figure 9.21 A partly prepared convenience food makes a meal, such as a salad, easier to prepare.

HANDS-ON

SCHOOL SMORGASBORD

PURPOSE

This activity gives you the opportunity to cater for and host a smorgasbord to celebrate Year 12 graduation using a variety of fresh and convenience foods. Make sure you follow safe and hygienic food-handling practices in your preparation and food presentation.

STEPS

1 Each student selects a suitable recipe.
2 As a class, design a suitable menu. Students can vote on each recipe.
3 Organise suitable quantities and fill out a food order.
4 Design the layout of the smorgasbord, including decorations.
5 Design an invitation for the event and give one to each Year 12 student.
6 Produce and prepare your smorgasbord in the allocated practical lesson.

ACTIVITIES

1 Rule up your notebook into three columns. List the fresh foods, partly prepared convenience foods, and pre-prepared convenience foods that you used in your smorgasbord.
2 Were there any problems in preparing these foods? Were the instructions clear?
3 In two or three paragraphs, give your opinion of using convenience foods to save time and effort. Would you use them again? Why or why not?

TIP

Organise your smorgasbord for lunchtime, as students will be unable to leave classes.

UNIT REVIEW

LOOKING BACK

1 How do convenience foods help a host?
2 Give two examples of partly prepared convenience foods.
3 Give two examples of pre-prepared convenience foods.
4 What is a smorgasbord?
5 Why would a host want to use a mixture of fresh and convenience foods?

FOR YOU TO DO

6 Create your own menu for a dinner party that uses a variety of fresh and convenience foods. Label which foods will be fresh and which will require some preparation. Type up your menu for display and include graphics.

TAKING IT FURTHER

7 Imagine you are catering for a cocktail party using the following menu:
   - bacon and mushroom tartlets filled with sliced button mushrooms and bacon cubes and topped with egg, milk and nutmeg
   - spinach and fetta filo pastry triangles
   - fruit platters with sliced pineapple, mango, kiwi fruit, strawberries, grapes and apples
   - cheese platters with brie, blue-vein and vegetarian cheddar cheese served with rice crackers and water crackers.

   What convenience products are available that can be used to save time when preparing this menu?

LOOKING BACK

1 What is a staple food?
2 List three types of bread.
3 Give an example of a social reason for celebrating with food.
4 List three religious celebrations that are associated with special foods.
5 What is 'Passover'?
6 List two historical events that we celebrate with food.
7 Give an example of a family celebration that involves special foods.
8 Outline five basic hygiene rules when preparing food for special occasions.
9 Refer to the caterers' quantity guide on page 305 and list the number of food items per person that should be served at a morning or afternoon tea.
10 List three things that need to be considered when planning a menu for a special event.

FOR YOU TO DO

11 Design, make and present a recipe card for your favourite bread recipe.
12 Interview someone with a different cultural or religious background from you. Investigate an event that you both participate in (such as a family barbecue or birthday) and find out the differences in foods served. Report your findings to the class.
13 Design a menu for an Australia Day barbecue and devise a list of convenience food ingredients that could be used to produce each recipe or meal.
14 Plan a 16th-birthday party celebration. Choose a location, set a suitable menu and justify your choices.
15 Explain why nutritional value is important when planning meals for special occasions.
16 In a table, list the advantages and disadvantages of using convenience foods when you are preparing food for special occasions.
17 Use your own words to describe a kitchen workflow plan. Explain why it is helpful and how it saves time.
18 What are the four main work areas in a commercial kitchen?
19 Select a complicated recipe and devise a workflow plan. Use a computer program to create a drawing/diagram of the kitchen work areas, and don't forget to include a key.
20 List five examples of garnishes and draw each of them.

TAKING IT FURTHER

21 Use a variety of resources to research a type of food that is eaten all over the world (such as milk, cheese or fish) and provide a report of at least two pages outlining the differences and similarities shared by various cultures or regions.
22 Design and produce an audio/visual presentation on a religious celebration of your choice. Present this to the class.
23 Produce an A3 list of guidelines that may be used by a small catering company to remind them of proper food preparation and production techniques.
24 Imagine you own a catering service that provides meals for the elderly in institutions such as nursing homes or retirement villages. Devise a list of menu planning considerations that will need to be determined before you set a seasonal menu. Then design a menu and justify your choices for each food.
25 Produce a career profile for a food stylist and a food photographer. Use a variety of resources (and interview a stylist and a photographer if possible) and list the education, training and experience needed.

9780170383417

RECIPES

Double corn bread

This quick and easy corn bread that originated in the corn-growing regions of America does not use yeast, so you don't have to wait for the bread to rise before baking it.

 **Preparation time** 15 minutes  **Cooking time** 25 minutes  **Makes** 1 large loaf

INGREDIENTS

- ¾ cup white flour
- 1½ cups yellow cornmeal (use polenta if unavailable)
- 1 teaspoon salt
- 1½ tablespoons baking powder
- 1 tablespoon caster sugar
- 50 grams butter, melted
- 1 cup milk
- 3 eggs
- 1¼ cups canned sweet corn, drained

METHOD

1. Preheat oven to 200°C. Grease and line a 22 cm round tin.
2. Sift the flour, cornmeal, salt and baking powder together in a large bowl. Stir in the sugar and make a well in the centre.
3. Mix the melted butter, milk and eggs together. Add this to the centre of the flour mixture and beat until just combined.
4. Using a wooden spoon, stir the sweet corn quickly into the mixture. Pour into the prepared tin and bake for 20–25 minutes, or until a metal skewer inserted into the middle comes out clean.
5. Turn the bread onto a wire rack and lift off the paper. Cool slightly and serve warm.

QUESTIONS

1. Where does this recipe originate?
2. List as many corn-based products you can think of.
3. Even though yeast is not used in this recipe, there is an ingredient that makes the bread rise. Do you know what it is and how it works?

Mark Fergus

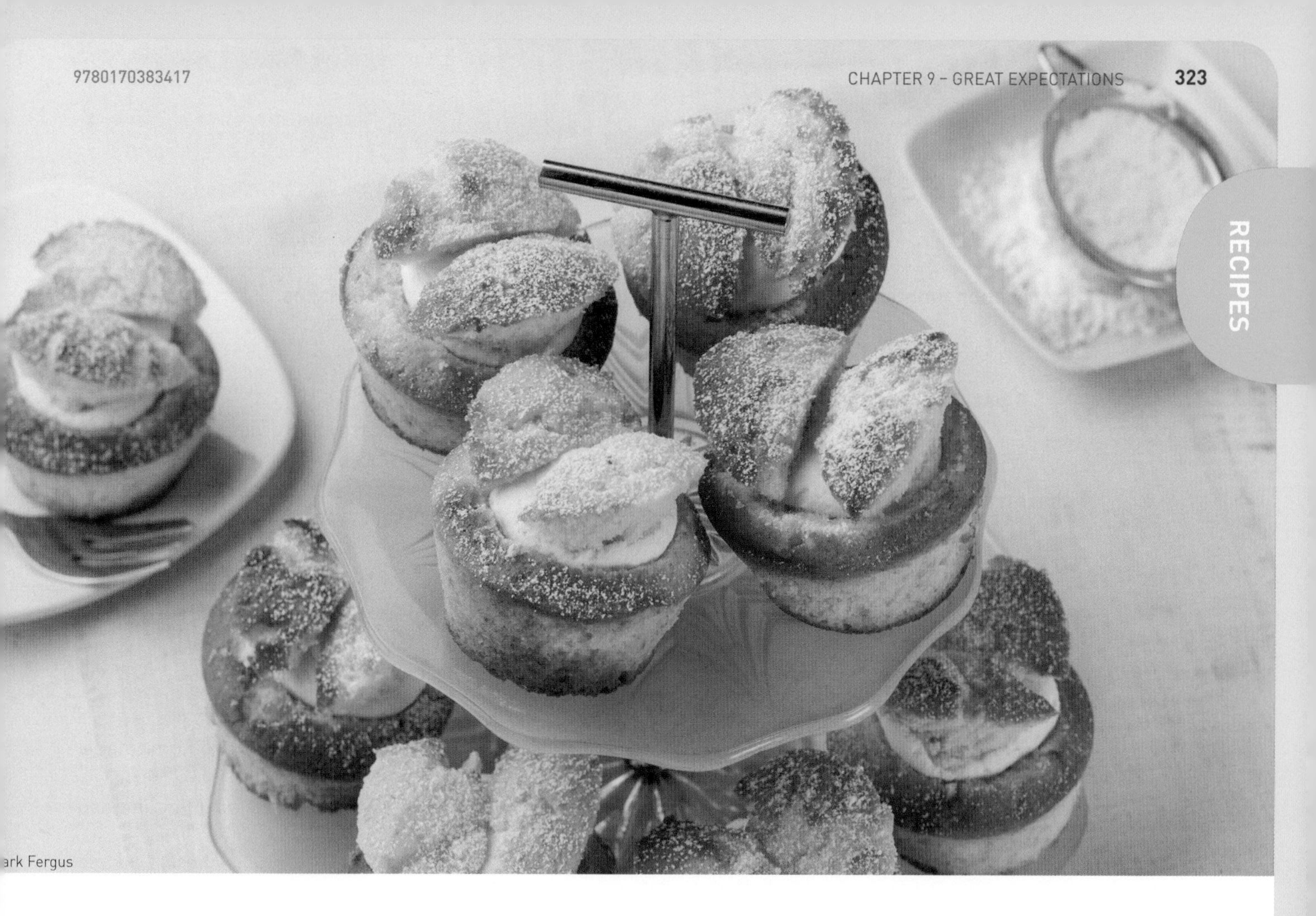

ark Fergus

Butterfly cakes

This recipe is an old-time favourite for children's birthday parties, and morning and afternoon teas.

 **Preparation time** 30 minutes  **Cooking time** 10 minutes  **Makes** 12 cakes

INGREDIENTS

- 125 grams soft butter
- ¾ cup caster sugar
- 1 teaspoon vanilla essence
- 2 eggs
- 1½ cups self-raising flour
- ½ cup milk
- 125 grams cream
- 2 teaspoons sugar
- 120 grams jam (any flavour)
- icing sugar to dust

METHOD

1. Preheat oven to 190°C. Grease 12 medium (three-quarter cup) muffin tins.
2. With electric beaters, mix the butter, sugar and vanilla essence in a large bowl until pale and creamy. Add the eggs one at a time, beating well after each addition. Fold in the flour alternatively with the milk until well combined.
3. Spoon the mixture into muffin pans and bake for about 10 minutes or until a skewer inserted into the centre comes out clean. Stand for 5 minutes before turning onto a wire rack to cool. Whip the cream and sugar with electric beaters until soft peaks form.
4. When cool, cut a shallow cone-shaped piece out of the top of each cake. Cut this cone in half. Fill the centre with jam and cream and arrange the two 'wings' on top of each cake. Dust with icing sugar and serve.

QUESTIONS

1. Describe the process of 'folding' flour into a mixture
2. What else could be used as a filling in the butterfly cakes?
3. List various garnishes and decorations that could be used to increase the appeal of this recipe at a party event.

RECIPES

Quick Christmas fruit mince tarts

This traditional Christmas recipe uses convenience foods to help save time and effort.

 **Preparation time** 15 minutes  **Cooking time** 15–20 minutes  **Makes** 12 tarts

INGREDIENTS

1 packet sweet shortcrust tart cases

1 cup bottled fruit mince

1 sheet frozen shortcrust pastry, thawed

2 tablespoons sugar

1 cup ready-made custard

METHOD

1. Preheat oven to 190°C.
2. Place the tart cases on a baking tray and divide the fruit mince equally between each case. Press down with back of spoon.
3. Using a 6 cm fluted or plain cutter, cut 12 rounds from the sheet of shortcrust pastry. Place a pastry round on top of each tart case.
4. Brush the tops with water and sprinkle with sugar.
5. Bake for 15–20 minutes or until golden. Dust tops with icing sugar if desired.
6. Serve warm or cold with ready-made custard.

QUESTIONS

1. What raw ingredients would be required to make homemade custard?
2. List two other ingredients that may be used for filling instead of fruit mince.

Mark Fergus

ark Fergus

Spicy roasted chick peas

This simple, healthy snack is a great starter to a special meal or occasion.

 **Preparation time** 5 minutes  **Cooking time** 25–35 minutes  **Serves** 2

INGREDIENTS

210 grams canned chick peas (half a large can)

1 tablespoon olive oil

Salt

Paprika or garlic salt (optional)

Pepper (optional)

METHOD

1. Preheat oven to 230°C
2. Blot the chick peas with a paper towel to dry. Place the chick peas, olive oil and seasoning in a bowl and toss to combine.
3. Place some baking paper on a tray, spread the chick peas over the paper and bake for 25–35 minutes until golden brown and crunchy.
4. Watch carefully for the last 10 minutes of cooking to avoid burning.

QUESTIONS

1. How do you 'blot' chick peas?
2. What other types of seasoning or spices could you use to flavour the chick peas?
3. List three other starters that could be served alongside these chick peas.

RECIPES

Chicken and ricotta sausage rolls

Complete this recipe in class with a partner and record how long it takes.

 **Preparation time** 30 minutes  **Cooking time** 20 minutes  **Makes** 10–12 rolls

INGREDIENTS

- 1 egg
- 190 grams lean chicken mince
- 100 grams ricotta cheese
- 1 tablespoon finely chopped fresh basil
- ½ tablespoon finely chopped fresh chives
- ¼ teaspoon salt
- Pepper (optional)
- 1 sheet frozen puff pastry, thawed

METHOD

1. Preheat oven to 210°C. Cover an oven tray with baking paper.
2. Lightly beat the egg and divide it into two equal portions. Set one half aside for glazing.
3. Place the egg, mince, ricotta, basil, chives and salt in a medium bowl. Season with pepper if desired. Mix well with clean hands. Halve the mixture and roll into two sausages 25 cm long.
4. Cut the sheet of pastry in half and brush the surface of each with water. Lay a roll of filling on the bottom third of each sheet across the widest part. Roll up like a sausage roll.
5. Use a sharp knife and cut each roll into 6 even pieces. Place these on an oven tray and glaze each one with lightly beaten egg. Score or pierce the surface with a sharp knife and bake for 20–25 minutes or until puffed and golden brown.

QUESTIONS

1. How long did it take you to prepare this recipe?
2. Can you think of any time-saving devices or ingredients that could be used to further speed up the process?

TIP: Use baking paper on the baking tray to help keep the tray clean.

Mark Fergus

Potato and corn chowder with warm bread rolls

This quick and hearty soup can be served as a snack or a meal.

 **Preparation time** 15 minutes  **Cooking time** 20 minutes  **Serves** 4–6

INGREDIENTS

- 4 small part-baked dinner rolls
- 2 brown onions, chopped
- 2 fresh corn cobs, with kernels cut off
- 350 grams potatoes, peeled and cut into small cubes
- 100 grams ham slices, finely chopped (optional)
- 50 grams butter
- 2 tablespoons plain flour
- 1 litre chicken stock
- 310 grams canned creamed corn
- ½ cup cream

TIP: Remember to use different chopping boards for meat and vegetable ingredients.

METHOD

1. Preheat oven for the bread rolls as directed on the packet. Prepare the onions, corn, potatoes and ham (optional) as directed.
2. Melt the butter in a large saucepan and cook the onion for 2–3 minutes or until soft. Add the flour and heat on a low temperature, stirring for 1 minute.
3. Add the stock, corn kernels, creamed corn, potato and ham to the pan.
4. Bring to the boil, reduce heat and simmer for 15 minutes or until the potatoes are tender and the mixture has thickened. Heat the bread rolls in the oven as directed on the packet.
5. Add the cream to the soup just before serving with warm, crusty bread rolls.

QUESTIONS

1. What is a kernel?
2. How could this recipe be modified to make a low-fat version?

RECIPES

Basic butter cake

This plain butter cake recipe makes a good base for decorating because it is firm and does not collapse easily. You may ice this cake with Vienna cream icing (see p. 329) for a delicious afternoon tea treat.

 **Preparation time** 20 minutes  **Cooking time** 40–50 minutes  **Makes** 1 cakes

INGREDIENTS

125 grams butter
½ cup caster sugar
½ teaspoon vanilla essence
2 eggs
1½ cups self-raising flour
⅓ cup milk

METHOD

1. Preheat oven to 180°C. Line a 20-centimetre round cake tin with baking paper.
2. Beat the butter and sugar until light and creamy. Add the vanilla essence and beat until light and fluffy.
3. Add the eggs one at a time and beat well after each addition.
4. Add half the flour and half the milk. Mix well. Add the remaining flour and milk. Beat until mixture is smooth.
5. Spread into greased tin and bake for 40–50 minutes or until a skewer inserted into the centre of the cake comes out clean. Turn onto a wire rack to cool.

TIP: Allow the cake to cool completely before turning it out of the tin to prevent it from sticking.

Mark Fergus

rk Fergus

Vienna cream icing

This icing is very easy to work with and is a great base for cake decorating.

 **Preparation time** 10 minutes  **Cooking time** nil  **Makes** enough icing for one cake

INGREDIENTS

125 grams butter, at room temperature

1½ cups sifted icing sugar

2 tablespoons milk

Food colouring (optional)

METHOD

1 Beat the butter with an electric mixer until it is almost white.
2 Gradually add about half the sifted icing sugar, beating constantly. Add the milk gradually and then beat in the remaining icing sugar.
3 Add food colouring if desired.
4 The mixture should be smooth and easy to spread with a spatula.

QUESTIONS

1 As well as greasing the tin, what could be used to prevent the cake from sticking?
2 Why are the butter and sugar beaten until the mixture is light and creamy?
3 Find another type of icing that may be suitable for this cake-decorating activity.

CHAPTER 10

Mark Fergus

What's in?

FOCUS AREA

FOCUS AREA: FOOD TRENDS

There are many reasons why a particular food, type of food service, cooking technique, recipe or food presentation technique becomes popular. A trend often reflects changes in society, technology or our beliefs about food. Today, with the power of social media, a food trend can become a worldwide phenomenon. In this chapter, you will investigate current food trends and the influences on these trends. You will, of course, also be preparing and presenting food in a modern style.

CHAPTER OUTCOMES

In this chapter you will learn about:

- trends in food, dining and food service
- trends in food preparation and food styling
- food styling and photography
- factors affecting acceptance of new food trends
- marketing and current food trends.

In this chapter you will learn to:

- identify and compare past and present food trends
- identify services offered by hospitality establishments
- plate food for service
- design, plan, prepare and present the latest food trends
- style and photograph food
- explain the influence of food styling and photography
- discuss what influences the acceptance of new food trends
- discuss how the media promote new food trends
- create an innovative marketing concept
- produce an image of styled food using computer technology.

RECIPES

- Lemonade scones, p. 352
- Chocolate brownie bites, p. 353
- Spinach gozleme, p. 354
- Custard horns, p. 355
- Breakfast wrap, p. 356
- Beef stroganoff, p. 357

Australia's first KFC was located in the Sydney suburb of Guildford in 1968, because it had a large population of young families. It caused Australia's chicken production to increase significantly.

You can regrow a spring onion after you have used it. Place the end of the spring onion that has the roots into soil a few centimetres from the top of the soil, water regularly and wait for it to grow.

Australian Aboriginal culture has always practised what is now termed **clean eating**: eating unprocessed food direct from the land.

Many cafes and restaurants have their own herbs and gardens in order to source fresh produce.

Genetically modified canola, modified for herbicide tolerance, was approved for commercial production in Australia in 2003. Canola oil is used in margarine-type spreads, dairy blends and as an ingredient in tinned and snack foods.

Restaurants are not required to state if they use genetically modified ingredients.

For quality control, wrap fresh herbs in a paper towel and place in a plastic bag in the fridge, or stand for up to one week in a jar of water covered with a plastic bag.

The colour of a chilli will not indicate its degree of spiciness. Usually, the smaller the chilli, the hotter it will be.

The tea bag was created by accident, as tea bags were originally sent as samples. Today, a wide variety of teas are available, including herbal teas. High Tea events have come back into vogue.

The most popular carrots used to be purple. These vintage carrots have made a recent comeback.

If the food displayed on a food label is not exactly the same as is in the package, then the term 'serving suggestion' must be used. Otherwise, it is false advertising.

FOOD WORDS

cafe informal restaurant, offering a range of hot meals

clean eating a diet of whole foods that have minimal processing

cuisine style of cooking

degustation menu appreciative tasting of various foods and focusing on the senses

herbs leafy plants, valued for their flavour and scent

spices aromatic seasonings obtained from the bark, buds, fruit, roots, seeds or stems of various plants

10.1 Trends in food

Food trends reflect popular changes that consumers have made in the way they purchase, prepare and serve food. Current trends indicate that Australians lead hectic lifestyles but they appreciate technology, the environment and good health. They also have an interest in new and exotic foods.

Organic ingredients and produce

Organic foods are grown without the use of artificial fertilisers, pesticides, herbicides, antibiotics, growth regulators or hormones. Instead, farmers rely on non-chemical methods to produce food.

The 'paddock to plate' concept reflects the recent trends towards eating less processed foods and conserving the environment. This has resulted in an expanding range of organic and free-range foods available to consumers. Animals kept free range are allowed to roam around for a majority of the day, rather than caged in a very small space.

Organically produced grains, meats, chicken, milk, eggs, fruit and vegetables are now commonly found in supermarkets, butchers, greengrocers and fresh food markets. The range of products using organic ingredients is also expanding; it now includes items such as bread, muesli and baby food.

Organic foods are generally more expensive than their non-organic counterparts. They can also be smaller, have more pest marks and have a shorter shelf life than non-organic foods. However, many consumers believe they are more tasty and nutritious and have had less impact on the environment. To certify foods as having been legitimately organically produced, many organic products carry a logo to state their authenticity.

10.1a Trends in food

Figure 10.1 The range of certified organic foods available in the marketplace is constantly expanding.

FOOD IN FOCUS

SADHANA KITCHEN

Maz Valcorza is founder of Sadhana Kitchen. Sadhana Kitchen opened Sydney's first organic wholefoods & raw foods cafe, which promotes the trend towards clean eating.

According to Sadhana Kitchen, raw foods are those that are prepared and processed without exceeding 40°C. It is believed that this allows the food's natural enzymes to stay intact, giving your body the most benefit and allows for easier digestion. The cafe uses organically grown ingredients in season and stays clear of sugars and flours. Being organic they are free from pesticides, chemicals, hormones and artificial additives. Their meals suit vegetarians, vegans and those requiring gluten-free and lactose-free diets.

Open seven days for lunch, they also offer lunch delivery service. Every Friday night they offer a seven-course raw, vegan, gluten-free **degustation menu**, and daily at 10 a.m. they offer a raw, vegan High Tea. They also offer catering services for functions and even cakes for special events, all gluten-free and organic.

ACTIVITIES

1 How does Sadhana Kitchen define raw foods?
2 Define an 'organic' food.
3 Who do Sadhana Kitchen meals suit?
4 Besides lunch, what other services does Sadhana Kitchen offer customers?
5 Find a raw food vegan recipe on the internet.
6 Find an example of a degustation menu on the internet.

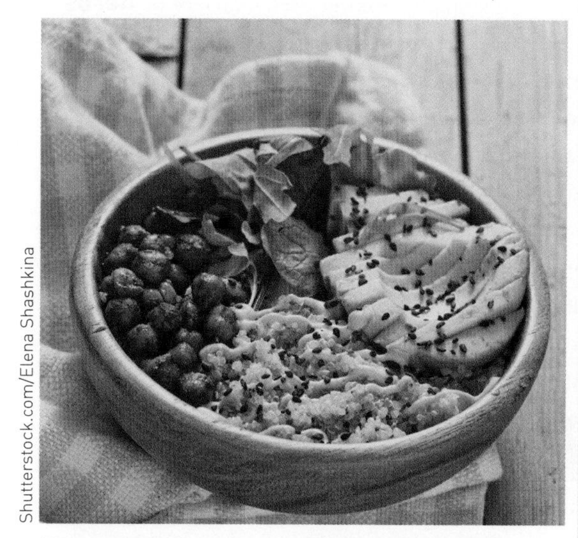

Shutterstock.com/Elena Shashkina

Figure 10.2 Organic, vegetarian menu option

Genetically modified foods

Genetic modification involves a gene from one plant or animal being transferred into another plant or animal. For example, the gene that makes a plant tolerate drought can be inserted into the gene of a plant that does not.

With genetic modification, the good features of one food can be incorporated into another food, which can help to improve the nutritional value, appearance, shelf life and processing of the food. It is claimed that genetic modification could also protect the environment, because fewer chemicals are needed to grow crops and combat pests.

The technology is still new but it is being used extensively overseas. Tomatoes with firmer skins and soy bean plants that are herbicide-resistant have been developed as a result of genetic modification. Genetically modified varieties of corn, potatoes, sugar beet, cotton oil and canola oil are now found in many food products overseas.

Australian authorities are uncertain of the long-term effects of genetic modification, so they have banned the growing of genetically modified crops in Australia except for canola and cotton, which can be used to make cooking oils. Imported products such as corn chips, taco shells, burritos, spaghetti sauces and soy sauce may contain genetically modified ingredients. Australian manufacturers are permitted to use imported genetically modified ingredients but they need to be identified on food labels by law.

Ingredients: Soy protein isolate (genetically modified); maltodextrin; vegetable oil; food acid (332); emulsifier (471); vegetable gum (407); water.

Figure 10.3 GM food labelling is required under Australian law.

10.1b Trends in food

Prepared fresh food products

Australians appreciate the flavours of fresh ingredients, but with many Australians leading busy lives, there has been a trend towards purchasing fresh foods already prepared for us in some way; for example, marinated chicken wings, pre-mixed salads, crumbed meat, prepared rissoles, dips and fresh pasta.

Fresh herbs and spices

Modern Australian cooking uses a variety of **herbs** and **spices** – fresh are preferred because of their flavour. Many of these herbs and spices stem from Mediterranean and Asian cooking but, increasingly, Australian native herbs and spices are being used in many contemporary recipes. Herbs and spices also provide a wide range of nutrients and phytochemicals (plant chemicals) that can help prevent disease and contribute to better health.

Table 10.1 Common herbs and spices

MEDITERRANEAN	ASIAN	AUSTRALIAN NATIVE
Basil	Allspice	Lemon ironbark
Bay leaves	Cardamom	Lemon myrtle
Chilli	Chilli	Mountain pepper
Chives	Chinese five-spice	Native mint
Cinnamon	Cinnamon	Riberry
Dill	Coriander	Wattleseed
Garlic	Cumin	
Marjoram	Fenugreek	
Mint	Garlic	
Oregano	Ginger	
Parsley	Lemongrass	
Pepper	Pepper	
Rosemary	Saffron	
Sage	Thai basil	
Tarragon	Turmeric	
Thyme		
Vanilla		

HANDS-ON

GROW YOUR OWN MICROGREENS

PURPOSE

Microgreens are greens, lettuces and herbs that are harvested young. They are packed with nutrients and add flavour and goodness to salads, soups and sandwiches.

MATERIALS

Specially designed microgreen seeds or rocket, amaranth, coriander, basil, fennel, radish and mizuna (Japanese mustard) seeds

Plant liquid fertiliser

Spray water bottle

5 cm–8 cm pots or deep berry containers with lids (Take-away plastic containers with lids are also good but you will need to punch drainage holes in the base.)

STEPS

1. Fill a container almost to the top with potting mix. Water the soil and allow it to drain. Sprinkle the seeds thickly over the top and cover with a thin layer of potting mix, then water with a mist sprayer. If the container has no lid, cover it with plastic wrap to encourage germination. Remove lid once seedlings are growing well.
2. Place the container in a sunny spot – bright, indirect light works best, like a window sill – and turn the container regularly for even growth. Apply half-strength liquid plant food every 3–4 days and ensure the potting mix does not dry out. Use a mist sprayer or sit the container in water.
3. Once microgreens are between 3 cm and 5 cm tall and showing their second leaves, they're ready to eat. This usually happens 10–14 days after planting. Cut the whole seedling with scissors just above soil level.
4. You should only get one harvest, but you can start a second crop straight after harvest simply by scattering fresh seed and repeating the sowing process, with the old roots providing a good source of organic matter. This process can be repeated many times.

ACTIVITIES

Produce a collage of images where microgreens have been used in food preparation.

iStock.com/Cokanson

Heat-and-serve meals

Heat-and-serve meals include canned soups, frozen pizzas and TV dinners. These types of meals were traditionally used in cases of emergency or as treats; today, with smaller household structures and lifestyle pressures, heat-and-serve meals have become popular solutions for preparing quick meals, particularly as many are easy to heat in the microwave.

Besides frozen and canned, many heat-and-serve meals are now sold chilled, bottled or in long-life form. Chilled soups, pasta dishes, frozen dinners and simmer sauces are big sellers on the Australian market.

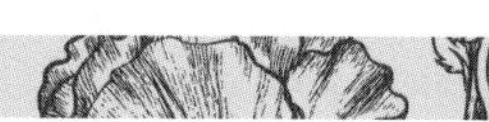

FOOD IN FOCUS

DELICIOUS NUTRITIOUS

Woolworths has launched a new range of frozen meals under the brand 'Delicious Nutritious'. This range has been designed and promoted by celebrity personal trainer Michelle Bridges. Woolworths said it decided to create the range after discovering that 90 per cent of Australians feel they do not eat enough vegetables.

Each meal is less than 450 calories, has three serves of vegetables and rates at least a four-star rating as part of the Federal Government's Health Star front-of-pack rating system for promoting healthier food.

Options within the 'Delicious Nutritious' range include:

- Chicken Pesto Pasta
- Beef & Tomato Casserole
- Asian Style Chicken
- Spicy Chimichurri Beef & Vegetables
- Salmon Fish Cakes
- Italian Style Chicken
- Moroccan Style Chickpea Tagine
- Mild Massaman Beef.

The recommended retail price for each of these meals at the time of launch is AUD$7.99.

According to Nielsen figures, Lean Cuisine currently has the largest brand volume in the health-related frozen meals sold in Australian supermarkets with 47.9 per cent share. Weight Watchers followed with 30.1 per cent and McCain had 19 per cent of the market, while other brands make up the remaining 3 per cent.

Source: Adapted from 'Woolworths launches new 'Delicious Nutritious' frozen meals', *Australian Food News*, 28 October 2015

ACTIVITIES

1 Who is Woolworths using to promote their new range of frozen meals?
2 Why did Woolworths develop the Delicious Nutritious range of frozen meals?
3 Identify the nutritional features of the Delicious Nutritious range.
4 Which option from the Delicious Nutritious range is suitable for vegetarians?
5 Identify two competitors for the Delicious Nutritious range of products.
6 Suggest another meal option for the Delicious Nutritious range.

Mark Fergus

Meal replacements

Breakfast beverages and bars

With so many Australians citing lack of time as a reason for not eating breakfast, breakfast bars and breakfast beverages have become a popular substitute. Unfortunately, breakfast bars and beverages usually contain more sugar and fat and less dietary fibre than packaged cereals, so dieticians recommend that they only be consumed occasionally. For young children, cereal with milk is preferred over breakfast bars, because milk contains the valuable nutrients that children need.

Weight control products

Some meal-replacement products, such as milkshakes, soups, bars and powders, promote weight loss. While they may initially lead to weight loss, these products can become boring and restrictive, and the weight lost may be regained. This is because the dieter may not have learned to make healthier food choices.

Many such diet products also recommend a minimum 30 minutes of exercise a day. For many people, increasing physical activity rather than dramatically changing food habits can help to control weight.

Some meal-replacement products are designed to increase body mass, particularly muscle mass, because of their high protein, vitamin and mineral content. These products are popular with males. It is important to note that excessive amounts of some vitamins and minerals may be dangerous, and excess protein is stored as body fat.

Snack bars

Snacking is an important part of the diet, but research indicates that Australians are replacing meals with snacks. Many food manufacturers have responded to this trend by producing snack bars.

While chocolate bars have always been popular snacks, they have become smaller and available in bulk packaging, making it tempting to always have one on hand for a treat.

Cereal, muesli and protein bars are also popular snacks. Care must be taken to examine the sugar content of these products, as many contain high quantities of glucose and honey, which can cling to the teeth and so contribute to tooth decay. Added ingredients such as chocolate and nuts can also make such bars high in fat. Bars that claimed to be low-carbohydrate and high-protein are believed to help to delay hunger, but any excess protein the body does not require for growth or maintenance is stored as fat.

Snack bars and drinks containing extract of guarana, a plant containing high levels of caffeine, have become popular with children and teenagers. They are marketed as energy boosters, as caffeine is a stimulant, providing a short, sharp boost of energy. Caffeine in high levels can also increase blood pressure, contribute to anxiety and dehydration, and disturb sleep.

Electrolyte replacement drinks

During strenuous exercise the body loses fluids, making it harder for a person to cool down and keep performing well. Electrolyte drinks help to replace and retain the fluids, electrolytes (sodium, potassium, chloride) and salts lost through sweating. These drinks also contain some carbohydrate, usually glucose, to provide energy.

While originally intended for athletes, electrolyte drinks are now marketed to the general public. They are advertised as thirst quenchers and energy boosters, and come in many flavours that appeal to the young. These drinks should be drunk in moderation, as large quantities of electrolytes such as sodium can contribute to health problems such as hypertension.

Functional foods and ingredients

Functional foods are those that have been modified to provide extra health benefits beyond the traditional nutrients they already contain. There are numerous varieties of functional foods on the market, each aiming to provide different health benefits.

FOOD IN FOCUS

DO MEAL REPLACEMENT SHAKES HELP YOU LOSE WEIGHT?

JOCELYN LOWINGE

The idea behind meal replacement shakes is to replace one or two meals a day to help reduce the chances of over-eating. According to Professor Tim Crowe, a dietitian and nutrition researcher at Deakin University, 'Shakes have an advantage in that they are very easy to use, but to lose weight you do need to eat sensibly around them and your other meals still have to be portion controlled. They are reasonably high in protein and lead to a feeling of fullness, but you will still be hungry.'

Crowe says replacing meals isn't going to help you learn how to eat in a healthy way so that you can maintain weight loss for the long-term. Evidence suggests they can help with weight loss, but most people regain weight. People tend to do better with weight loss if they get support from a dietitian, he says.

Most meal replacements come as powders that you make into shakes, but some come in the form of soups or bars. Crowe says there are regulations around what can be called a meal replacement; for instance, they must provide adequate amounts of 16 vitamins and minerals.

Crowe says you should ensure you know what you're buying as other protein and energy bars are likely to be high in sugar and kilojoules and not suitable if you're trying to lose weight. 'Meal replacements also don't contain enough fibre and they lack phytonutrients, which are beneficial nutrients from fruit and vegetables.'

'Shakes also don't address the social aspect of eating, and people using them will have a hard time eating out,' Crowe says. 'You can't eat like that forever. Taste fatigue can be quite common.'

Dr Priya Sumithran an endocrinologist from the University of Melbourne, says 'Using meal replacements as part of a very-low-energy diet should only be done under medical supervision. This way they can have their health monitored.'

Source: Adapted from 'Do meal replacement shakes help you lose weight?' by Jocelyn Lowinge, *Australian Food News*, 8 July 2015. Republished with permission of 'Australian Food News Media'

ACTIVITIES

1 How often should you use meal replacement shakes?
2 Besides shakes, what other meal replacement products can you purchase?
3 Give an example of what is required before a product could be labelled as a meal replacement.
4 Outline the advantages of meal replacements.
5 Outline the disadvantages of meal replacements

Shutterstock.com/Africa Studio

Figure 10.4 Many brands of diet shakes are available, but in the long term they may not help you to maintain weight loss.

Table 10.2 Food and functional variety

FOOD	FUNCTIONAL VARIETY
Milks, yoghurts and cheeses	Low-fat, high-calcium, cholesterol-free, iron-enriched, fermented
Breads and cereals	High-fibre, high-protein, low-carbohydrate, cholesterol-reducing, vitamin- and iron-enriched
Margarine, butters. cream	Cholesterol-reducing, fat-reduced, salt-reduced, rich in omega 3
Jams, jellies, spreads	Low-kilojoule, fat-reduced, sugar-reduced
Drinks	Low-kilojoule, low-carbohydrate, low-sugar

UNIT REVIEW

LOOKING BACK

1 Define the term 'organic food'.
2 Briefly explain what is involved in genetically modifying food.
3 List two examples of prepared fresh foods.
4 Name three herbs and/or spices used in Thai cuisine.
5 Identify three meal-replacement products available on the market.
6 Outline the role of electrolyte replacement drinks.
7 Define the term 'functional food'.

FOR YOU TO DO

8 Find an image of a functional food and outline the health benefits that may be gained from using the product.
9 Design your ideal snack bar by sketching and annotating its features. Justify your reasoning for the ingredients you use.
10 Find a recipe using an Australian native herb or spice.

TAKING IT FURTHER

11 a Interview an older member of your community, such as a grandparent, your parents, a senior from the local retirement village or a neighbour, to discover the foods they ate when they were your age.
b Write a one- to two-page report on your findings, comparing their food intake to yours today.
c Prepare a dish from the era of the person you interviewed.

12 a Purchase a heat-and-serve meal, such as chilled lasagne or a prepared soup.
b Find and prepare a recipe that makes the same product but from raw ingredients.
c Compare and evaluate the:
- time taken
- ingredients used
- quantity produced
- flavour
- appearance
- cost.

10.2 Trends in dining

Despite the huge array of cooking books, cooking shows and cooking appliances available today, Australians are eating out and ordering in more than ever before.

Establishments and levels of service

Restaurants, **cafes**, canteens, bistros and fast food and take-away stores are examples of establishments providing the service of preparing, cooking and serving food to customers. These form part of the hospitality industry and each provides different types of foods and level of service.

Take-away service

Take-away establishments offer quick service and relatively inexpensive meals and snacks in a throwaway container. Some provide dine-in seating, while others provide 'drive-thru' or home deliveries.

WS 10.2 Trends in dining and food service

Prior to the 1960s, the milk bar and the fish and chip shop were the most common take-away establishments, but today there are many different types of take-aways selling a variety of foods such as hamburgers, pizza, salads, sushi and wraps.

Most take-away outlets are located at convenient locations near workplaces and entertainment venues, in shopping centres and even in service stations. Many restaurants and cafes also offer take-away and home delivery services. Current trends also include serving restaurant-quality food from food trucks.

Dining out

Cafes are very popular in Australia. They offer casual dining where a customer can enjoy a cup of coffee, a sweet or a light meal. They have become a social meeting place, with many located near workplaces, entertainment venues, tourist attractions and in shopping centres. Most cafes open early to cater for breakfast and stay open until late. Many offer 'modern Australian' menus that include updated versions of Australian classics, as well as dishes from other cultures.

The bistro is another popular dining venue. Often located in clubs and pubs, bistros offer a variety of food choices and relaxed, inexpensive dining.

Restaurants provide menus consisting of several courses. A variety of restaurants tend to be commonly found on main streets and in shopping complexes in towns and cities across Australia. Modern Australian, Chinese, Italian, Thai, Mexican and Indian are the most popular **cuisines** but others, such as Japanese, Vietnamese, Korean, Lebanese and Moroccan, as well as vegetarian restaurants, are increasing in popularity. Many of these dining venues also offer alfresco dining, where customers can enjoy eating outdoors.

FOOD IN FOCUS

CRUST PIZZA

When Costa Anastasiadis and Michael Logos entered the pizza business over a decade ago, everyone said they were crazy, particularly with established brands like Domino's and Pizza Hut where you could get a pizza for a few dollars. They noticed the pizza industry was either low-cost take-away and home delivery chains, or quality individual stores that mostly didn't do deliveries. So they decided to offer high-quality pizzas and allow take-away customers to see the ingredients going into the pizzas.

They piloted their idea at their first Crust Gourmet Pizza Bar at Annandale (in Sydney), and now Crust has over 100 stores in Australia. They've also expanded the product range to offer healthy pizzas as well as dessert pizzas.

Most of the stores are franchised, where franchisees pay around $400 000 to set up a store and pay Crust a royalty on turnover. Anastasiadis says being flexible and keeping up with change and trends has been one of the reasons for Crust's success. The use of social media as a marketing platform has been a good way of keeping up with change.

Source: Adapted from 'Not so crazy: how Crust cracked 100 stores' by Rod Myer, *Sydney Morning Herald*, 9 November 2011.

ACTIVITIES

1 What did Crust decide to offer customers when they first opened?
2 How have they expanded their product range since opening?
3 How much does it cost to set up and buy a Crust franchise?
4 How do you think Crust uses social media to market their business?

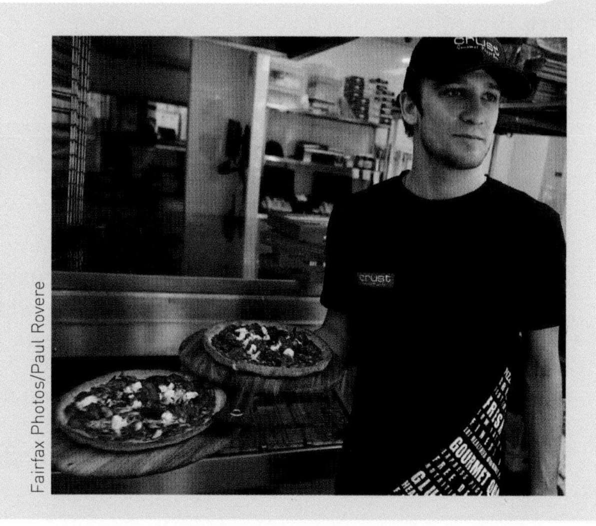

Fairfax Photos/Paul Rovere

Formal and buffet service

In the past, tables at home and at restaurants were set in a formal style. Tables would be set with appropriate cutlery and glassware, and menu courses would be individually presented to each diner.

Today, while formal service is still used, less formal service is commonly used at home and by many food service providers. It is not uncommon in many dining venues to use paper serviettes and dinner knives and forks for most courses, rather than specific cutlery such as dessert forks and fish knives and forks. There is also a trend to share food, such as Spanish Tapas, pizza and platters of pasta.

Many food service providers, including function centres, caterers and bistros, commonly use buffet service, particularly when feeding large numbers of people. This type of service is again very casual as prepared foods are set on a table and guests help themselves.

Trends in table setting

A few decades ago Australians would set their fine bone china dinner set, crystal glassware and silver cutlery on a crisp white linen tablecloth for their formal dinners. These were served in the dining room and always ended with a cup of tea.

Technology and social changes have now resulted in a wider selection of table setting options.

Crockery and cutlery that are microwave- and dishwasher-safe are now available, and plastic items are commonly used for outdoor dining. Table-setting items from different cultures, such as chopsticks and pasta plates, are common in most households, while the coffee mug has replaced the teacup in many cupboards. Wine glasses are no longer reserved for formal dinners, and while white china and white tablecloths are still popular, so are bright colours, prints and placemats.

Even the venue for family meals has changed. The formal dining room has been replaced by a more open-plan design combining the kitchen, dining and lounge rooms, and often opening to an outdoor dining area. Family sit-down meals are increasingly being replaced by meals eaten individually at the breakfast bar or in front of devices such as televisions or computer tablets.

Alamy Stock Photo/Rob Walls

Figure 10.5 Alfresco dining suits the Australian lifestyle.

iStock.com/Kris Vandereycken

Figure 10.6 Formal dining setting is still commonly used for formal occasions.

UNIT REVIEW

LOOKING BACK

1 List five different types of food service establishments.
2 Explain what take-away establishments offer customers.
3 Name three types of dining out venues popular with Australians.

FOR YOU TO DO

4 You are about to move out of home into your own place. Make a list of the table setting equipment you would need to purchase.
5 Name your favourite eating out establishment and explain why you like the venue.

>

TAKING IT FURTHER

6 Design your own modern food service establishment, including the following tasks:
- Choose a type of establishment.
- Name your establishment.
- Describe and justify its location.
- Describe and justify the services it will provide.
- Produce a professional-looking set menu.
- Design a computer-generated logo that would help to identify your establishment.
- Prepare and present a dish of your menu in a practical lesson.

7 Conduct a survey by asking each student in another class to list the favourite menu item from a take-away venue.
- Enter the results on a spreadsheet.
- Produce a suitable graph that best reflects the result.

10.3 Trends in food presentation and food styling

We are drawn to food that looks interesting and appealing. Various techniques are often applied to help make food look more enticing.

Garnishing and decorating

Garnishes and decorations are ingredients that are usually placed last on a dish to provide a special finishing touch. Garnishes are used for savoury dishes and decorations for sweet dishes.

In the 1970s, a sprig of parsley garnished most savoury dishes, while whipped cream, strawberries, dusted icing sugar, canned fruits and chocolate shavings were common decorations for desserts. In the 1990s, garnishes became more dramatic with the use of watercress, dill, and continental parsley, while desserts were decorated with fruit sauces and spun toffee.

Today, garnishes and decorations have become less dramatic. Chefs use simple, fresh ingredients, such as fruit, herbs and liquids, including oils, vinegars and fruit sauces, to highlight dishes and add flavour, colour and texture to a dish.

Shutterstock.com/Yulia Davidovich

Figure 10.7 The fresh fruit on this dessert adds flavour, texture, colour and height to the berry cheesecake.

Shutterstock.com/Losangela

Figure 10.8 The use of garnish on a soup adds flavour, texture and colour.

HANDS-ON

DECORATING WITH SAUCES

PURPOSE

Sauces can add not only favour to a dish but also visual appeal. Practise these techniques of plate decoration using sauces.

MATERIALS

White plates
Strawberry or chocolate topping in a squeeze bottle
Skewers

STEPS

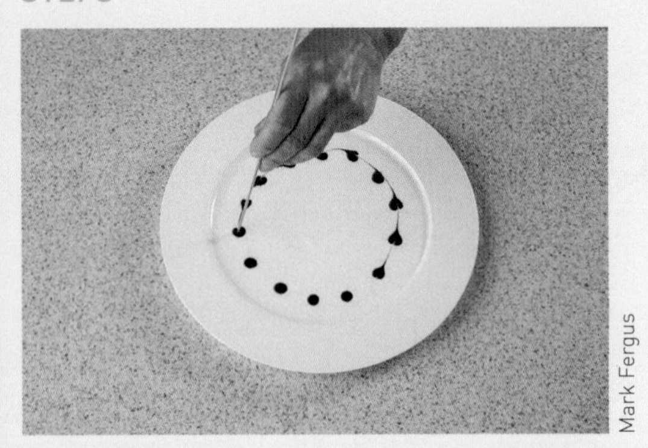

Figure 10.9 Pulling a skewer through chocolate sauce dots creates a border of hearts.

Figure 10.10 Draw circles of topping on a plate and then use a skewer to produce a spider's web.

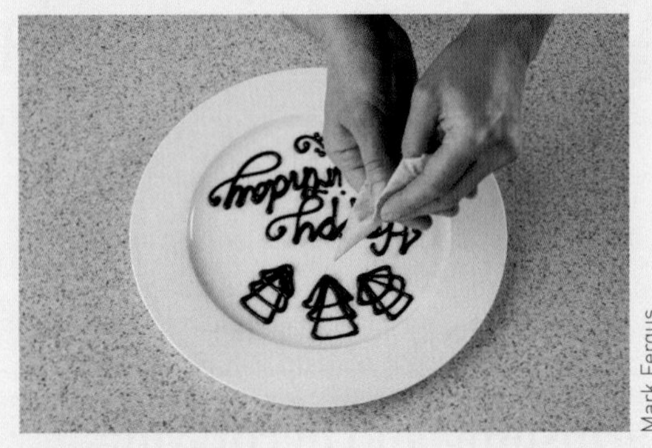

Figure 10.11 Design your own freehand signature design.

Plating styles

The classical plating technique uses the three basic food items of starch (usually a white vegetable), vegetables and protein in a specific arrangement. A simple guide to classical plating is to think of the plate as the face of a clock. The meat is placed at the 6 o'clock position and the vegetables arranged from 10 o'clock through to 2 o'clock. Plating is an art form that is always evolving. Today, more freeform plating styling is a trend. White plates are still popular because they contrast food well, but more natural objects such as stone, and recycled items such as jam jars and enamel plates, have become popular to serve food and drinks.

The basics of plating

Create a framework

Start with drawings and sketches to visualise the plate. Find inspiration from a picture or object and also do a practice plate.

Keep it simple

Select one ingredient to focus on and use space to simplify the presentation. Clutter distracts from the main elements of a dish.

Balance the dish

Use a range of colours, shapes and textures, but remember that too many colours, shapes and textures can become confusing. Presentation should never overpower the flavour.

Get the right portion size

Ensure there is the right amount of ingredients and the plate complements the dish, not too big or too small. Make sure each diner receives the same amount and ensure the right proportion of protein, carbohydrates and vegetables to create a nutritionally balanced meal.

Highlight the key ingredient

Ensure the main ingredient stands out and pay equal attention to the other elements on the plate, such as garnishes, sauces and even the plate itself.

Cleverly cut or sculpted ingredients can enhance the visual appeal of dishes. Sculpted food also provides height and structure and keeps the plate neat and clean. Sauces and garnishes should complement and highlight the main ingredients.

Figure 10.12 Plating styles

The classic clock-face plating
This is a simple method of plating food for meat and vegetables.

Freeform plating
Just like modern paintings, freeform plating involves placing the dish in an abstract yet intriguing set-piece on a plate.

Landscape plating
Taking inspiration from landscape gardens, this arrangement of food components is laid out flat and 'landscaped'.

Food art
Decorating food into a work of art, this is commonly used as a centre piece.

Left to right: Alamy Stock Photo/OJO Images Ltd; Shutterstock.com/Hvoenok; Shutterstock.com/Shebeko; iStock.com/serg78

UNIT REVIEW

LOOKING BACK

1 What is the role of a garnish or decoration?
2 Distinguish between a garnish and a decoration.
3 List two examples of garnishes.
4 Give two examples of decorations.
5 Describe how to position food in a classical plating style.

FOR YOU TO DO

6 Produce a one-page collection of food garnishes and decorations ideas.
7 Research the sauces that traditionally accompany:
- **a** fish and chips
- **b** roast beef
- **c** roast pork
- **d** spring rolls.

8 Justify why each garnish or decoration might have been used on each of the following foods:
- **a** a lemon wedge on fried calamari
- **b** a strawberry fan on vanilla ice-cream
- **c** croutons on pumpkin soup.

TAKING IT FURTHER

9 Use a slice of plain cheesecake and plate it in a modern style.
10 Sketch and annotate a design idea for the traditional 'Australian dinner' of meat and three vegetables.

10.4 Food styling and photography

Preparing a visual display of food is known as food styling. A food stylist often works in consultation with a photographer in order to get a perfect shot to meet the client's brief. Food styling is time-consuming but it creates an atmosphere that stimulates the appetite to encourage consumers to try the food. You see examples of food styling and photography in:

- recipe books and magazines
- newspaper articles
- television, billboard and magazine advertisements
- brochures and leaflets
- posters and framed pictures
- webpages and social media
- food labels and packages
- menu boards
- television shows.

When presenting foods, food stylists employ the basic design principles of variety in colour, texture and flavour. They also like to create themes; for example, food suitable for a children's birthday party may be displayed with streamers, balloons and a bright background. In food advertisements, food stylists usually repeat the colours of the food package being promoted in the background to help reinforce the product.

Getting the perfect shot when photographing food can be difficult. Lighting causes foods to dry out, shrink, sag and discolour, so food stylists employ some interesting techniques. Computer programs and digital cameras assist work of the photographer.

A stylist's tool kit will often consist of tapes, a tape measure, tacks, pins, clips, elastic bands, safety pins, glue, matches, scissors, scalpels, eye droppers, syringes, wire, brushes, touch-up paints and other items to help in setting up the food.

Tricks of the trade

- A whole roast chicken is often only partially cooked to keep it plump and juicy. It is then browned with a coat of soy sauce and a blowtorch crisps up the edges. A final coat of oil or liquid glucose gives a hot, freshly cooked shine. Sometimes mashed potato is placed under the skin to help plump the chicken up further.
- A hamburger shoot involves treating the meat in the same manner as chicken and then finding the perfect bun, tomato slices and lettuce leaves. Sometimes extra sesame seeds are glued onto the bun. Layers of plastic may be used to separate the layers and pins to hold everything in place. Tomato sauce is mixed with tomato paste to stop it running and placed exactly where desired with a syringe.
- Ingredients for stir-fries are usually individually cooked, then tossed in a sauce and painstakingly arranged with tweezers to look random.
- In breakfast cereal advertisements, cream often replaces milk to prevent cereal looking soggy, while plastic ice cubes are used instead of real ice in soft drink commercials.
- Ice-cream melts under lights, so coloured mashed potato or a mixture of corn syrup, margarine and icing sugar may be used.
- Meat pies are often cooked with a filling of mashed potato to make them easier to slice (usually with a scalpel to get a perfect cut). Once the filling is removed, chunks of meat are arranged neatly in place.
- Swiss cheese is usually given extra holes with little round cutters.

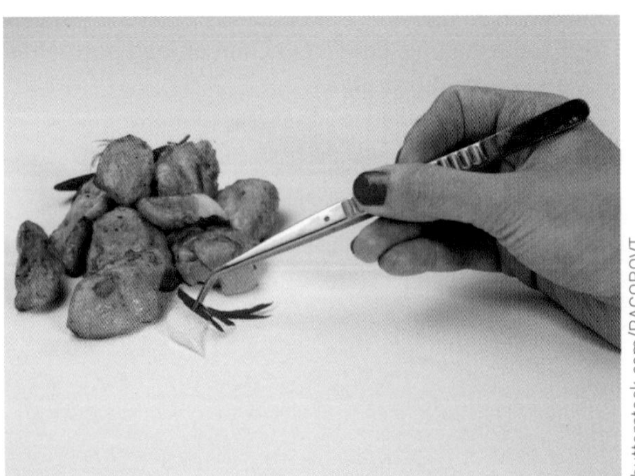

Shutterstock.com/RACOBOVT

Figure 10.13 A food stylist at work

UNIT REVIEW

LOOKING BACK

1 Define the term 'food styling'.
2 With whom does a food stylist normally work in consultation?
3 List six situations in which food styling is used.
4 List tools you would have in your food stylist's tool box.

FOR YOU TO DO

5 Select a photo of a food that has been styled for an advertisement, recipe book or webpage. Analyse:
   a the theme or mood that has been created
   b the styling techniques employed
   c how the styling may influence the viewer.
6 You have been employed as a food stylist for a photo shoot for an advertisement for a tropical fruit juice. Sketch a design idea for the advertisement. Describe your theme and label your props.

TAKING IT FURTHER

7 You have been commissioned to produce a recipe card to promote the use of fresh fruits and vegetables.
   a Find a suitable recipe.
   b Prepare the recipe.
   c Style and photograph the dish.
   d Produce the recipe with the photo at the front and a recipe at the back.

10.5 Accepting new food trends

There are a number of factors that affect whether individuals will accept and adopt a new food trend.

Personal experiences

Food habits are established early in life and can be difficult to change. Bad experiences with food, such as food poisoning or being forced to eat a food, can mean that you will not even try a particular food no matter how appetising it looks. Good experiences may encourage you to adopt new food trends. For instance, most people are willing to try new chocolate bars, because chocolate is often associated with good times.

Cultural taboos and beliefs

Australians enjoy foods of many different cultures. However, there is still resistance to some cultural foods, such as Japanese raw fish or French frogs' legs and snails, and we tend to strongly believe that some meat, such as that of whales, should not be consumed.

In some cultural groups, certain foods may be restricted or forbidden. Examples include the following.

- Muslims do not eat pork or drink alcohol.
- Seventh-day Adventists are vegetarian and avoid drinking tea, coffee and alcohol.
- Many Jews follow a kosher diet and therefore do not eat shellfish or pork products, and also have strict laws in relation to food preparation.

Christmas

Traditional Christmas foods are changing. It is not uncommon to find Christmas cakes of other cultures, such as the Italian panettone or the German stollen, being sold alongside mince tarts and fruitcakes. Even the traditional hot Christmas lunch of turkey or ham and hot plum pudding with custard is being replaced by cooler, lighter foods such as cold meats, seafood, salads, trifle and pavlova. Lunch is often eaten outdoors to take advantage of the Australian summer. Traditional hot and heavy Christmas foods are now commonly enjoyed during July, when many 'Christmas in July' functions are held.

For many families, occasions such as Christmas mean combining the traditional foods from different

Figure 10.14 Examples of a traditional and b non-traditional wedding cakes
Left: iStock.com/Neustockimages. Right: iStock.com/David Freund.

cultures. For example, many families with Italian backgrounds have lasagne followed by the Anglo-Saxon tradition of a roast as a main course.

Weddings

The Anglo-Saxon tradition is to celebrate weddings with an iced fruitcake. Today, many couples like to demonstrate individuality by selecting non-traditional cakes, such as mud cakes, sponge cakes or towers of cupcakes or profiteroles.

Menus for weddings have also become less traditional, with some couples preferring formal sit-down meals while others prefer finger foods. Many couples now like to give their guests the traditional European gift of a bomboniere with sugared almonds as a gesture of gratitude.

FOOD IN FOCUS

CULTURAL TRADITIONS

Josephine, Tom, Ali and Steve are all in Year 10 at the same school. All were born in Australia but each comes from a different cultural and religious background.

Josephine's background is Greek, and Easter is taken very seriously in her family. They avoid meat and dairy products during Lent, and on Good Friday they eat fish. On Easter Sunday they eat chocolate eggs as well as coloured boiled eggs and Greek Easter biscuits. Lamb is usually on the menu at the big family lunch.

Ali has a Lebanese Muslim background. During Ramadan he does not eat during daylight to remind him what it would be like to go hungry. The family all get up before daylight to have breakfast and then gather at night for dinner. They break their fast with a fresh date. At the end of Ramadan, his family enjoys the company of others and hold several parties where a lot of food is prepared, such as barbecued meat and Lebanese sweets.

Tom is of Chinese origin. For Chinese New Year, his family like to eat and enjoy the entertainment provided by a Chinese restaurant, where they enjoy a banquet that includes a variety of dishes such as duck and pork. Of course, the whole family read their fortune cookies and the children wait for their red envelopes of money.

Steve has a Macedonian background. In his Orthodox Christian religion, Christmas is celebrated on the 7th of January. On this day, a cake is baked with a coin and whoever finds the coin will receive good luck for the year.

>

FOOD IN FOCUS

ACTIVITIES

1 What are the four cultural groups represented in the text above?
2 Identify two religious events in the text above.
3 Identify a traditional food that Tom eats.
4 Why would pork not be on the menu at Ali's home?
5 What is the significance of placing a coin in a cake?

Figure 10.15 Special occasions are often celebrated with an abundance of food.

UNIT REVIEW

LOOKING BACK

1 Identify one example of a religion and a food that it forbids.
2 What is the traditional wedding cake of the Anglo-Saxon culture?
3 What is a bomboniere?

FOR YOU TO DO

4 Choose a festive celebration your family celebrates. Describe the traditional foods served and the roles that family members undertake in preparing the feast.
5 Sketch an idea for a modern wedding cake.
6 Make up a quote for Chinese fortune cookies.

TAKING IT FURTHER

7 Research the history of a festive tradition of a particular culture and produce a story book to teach young children about this tradition.

10.6 Marketing and food trends

Marketing is more than just advertising a product. For a new product to become a new food trend, it must be:

- aimed at particular consumers
- a product that consumers need or want
- easily available
- affordable
- suitably promoted.

Promotion is used to inform consumers about a new product and to persuade them to purchase it. Promotion is planned to gain the attention of the consumers most likely to buy their products. For example, a women's breakfast cereal is often advertised in a women's magazine or on websites that appeal to women. Techniques such as jingles, slogans, celebrity endorsements and references to happier lives are common in food advertisements.

Most companies promote new foods through several of the different forms listed below. This allows consumers to hear about products in a number of different ways.

Forms of promotion include:

- advertisements in newspapers and magazines, and on billboards, radio, television and the internet
- leaflets and brochures
- point-of-sale displays
- sponsorship
- sample giveaways
- in-store demonstrations
- competitions.

The media and food trends

The media greatly influence food trends. There are many food and cooking shows both online and on television. There are also many articles on food and nutrition online, on radio and in the print media.

Celebrities and celebrity chefs are commonly used to demonstrate how easy it is to prepare modern foods. They introduce people to new recipes, ingredients, cooking and food products. Sometimes they also discuss health issues. Today, with the availability of the internet, home cooks and food bloggers also promote food trends via cooking channels and social media.

Newspix/Richard Jupe

Figure 10.16 Celebrity chefs (such as Curtis Stone, pictured) and competitions show how easy it is to cook modern food.

FOOD IN FOCUS

THE ONE POT CHEF

David Chilcott began making cooking videos on YouTube as a hobby in 2007. Having no previous experience with video making, he had to learn everything from scratch. Starting out with an old camcorder and setting up in his home kitchen, the OnePotChefShow YouTube channel was born.

David's passion for cooking and his love of simple, home-style food was an instant recipe for success, drawing viewers from all over the world. Within 12 months, David was offered the chance to join YouTube's Partner Program, allowing him to start making money from the video content he uploaded to the site.

Using the money earned from his videos, David invested in new camera equipment, professional editing software and updating his kitchen supplies. Taking the channel to the next level, the quality of videos improved and the audience started growing rapidly.

Since then, David runs the OnePotChefShow channel as a full time business. He has also created a series of cookbooks based on his recipes and he even filmed a studio-based version of his popular cooking show in a real-life film studio in Los Angeles.

Source: Adapted from Onepotchefshow.com with permission David Chilcott

ACTIVITIES

As a team, produce your own short video clip on how to cook a recipe. Make sure everyone in your team is clear about their job. For example, presenter/cook, camera operator, video editor, sound technician/editor, researcher and food stylist. Each team member may need to take on more than one role.

David Chilcott

Figure 10.17 David Chilcott, the One Pot Chef

UNIT REVIEW

LOOKING BACK

1 Identify four different forms of media.
2 List four forms of promotions used in marketing foods, apart from advertising.
3 What do you think Australians gain from watching cooking shows?

FOR YOU TO DO

4 You want to take the country by storm with your healthy ice-cream aimed at the teen demographic. Market this product by:
- describing it
- listing places where it would be sold
- identifying the recommended retail price
- developing a slogan
- selecting and justifying a celebrity to endorse it
- justifying the forms of media you would employ to advertise it.

TAKING IT FURTHER

5 Observe a current television cooking show or video and discuss the food trends portrayed.
6 Collect two advertisements for food from magazines or newspapers. Assess the:
- product being advertised
- audience the advertisement is aimed at
- message(s) the advertising is trying to communicate.

10.7 Additional content: marketing and current food trends

Marketing teams research the market in order to develop and promote appropriate products for particular consumers. The 'four Ps' of marketing are product, price, promotion and place. In the case of heat-and-serve meals, they are:

1 **product**: targeted at Australia's rising population of single adults and busy working families
2 **price**: commonly discounted with regular specials and competitive with take-away meals
3 **promotion**: samples and taste testing followed up with prominent advertising in the media
4 **place**: easily available at supermarkets, convenience stores and service stations.

Marketing, the media and food styling

The print, electronic and social media use food styling and photography to market products. Food styling tempts customers to buy the product that is being promoted. Many cooking shows are sponsored by food companies, and many magazines and websites feature enticing colour photographs of food dishes promoting a food product. Celebrities also often become involved in marketing campaigns. Social media in particular can create worldwide food trends in a short space of time.

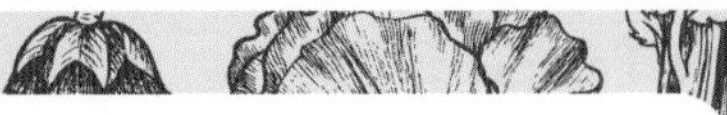

FOOD IN FOCUS

SKI D'LITE

Ski D'Lite yoghurt has been relaunched with 25 per cent less added sugar.

The yoghurts will continue to be low-fat and come in a variety of flavours including strawberry, smashed berry and peach & mango.

A new marketing campaign will accompany the relaunch. Titled 'Practically Perfect' it will focus in on Australian mothers trying to do their best in raising their children.

Advertising efforts will include a video in which children recognise their Mum as being perfect despite not always getting the little things right.

Public relation efforts and influencer relations (such as bloggers, journalists, celebrities) and Facebook advertising will also be part of the campaign.

Ski D'Lite is currently available for purchase in supermarkets across Australia.

Source: 'Ski D'Lite relaunches with 25 per cent less sugar', *Australian Food News*, 5 October 2015. Republished with permission of 'Australian Food News Media'

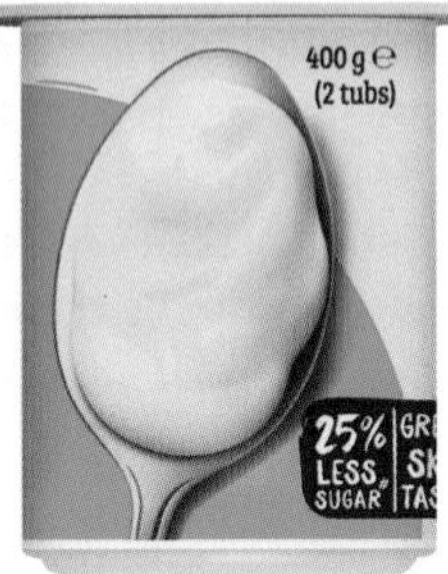

dLibrary.com.au

Figure 10.18 Part of the relaunched range of Ski D'Lite yoghurts with 25 per cent less sugar

ACTIVITIES

1 Is the Ski D'Lite with 25 per cent less sugar a new product? Explain your answer.
2 What is the slogan used in the yoghurt in the new marketing campaign?
3 Where is the product to be placed for sale?
4 Who is the product being promoted towards in new marketing campaign?
5 What strategies will be employed to promote the product?

UNIT REVIEW

LOOKING BACK

1 What are the 'four Ps' of marketing?
2 Identify three examples of print media.
3 What is the purpose of food styling?

FOR YOU TO DO

4 a Design your own new food product.
  b Address the 'four Ps' of marketing to ensure your food trend will be a success.
  c Explain how you can determine whether your marketing plan has been successful.

TAKING IT FURTHER

5 Use a camera and related computer software to produce a finished visual image of styled foods; for example:
- an advertisement for a new food product
- a supermarket catalogue with sale prices
- a webpage with images of recipe directions
- a calendar with different recipes and health hints for each month.

Chapter review

LOOKING BACK

1 Define the term 'organic food'.
2 Outline one advantage and disadvantage of organic foods.
3 Define the term 'genetically modified food'.
4 Outline one advantage and disadvantage of genetically modified foods.
5 List two examples of meal replacement products.
6 Name two herbs.
7 Define the term 'functional food'.
8 Identify an example of a functional food.
9 Distinguish between a garnish and a decoration and provide an example of each.
10 List four places where you could expect to see the work of a food stylist.
11 List five items you would expect to see in a food stylist's tool box.
12 Identify and describe three techniques a food stylist may use when shooting a television commercial.
13 Identify four different forms of media.
14 List five examples of how food products can be promoted.

FOR YOU TO DO

15 You are planning to go away for a week with your friends. Your accommodation has a microwave and a stove. List some heat-and-serve foods that you can purchase at the local supermarket. Justify your reasoning for purchasing heat-and-serve foods.
16 You notice a friend frequently drinking an electrolyte replacement drink when she is thirsty throughout the day. Discuss the advantages and disadvantages of consuming electrolyte replacement drinks in this situation.
17 What advice would you give a parent or caregiver when choosing muesli bars for their children to take to school?
18 Compare and contrast the benefits of take-away service with restaurant dining.
19 Discuss how table setting has changed by comparing the past few decades.

TAKING IT FURTHER

20 Plan a wedding menu for a sit-down dinner.
21 Make a list of celebrities and the products they promote.

RECIPES

Lemonade scones

Scones are suitable for High Teas and morning teas.

 **Preparation time** 15 minutes  **Cooking time** 15 minutes  **Makes** 12

INGREDIENTS

2 cups self-raising flour
½ teaspoon salt
¼ cup caster sugar
½ cup thick cream
½ cup lemonade
2 tablespoons milk

METHOD

1. Preheat the oven to 220°C.
2. Lightly grease a baking tray.
3. Sift the flour and salt.
4. Add the sugar.
5. Add the cream and lemonade and mix to form a soft dough in the shape of a ball.
6. Turn out onto a lightly floured board and knead lightly until combined.
7. Press the dough with your hands to a thickness of about 2centimetre.
8. Use a 4 centimetre round cutter to cut out 12 scones.
9. Place on baking tray, with scones just touching each other.
10. Brush the tops with some milk.
11. Bake for 10–15 minutes until lightly browned.

QUESTIONS

1. Why does the method instruct you to glaze the scones with milk?
2. What ingredients in the recipe help to make the scones rise?
3. What other ingredients could be added to the scones to vary the flavour?
4. If there are 55 grams in a ¼ cup of caster sugar, how many grams are there in 1 cup of caster sugar?
5. If there are 300 grams in two cups of flour, how many cups are there in 1.5 kilograms of flour?

Mark Fergus

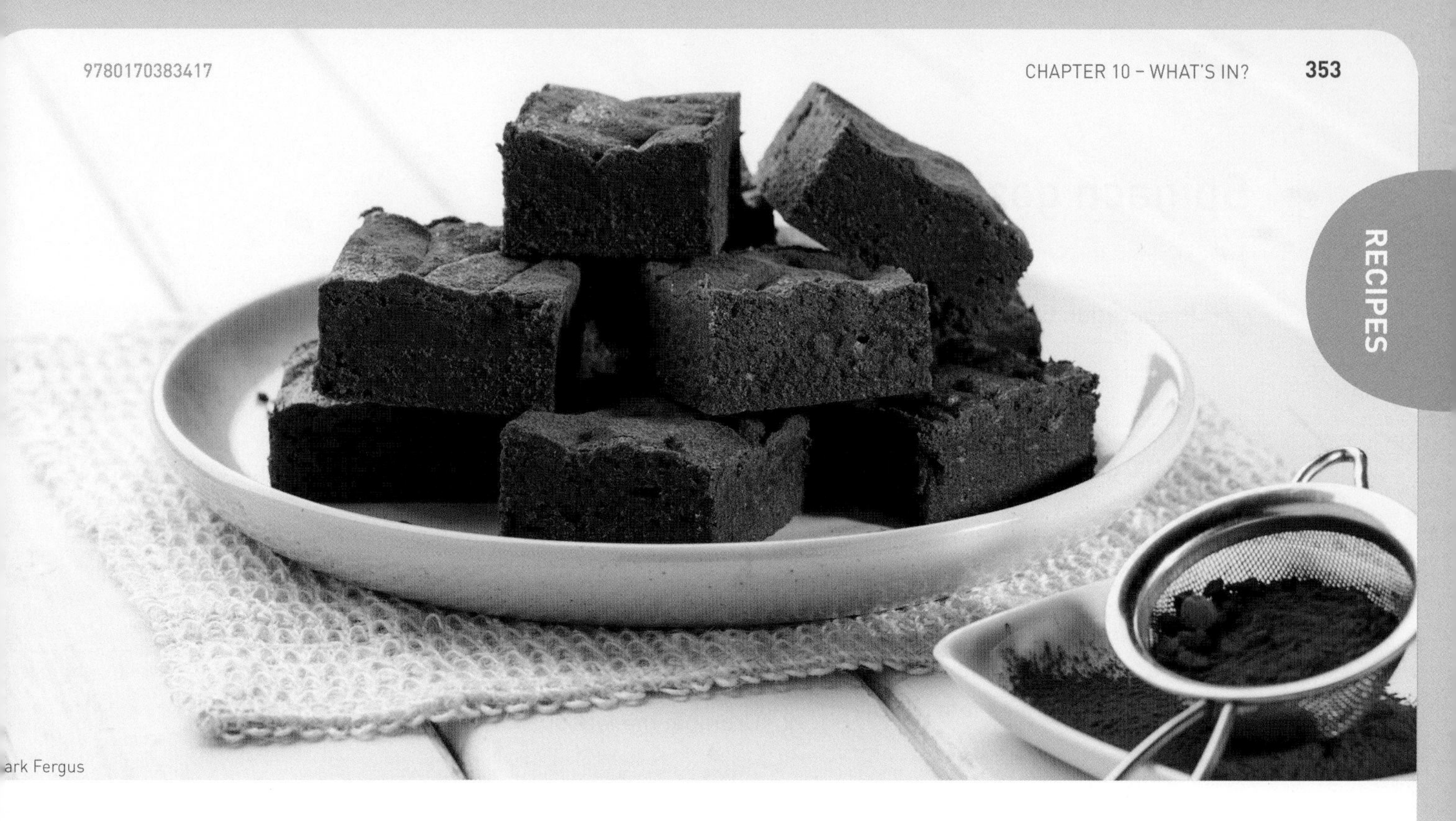

ark Fergus

Chocolate brownie bites

Many menus now offer gluten-free options for customers.

 **Preparation time** 15 minutes  **Cooking time** 20 minutes  **Makes** 16 slices

INGREDIENTS

- **200 grams unsalted butter, chopped**
- **200 grams dark chocolate, broken into pieces**
- **1 cup brown sugar**
- **3 eggs, lightly beaten**
- **1 teaspoon vanilla extract**
- **¾ cup gluten-free plain flour**
- **2 tablespoons cocoa powder**

HINT: It is best to cut the brownies the day after baking.

METHOD

1. Preheat oven to 190°C.
2. Lightly grease and line a 5 cm deep, 18 cm square baking pan.
3. Melt the butter, sugar and chocolate in a saucepan over a low heat, stirring constantly until melted and smooth. Transfer to a heatproof bowl and cool slightly.
4. Add the eggs and vanilla to the chocolate mixture, and mix well.
5. Sift the flour and cocoa over the chocolate mixture. Stir to combine.
6. Pour the brownie mixture into the pan. Bake 20 minutes or until just set. Cool and place in an airtight container.

QUESTIONS

1. Predict the equipment needed to prepare the dish.
2. What preparation is required for the ingredients before you start the method?
3. Why do you need to cool the mixture before adding in the eggs?
4. Determine the duties of each partner if you were to work in pairs when preparing this recipe.
5. What is gluten?
6. Why is low-gluten flour recommended for use in cake making?

9780170383417

Spinach gozleme

This Turkish dish is often prepared on site at fetes and festivals.

 **Preparation time** 15 minutes  **Cooking time** 20 minutes  **Serves** 2

INGREDIENTS

DOUGH

100 grams Greek-style plain yoghurt

125 grams self-raising flour

2 tablespoons olive oil

SPINACH FILLING

50 grams baby spinach

100 grams feta cheese

METHOD

Dough

1 Combine the yoghurt with the self-raising flour and mix with a fork until a ball is formed in the bowl.
2 Knead the dough on a floured board for 2–3 minutes.
3 Cut into 2–4 portions. Cover with a damp tea towel until ready to use.

Filling

1 Pour boiling water over the spinach and then drain.
2 Crumble the feta up and stir through the spinach.

Cooking

1 Roll each dough ball into a 25 cm round and place 4 tablespoons of filling on one half.
2 Fold each dough round in half and press together the edges.
3 Heat a frypan or barbecue hotplate to medium heat and brush each gozleme with olive oil.
4 Cook on each side until golden 'eyes' appear.

QUESTIONS

1 Identify the ingredients in the dish that are high in both calcium and protein.
2 What are the nutrients that are found in spinach?
3 Why do you cover the dough with a tea towel?
4 Identify some other alternative fillings you could use in the gozleme.

Mark Fergus

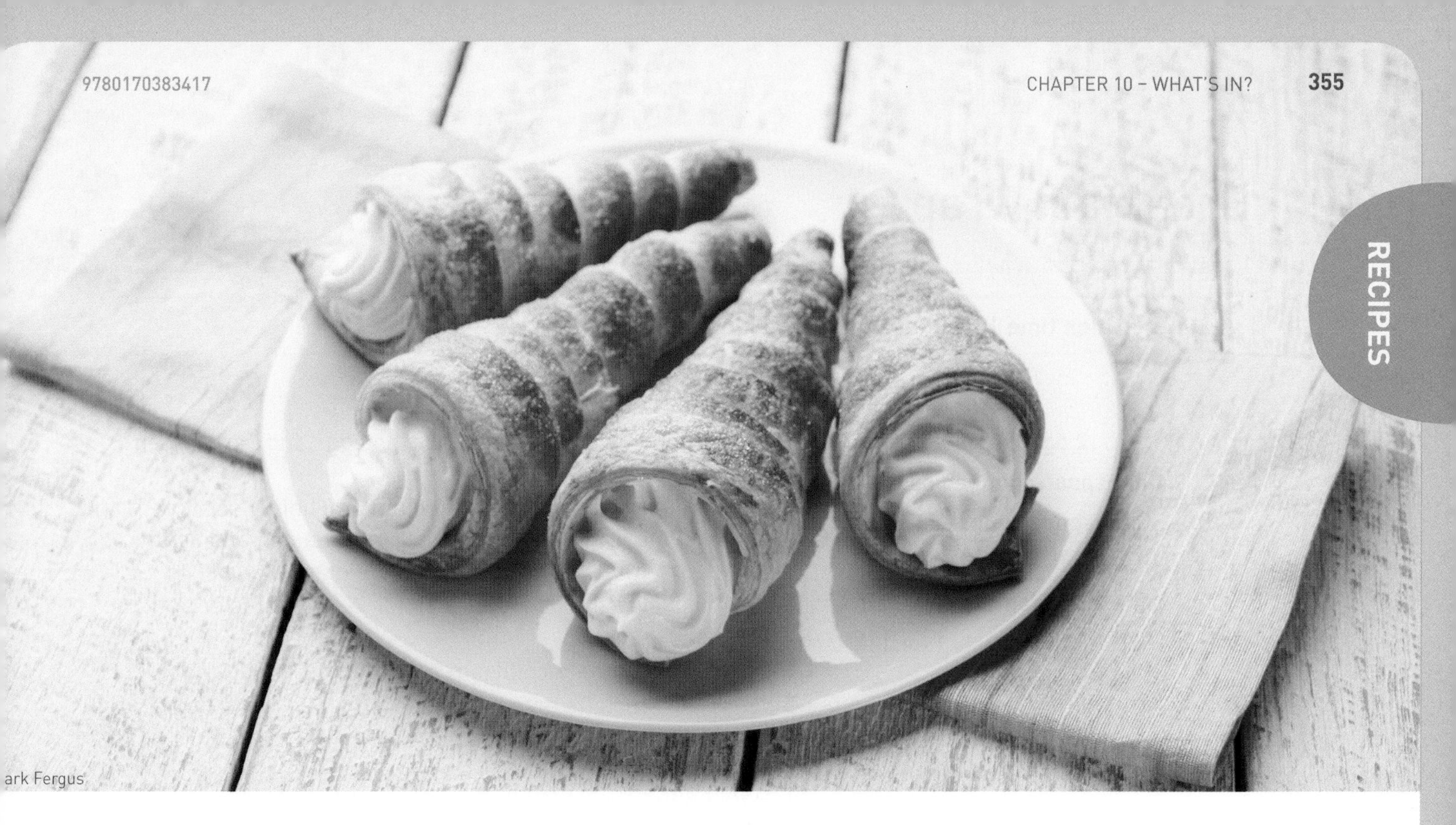
ark Fergus

Custard horns

This old-time favourite is good finger food for special occasions.

 **Preparation time** 15 minutes  **Cooking time** 15 minutes  **Serves** 2

INGREDIENTS

- 1 sheet of puff pastry
- spray oil
- 1 tablespoon of whisked egg or milk
- 1 tablespoon sugar
- ½ cup thickened cream
- 1 cup thick chilled custard

METHOD

1. Preheat oven to 200°C. Line a baking tray with baking paper.
2. Grease the outside of 6 cream horn moulds.
3. Cut one of the pastry sheets into 2 cm thick strips.
4. Starting at the pointed end of the mould, wrap the pastry strips around in a spiral manner to cover completely, slightly overlapping the strips.
5. Brush the pastry with beaten egg or milk, being careful not to get any on the top of the mould. Sprinkle with sugar and lay onto the tray.
6. Repeat with the remaining moulds.
7. Bake for 12–15 minutes, or until golden.
8. Cool for 10 minutes and then carefully remove the moulds.
9. Fill with custard when horns are cold or store in an airtight container until required for filling.

Filling

1. Whisk the cream and custard together until thick.
2. Place into a piping bag fitted with a 13 mm star-shaped nozzle.
3. Fill the horns with custard cream and serve.

QUESTIONS

1. What other filling can be used?
2. How is custard made?
3. Why should you not glaze the metal mould?
4. How can you change this recipe into a savoury food?

9780170383417

Breakfast wrap

Breakfast menus are now common in many dining establishments.

 **Preparation time** 10 minutes  **Cooking time** 5 minutes  **Serves** 1

INGREDIENTS

2 eggs, lightly beaten
1 tablespoon chopped parsley
1 tablespoon chopped chives
50 grams sliced ham, chopped
1 tablespoon tomato chilli jam
1 tortilla

METHOD

1 In a small bowl, whisk together the eggs and herbs, then season to taste.
2 Heat a non-stick frypan on medium and lightly spray with cooking oil. Pour the egg mixture into the pan and gently swirl, evenly distributing the mixture like a pancake.
3 Sprinkle over the ham and cook for 2–3 minutes or until the egg is just set.
4 Spread the tomato chilli jam over the flour tortilla.
5 Slide the omelette onto the tortilla, and then roll up like a wrap.

QUESTIONS

1 What equipment would you require to prepare this dish?
2 What could you substitute for the ham?
3 What are some common breakfast dishes on cafe menus?
4 From which country does the tortilla originate?
5 Describe the flavours you would expect from the jam.
6 What does it mean to 'season'?
7 What is the role of the herbs in this recipe?

Mark Fergus

ark Fergus

Beef stroganoff

A good dish for buffet menus because it is easy to serve and make in bulk quantities.

 **Preparation time** 15 minutes  **Cooking time** 25 minutes  **Serves** 2

INGREDIENTS

- 1 teaspoon oil
- 200 grams beef
- 2 tablespoons plain flour
- ½ onion, finely sliced
- ½ cup water
- 1 mushroom, sliced finely
- 1 beef stock cube
- 1 tablespoon tomato paste
- ¼ cup sour cream
- ½ cup long grain rice
- 4 sprigs parsley, chopped finely

METHOD

1. Half-fill a medium saucepan with water and bring to boil. Add the rice and boil for 15 minutes or until tender. Drain.
2. Cut the beef into thin strips and then coat in flour.
3. Heat the oil in frypan, add the meat and cook until brown. Remove from pan and keep warm.
4. Fry the onion and mushrooms for 2–3 minutes. Return the meat to the pan. Add the water, stock cube and tomato paste.
5. Bring mixture to the boil and then reduce heat to simmer for 5 minutes.
6. Mix the sour cream into the meat mixture to allow the sauce to thicken. Simmer for another 5 minutes.
7. Serve stroganoff on a bed of rice, garnished with parsley.

QUESTIONS

1. List the main nutrients found in red meat.
2. Historically, why did Australians eat large quantities of lamb and beef?
3. How can you test if rice is cooked?
4. What is the purpose of coating the meat in flour?
5. What does 'simmering' mean?

9780170383417

index

A

B

C

G

H

I

J

K

L

M

N

9780170383417

recipe index

Ex

Bonus recipes